INTERNATIONAL SERIES OF MONOGRAPHS ON PHYSICS

INTERNATIONAL SERIES OF MONOGRAPHS ON PHYSICS

Statistical Physics of
Spin Glasses and
Information Processing

An Introduction

HIDETOSHI NISHIMORI

Department of Physics
Tokyo Institute of Technology

OXFORD
UNIVERSITY PRESS

OXFORD

UNIVERSITY PRESS

Great Clarendon Street, Oxford OX2 6DP

Oxford University Press is a department of the University of Oxford.
It furthers the University's objective of excellence in research, scholarship,
and education by publishing worldwide in

Oxford New York

Athens Auckland Bangkok Bogotá Buenos Aires Cape Town
Chennai Dar es Salaam Delhi Florence Hong Kong Istanbul Karachi
Kolkata Kuala Lumpur Madrid Melbourne Mexico City Mumbai Nairobi
Paris São Paulo Shanghai Singapore Taipei Tokyo Toronto Warsaw
with associated companies in Berlin Ibadan

Oxford is a registered trade mark of Oxford University Press
in the UK and in certain other countries

Published in the United States
by Oxford University Press Inc., New York

British Library Cataloguing in Publication Data

Data available

Library of Congress Cataloging in Publication Data

ISBN 0 19 850941 3 paperback
ISBN 0 19 850940 5 hardback

1 3 5 7 9 10 8 6 4 2

Typeset by the author using LaTex
Printed in Great Britain
on acid-free paper by Biddles Ltd, Guildford and King's Lynn

PREFACE

The scope of the theory of spin glasses has been expanding well beyond its original goal of explaining the experimental facts of spin glass materials. For the first time in the history of physics we have encountered an explicit example in which the phase space of the system has an extremely complex structure and yet is amenable to rigorous, systematic analyses. Investigations of such systems have opened a new paradigm in statistical physics. Also, the framework of the analytical treatment of these systems has gradually been recognized as an indispensable tool for the study of information processing tasks.

One of the principal purposes of this book is to elucidate some of the important recent developments in these interdisciplinary directions, such as error-correcting codes, image restoration, neural networks, and optimization problems. In particular, I would like to provide a unified viewpoint traversing several different research fields with the replica method as the common language, which emerged from the spin glass theory. One may also notice the close relationship between the arguments using gauge symmetry in spin glasses and the Bayesian method in information processing problems. Accordingly, this book is not necessarily written as a comprehensive introduction to single topics in the conventional classification of subjects like spin glasses or neural networks.

In a certain sense, statistical mechanics and information sciences may have been destined to be directed towards common objectives since Shannon formulated information theory about fifty years ago with the concept of entropy as the basic building block. It would, however, have been difficult to envisage how this actually would happen: that the physics of disordered systems, and spin glass theory in particular, at its maturity naturally encompasses some of the important aspects of information sciences, thus reuniting the two disciplines. It would then reasonably be expected that in the future this cross-disciplinary field will continue to develop rapidly far beyond the current perspective. This is the very purpose for which this book is intended to establish a basis.

The book is composed of two parts. The first part concerns the theory of spin glasses. Chapter 1 is an introduction to the general mean-field theory of phase transitions. Basic knowledge of statistical mechanics at undergraduate level is assumed. The standard mean-field theory of spin glasses is developed in Chapters 2 and 3, and Chapter 4 is devoted to symmetry arguments using gauge transformations. These four chapters do not cover everything to do with spin glasses. For example, hotly debated problems like the three-dimensional spin glass and anomalously slow dynamics are not included here. The reader will find relevant references listed at the end of each chapter to cover these and other topics not treated here.

The second part deals with statistical-mechanical approaches to information processing problems. Chapter 5 is devoted to error-correcting codes and Chapter 6 to image restoration. Neural networks are discussed in Chapters 7 and 8, and optimization problems are elucidated in Chapter 9. Most of these topics are formulated as applications of the statistical mechanics of spin glasses, with a few exceptions. For each topic in this second part, there is of course a long history, and consequently a huge amount of knowledge has been accumulated. The presentation in the second part reflects recent developments in statistical-mechanical approaches and does not necessarily cover all the available materials. Again, the references at the end of each chapter will be helpful in filling the gaps. The policy for listing the references is, first, to refer explicitly to the original papers for topics discussed in detail in the text, and second, whenever possible, to refer to review articles and books at the end of a chapter in order to avoid an excessively long list of references. I therefore have to apologize to those authors whose papers have only been referred to indirectly via these reviews and books.

The reader interested mainly in the second part may skip Chapters 3 and 4 in the first part before proceeding to the second part. Nevertheless it is recommended to browse through the introductory sections of these chapters, including replica symmetry breaking (§§3.1 and 3.2) and the main part of gauge theory (§§4.1 to 4.3 and 4.6), for a deeper understanding of the techniques relevant to the second part. It is in particular important for the reader who is interested in Chapters 5 and 6 to go through these sections.

The present volume is the English edition of a book written in Japanese by me and published in 1999. I have revised a significant part of the Japanese edition and added new material in this English edition. The Japanese edition emerged from lectures at Tokyo Institute of Technology and several other universities. I would like to thank those students who made useful comments on the lecture notes. I am also indebted to colleagues and friends for collaborations, discussions, and comments on the manuscript: in particular, to Jun-ichi Inoue, Yoshiyuki Kabashima, Kazuyuki Tanaka, Tomohiro Sasamoto, Toshiyuki Tanaka, Shigeru Shinomoto, Taro Toyoizumi, Michael Wong, David Saad, Peter Sollich, Ton Coolen, and John Cardy. I am much obliged to David Sherrington for useful comments, collaborations, and a suggestion to publish the present English edition. If this book is useful to the reader, a good part of the credit should be attributed to these outstanding people.

H. N.

Tokyo
February 2001

CONTENTS

1
MEAN-FIELD THEORY OF PHASE TRANSITIONS

Methods of statistical mechanics have been enormously successful in clarifying the macroscopic properties of many-body systems. Typical examples are found in magnetic systems, which have been a test bed for a variety of techniques. In the present chapter, we introduce the Ising model of magnetic systems and explain its mean-field treatment, a very useful technique of analysis of many-body systems by statistical mechanics. Mean-field theory explained here forms the basis of the methods used repeatedly throughout this book. The arguments in the present chapter represent a general mean-field theory of phase transitions in the Ising model with uniform ferromagnetic interactions. Special features of spin glasses and related disordered systems will be taken into account in subsequent chapters.

1.1 Ising model

A principal goal of statistical mechanics is the clarification of the macroscopic properties of many-body systems starting from the knowledge of interactions between microscopic elements. For example, water can exist as vapour (gas), water (liquid), or ice (solid), any one of which looks very different from the others, although the microscopic elements are always the same molecules of H_2O. Macroscopic properties of these three phases differ widely from each other because intermolecular interactions significantly change the macroscopic behaviour according to the temperature, pressure, and other external conditions. To investigate the general mechanism of such sharp changes of macroscopic states of materials, we introduce the Ising model, one of the simplest models of interacting many-body systems. The following arguments are not intended to explain directly the phase transition of water but constitute the standard theory to describe the common features of phase transitions.

Let us call the set of integers from 1 to N, $V = \{1, 2, \ldots, N\} \equiv \{i\}_{i=1,\ldots,N}$, a *lattice*, and its element i a *site*. A site here refers to a generic abstract object. For example, a site may be the real lattice point on a crystal, or the pixel of a digital picture, or perhaps the neuron in a neural network. These and other examples will be treated in subsequent chapters. In the first part of this book we will mainly use the words of models of magnetism with sites on a lattice for simplicity. We assign a variable S_i to each site. The *Ising spin* is characterized by the binary value $S_i = \pm 1$, and mostly this case will be considered throughout this volume. In the problem of magnetism, the Ising spin S_i represents whether the microscopic magnetic moment is pointing up or down.

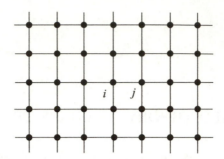

FIG. 1.1. Square lattice and nearest neighbour sites $\langle ij \rangle$ on it

A *bond* is a pair of sites (ij). An appropriate set of bonds will be denoted as $B = \{(ij)\}$. We assign an *interaction energy* (or an interaction, simply) $-JS_iS_j$ to each bond in the set B. The interaction energy is $-J$ when the states of the two spins are the same ($S_i = S_j$) and is J otherwise ($S_i = -S_j$). Thus the former has a lower energy and is more stable than the latter if $J > 0$. For the magnetism problem, $S_i = 1$ represents the up state of a spin ($\uparrow$) and $S_i = -1$ the down state ($\downarrow$), and the two interacting spins tend to be oriented in the same direction ($\uparrow\uparrow$ or $\downarrow\downarrow$) when $J > 0$. The positive interaction can then lead to macroscopic magnetism (ferromagnetism) because all pairs of spins in the set B have the tendency to point in the same direction. The positive interaction $J > 0$ is therefore called a *ferromagnetic interaction*. By contrast the negative interaction $J < 0$ favours antiparallel states of interacting spins and is called an *antiferromagnetic interaction*.

In some cases a site has its own energy of the form $-hS_i$, the Zeeman energy in magnetism. The total energy of a system therefore has the form

$$H = -J \sum_{(ij)\in B} S_iS_j - h \sum_{i=1}^{N} S_i. \tag{1.1}$$

Equation (1.1) is the *Hamiltonian* (or the energy function) of the *Ising model*.

The choice of the set of bonds B depends on the type of problem one is interested in. For example, in the case of a two-dimensional crystal lattice, the set of sites $V = \{i\}$ is a set of points with regular intervals on a two-dimensional space. The bond $(ij)\,(\in B)$ is a pair of *nearest neighbour* sites (see Fig. 1.1). We use the notation $\langle ij \rangle$ for the pair of sites $(ij) \in B$ in the first sum on the right hand side of (1.1) if it runs over nearest neighbour bonds as in Fig. 1.1. By contrast, in the infinite-range model to be introduced shortly, the set of bonds B is composed of all possible pairs of sites in the set of sites V.

The general prescription of statistical mechanics is to calculate the thermal average of a physical quantity using the probability distribution

$$P(\boldsymbol{S}) = \frac{\mathrm{e}^{-\beta H}}{Z} \tag{1.2}$$

for a given Hamiltonian H. Here, $\boldsymbol{S} \equiv \{S_i\}$ represents the set of spin states, the *spin configuration*. We take the unit of temperature such that Boltzmann's constant k_{B} is unity, and β is the inverse temperature $\beta = 1/T$. The normalization factor Z is the *partition function*

$$Z = \sum_{S_1=\pm1} \sum_{S_2=\pm1} \cdots \sum_{S_N=\pm1} \mathrm{e}^{-\beta H} \equiv \sum_{\boldsymbol{S}} \mathrm{e}^{-\beta H}. \qquad (1.3)$$

One sometimes uses the notation Tr for the sum over all possible spin configurations appearing in (1.3). Hereafter we use this notation for the sum over the values of Ising spins on sites:

$$Z = \mathrm{Tr}\, \mathrm{e}^{-\beta H}. \qquad (1.4)$$

Equation (1.2) is called the *Gibbs–Boltzmann distribution*, and $\mathrm{e}^{-\beta H}$ is termed the *Boltzmann factor*. We write the expectation value for the Gibbs–Boltzmann distribution using angular brackets $\langle \cdots \rangle$.

Spin variables are not necessarily restricted to the Ising type ($S_i = \pm1$). For instance, in the *XY model*, the variable at a site i has a real value θ_i with modulo 2π, and the interaction energy has the form $-J\cos(\theta_i - \theta_j)$. The energy due to an external field is $-h\cos\theta_i$. The Hamiltonian of the *XY* model is thus written as

$$H = -J \sum_{(ij)\in B} \cos(\theta_i - \theta_j) - h \sum_i \cos\theta_i. \qquad (1.5)$$

The *XY* spin variable θ_i can be identified with a point on the unit circle. If $J > 0$, the interaction term in (1.5) is ferromagnetic as it favours a parallel spin configuration ($\theta_i = \theta_j$).

1.2 Order parameter and phase transition

One of the most important quantities used to characterize the macroscopic properties of the Ising model with ferromagnetic interactions is the *magnetization*. Magnetization is defined by

$$m = \frac{1}{N} \left\langle \sum_{i=1}^{N} S_i \right\rangle = \frac{1}{N} \mathrm{Tr}\left((\sum_i S_i) P(\boldsymbol{S}) \right), \qquad (1.6)$$

and measures the overall ordering in a macroscopic system (i.e. the system in the thermodynamic limit $N \to \infty$). Magnetization is a typical example of an *order parameter* which is a measure of whether or not a macroscopic system is in an ordered state in an appropriate sense. The magnetization vanishes if there exist equal numbers of up spins $S_i = 1$ and down spins $S_i = -1$, suggesting the absence of a uniformly ordered state.

At low temperatures $\beta \gg 1$, the Gibbs–Boltzmann distribution (1.2) implies that low-energy states are realized with much higher probability than high-energy

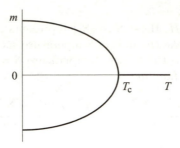

FIG. 1.2. Temperature dependence of magnetization

states. The low-energy states of the ferromagnetic Ising model (1.1) without the external field $h = 0$ have almost all spins in the same direction. Thus at low temperatures the spin states are either up $S_i = 1$ at almost all sites or down $S_i = -1$ at almost all sites. The magnetization m is then very close to either 1 or -1, respectively.

As the temperature increases, β decreases, and then the states with various energies emerge with similar probabilities. Under such circumstances, S_i would change frequently from 1 to -1 and vice versa, so that the macroscopic state of the system is disordered with the magnetization vanishing. The magnetization m as a function of the temperature T therefore has the behaviour depicted in Fig. 1.2. There is a critical temperature T_c; $m \neq 0$ for $T < T_c$ and $m = 0$ for $T > T_c$.

This type of phenomenon in a macroscopic system is called a *phase transition* and is characterized by a sharp and singular change of the value of the order parameter between vanishing and non-vanishing values. In magnetic systems the state for $T < T_c$ with $m \neq 0$ is called the *ferromagnetic phase* and the state at $T > T_c$ with $m = 0$ is called the *paramagnetic phase*. The temperature T_c is termed a *critical point* or a *transition point*.

1.3   Mean-field theory

In principle, it is possible to calculate the expectation value of any physical quantity using the Gibbs–Boltzmann distribution (1.2). It is, however, usually very difficult in practice to carry out the sum over 2^N terms appearing in the partition function (1.3). One is thus often forced to resort to approximations. *Mean-field theory* (or the mean-field approximation) is used widely in such situations.

1.3.1   *Mean-field Hamiltonian*

The essence of mean-field theory is to neglect fluctuations of microscopic variables around their mean values. One splits the spin variable S_i into the mean $m = \sum_i \langle S_i \rangle / N = \langle S_i \rangle$ and the deviation (fluctuation) $\delta S_i = S_i - m$ and assumes that the second-order term with respect to the fluctuation δS_i is negligibly small in the interaction energy:

$$H = -J \sum_{(ij) \in B} (m + \delta S_i)(m + \delta S_j) - h \sum_i S_i$$

$$\approx -Jm^2 N_B - Jm \sum_{(ij) \in B} (\delta S_i + \delta S_j) - h \sum_i S_i. \tag{1.7}$$

To simplify this expression, we note that each bond (ij) appears only once in the sum of $\delta S_i + \delta S_j$ in the second line. Thus δS_i and δS_j assigned at both ends of a bond are summed up z times, where z is the number of bonds emanating from a given site (the *coordination number*), in the second sum in the final expression of (1.7):

$$H = -Jm^2 N_B - Jmz \sum_i \delta S_i - h \sum_i S_i$$

$$= N_B Jm^2 - (Jmz + h) \sum_i S_i. \tag{1.8}$$

A few comments on (1.8) are in order.

1. N_B is the number of elements in the set of bonds B, $N_B = |B|$.
2. We have assumed that the coordination number z is independent of site i, so that N_B is related to z by $zN/2 = N_B$. One might imagine that the total number of bonds is zN since each site has z bonds emanating from it. However, a bond is counted twice at both its ends and one should divide zN by two to count the total number of bonds correctly.
3. The expectation value $\langle S_i \rangle$ has been assumed to be independent of i. This value should be equal to m according to (1.6). In the conventional ferromagnetic Ising model, the interaction J is a constant and thus the average order of spins is uniform in space. In spin glasses and other cases to be discussed later this assumption does not hold.

The effects of interactions have now been hidden in the magnetization m in the mean-field Hamiltonian (1.8). The problem apparently looks like a non-interacting case, which significantly reduces the difficulties in analytical manipulations.

1.3.2 *Equation of state*

The mean-field Hamiltonian (1.8) facilitates calculations of various quantities. For example, the partition function is given as

$$Z = \mathrm{Tr} \exp \left[\beta \left\{ -N_B Jm^2 + (Jmz + h) \sum_i S_i \right\} \right]$$

$$= \mathrm{e}^{-\beta N_B Jm^2} \left\{ 2 \cosh \beta (Jmz + h) \right\}^N. \tag{1.9}$$

A similar procedure with S_i inserted after the trace operation Tr in (1.9) yields the magnetization m,

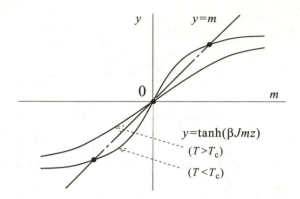

FIG. 1.3. Solution of the mean-field equation of state

$$m = \frac{\text{Tr} S_i e^{-\beta H}}{Z} = \tanh \beta (Jmz + h). \tag{1.10}$$

This equation (1.10) determines the order parameter m and is called the *equation of state*. The magnetization in the absence of the external field $h = 0$, the *spontaneous magnetization*, is obtained as the solution of (1.10) graphically: as one can see in Fig. 1.3, the existence of a solution with non-vanishing magnetization $m \neq 0$ is determined by whether the slope of the curve $\tanh(\beta Jmz)$ at $m = 0$ is larger or smaller than unity. The first term of the expansion of the right hand side of (1.10) with $h = 0$ is βJzm, so that there exists a solution with $m \neq 0$ if and only if $\beta Jz > 1$. From $\beta Jz = Jz/T = 1$, the critical temperature is found to be $T_c = Jz$. Figure 1.3 clearly shows that the positive and negative solutions for m have the same absolute value ($\pm m$), corresponding to the change of sign of all spins ($S_i \rightarrow -S_i, \forall i$). Hereafter we often restrict ourselves to the case of $m > 0$ without loss of generality.

1.3.3 *Free energy and the Landau theory*

It is possible to calculate the specific heat C, magnetic susceptibility χ, and other quantities by mean-field theory. We develop an argument starting from the free energy. The general theory of statistical mechanics tells us that the free energy is proportional to the logarithm of the partition function. Using (1.9), we have the mean-field free energy of the Ising model as

$$F = -T \log Z = -NT \log\{2 \cosh \beta (Jmz + h)\} + N_B Jm^2. \tag{1.11}$$

When there is no external field $h = 0$ and the temperature T is close to the critical point T_c, the magnetization m is expected to be close to zero. It is then possible to expand the right hand side of (1.11) in powers of m. The expansion to fourth order is

$$F = -NT \log 2 + \frac{JzN}{2}(1 - \beta Jz)m^2 + \frac{N}{12}(Jzm)^4 \beta^3. \tag{1.12}$$

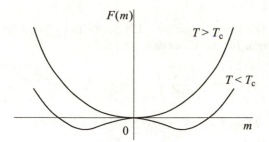

FIG. 1.4. Free energy as a function of the order parameter

It should be noted that the coefficient of m^2 changes sign at T_c. As one can see in Fig. 1.4, the minima of the free energy are located at $m \neq 0$ when $T < T_c$ and at $m = 0$ if $T > T_c$. The statistical-mechanical average of a physical quantity obtained from the Gibbs–Boltzmann distribution (1.2) corresponds to its value at the state that minimizes the free energy (*thermal equilibrium state*). Thus the magnetization in thermal equilibrium is zero when $T > T_c$ and is non-vanishing for $T < T_c$. This conclusion is in agreement with the previous argument using the equation of state. The present theory starting from the Taylor expansion of the free energy by the order parameter is called the *Landau theory* of phase transitions.

1.4 Infinite-range model

Mean-field theory is an approximation. However, it gives the exact solution in the case of the *infinite-range model* where all possible pairs of sites have interactions. The Hamiltonian of the infinite-range model is

$$H = -\frac{J}{2N} \sum_{i \neq j} S_i S_j - h \sum_i S_i. \tag{1.13}$$

The first sum on the right hand side runs over all pairs of different sites (i, j) $(i = 1, \ldots, N; j = 1, \ldots, N; i \neq j)$. The factor 2 in the denominator exists so that each pair (i, j) appears only once in the sum, for example $(S_1 S_2 + S_2 S_1)/2 = S_1 S_2$. The factor N in the denominator is to make the Hamiltonian (energy) extensive (i.e. $\mathcal{O}(N)$) since the number of terms in the sum is $N(N-1)/2$.

The partition function of the infinite-range model can be evaluated as follows. By definition,

$$Z = \text{Tr} \exp\left(\frac{\beta J}{2N}\left(\sum_i S_i\right)^2 - \frac{\beta J}{2} + \beta h \sum_i S_i\right). \tag{1.14}$$

Here the constant term $-\beta J/2$ compensates for the contribution $\sum_i (S_i^2)$. This term, of $\mathcal{O}(N^0 = 1)$, is sufficiently small compared to the other terms, of $\mathcal{O}(N)$, in the thermodynamic limit $N \to \infty$ and will be neglected hereafter. Since we cannot carry out the trace operation with the term $(\sum_i S_i)^2$ in the exponent, we decompose this term by the *Gaussian integral*

$$e^{ax^2/2} = \sqrt{\frac{aN}{2\pi}} \int_{-\infty}^{\infty} dm\, e^{-Nam^2/2+\sqrt{N}amx}. \qquad (1.15)$$

Substituting $a = \beta J$ and $x = \sum_i S_i/\sqrt{N}$ and using (1.9), we find

$$\mathrm{Tr}\, \sqrt{\frac{\beta J N}{2\pi}} \int_{-\infty}^{\infty} dm\, \exp\left(-\frac{N\beta J m^2}{2} + \beta J m \sum_i S_i + \beta h \sum_i S_i\right) \qquad (1.16)$$

$$= \sqrt{\frac{\beta J N}{2\pi}} \int_{-\infty}^{\infty} dm\, \exp\left(-\frac{N\beta J m^2}{2} + N \log\{2\cosh\beta(Jm+h)\}\right). \,(1.17)$$

The problem has thus been reduced to a simple single integral.

We can evaluate the above integral by steepest descent in the thermodynamic limit $N \to \infty$: the integral (1.17) approaches asymptotically the largest value of its integrand in the thermodynamic limit. The value of the integration variable m that gives the maximum of the integrand is determined by the saddle-point condition, that is maximization of the exponent:

$$\frac{\partial}{\partial m}\left(-\frac{\beta J}{2}m^2 + \log\{2\cosh\beta(Jm+h)\}\right) = 0 \qquad (1.18)$$

or

$$m = \tanh\beta(Jm+h). \qquad (1.19)$$

Equation (1.19) agrees with the mean-field equation (1.10) after replacement of J with J/N and z with N. Thus mean-field theory leads to the exact solution for the infinite-range model.

The quantity m was introduced as an integration variable in the evaluation of the partition function of the infinite-range model. It nevertheless turned out to have a direct physical interpretation, the magnetization, according to the correspondence with mean-field theory through the equation of state (1.19). To understand the significance of this interpretation from a different point of view, we write the saddle-point condition for (1.16) as

$$m = \frac{1}{N}\sum_i S_i. \qquad (1.20)$$

The sum in (1.20) agrees with the average value m, the magnetization, in the thermodynamic limit $N \to \infty$ if the law of large numbers applies. In other words, fluctuations of magnetization vanish in the thermodynamic limit in the infinite-range model and thus mean-field theory gives the exact result.

The infinite-range model may be regarded as a model with nearest neighbour interactions in infinite-dimensional space. To see this, note that the coordination number z of a site on the d-dimensional hypercubic lattice is proportional to d. More precisely, $z = 4$ for the two-dimensional square lattice, $z = 6$ for the three-dimensional cubic lattice, and $z = 2d$ in general. Thus a site is connected to very many other sites for large d so that the relative effects of fluctuations diminish in the limit of large d, leading to the same behaviour as the infinite-range model.

1.5 Variational approach

Another point of view is provided for mean-field theory by a variational approach. The source of difficulty in calculations of various physical quantities lies in the non-trivial structure of the probability distribution (1.2) with the Hamiltonian (1.1) where the degrees of freedom $\boldsymbol{S}$ are coupled with each other. It may thus be useful to employ an approximation to decouple the distribution into simple functions. We therefore introduce a single-site distribution function

$$P_i(\sigma_i) = \operatorname{Tr} P(\boldsymbol{S})\delta(S_i, \sigma_i) \tag{1.21}$$

and approximate the full distribution by the product of single-site functions:

$$P(\boldsymbol{S}) \approx \prod_i P_i(S_i). \tag{1.22}$$

We determine $P_i(S_i)$ by the general principle of statistical mechanics to minimize the free energy $F = E - TS$, where the internal energy E is the expectation value of the Hamiltonian and S is the entropy (not to be confused with spin). Under the above approximation, one finds

$$F = \operatorname{Tr}\left\{H(\boldsymbol{S})\prod_i P_i(S_i)\right\} + T\operatorname{Tr}\left\{\prod_i P_i(S_i)\sum_i \log P_i(S_i)\right\}$$

$$= -J\sum_{(ij)\in B} \operatorname{Tr} S_i S_j P_i(S_i)P_j(S_j) - h\sum_i \operatorname{Tr} S_i P_i(S_i)$$

$$+ T\sum_i \operatorname{Tr} P_i(S_i)\log P_i(S_i), \tag{1.23}$$

where we have used the normalization $\operatorname{Tr} P_i(S_i) = 1$. Variation of this free energy by $P_i(S_i)$ under the condition of normalization gives

$$\frac{\delta F}{\delta P_i(S_i)} = -J\sum_{j\in I} S_i m_j - hS_i + T\log P_i(S_i) + T + \lambda = 0, \tag{1.24}$$

where λ is the Lagrange multiplier for the normalization condition and we have written m_j for $\operatorname{Tr} S_j P_j(S_j)$. The set of sites connected to i has been denoted by I. The minimization condition (1.24) yields the distribution function

$$P_i(S_i) = \frac{\exp\left(\beta J\sum_{j\in I} S_i m_j + \beta h S_i\right)}{Z_{\mathrm{MF}}}, \tag{1.25}$$

where Z_{MF} is the normalization factor. In the case of uniform magnetization $m_j (= m)$, this result (1.25) together with the decoupling (1.22) leads to the distribution $P(S) \propto e^{-\beta H}$ with H identical to the mean-field Hamiltonian (1.8) up to a trivial additive constant.

The argument so far has been general in that it did not use the values of the Ising spins $S_i = \pm 1$ and thus applies to any other cases. It is instructive to use the values of the Ising spins explicitly and see its consequence. Since S_i takes only two values ± 1, the following is the general form of the distribution function:

$$P_i(S_i) = \frac{1 + m_i S_i}{2}, \tag{1.26}$$

which is compatible with the previous notation $m_i = \mathrm{Tr}\, S_i P_i(S_i)$. Substitution of (1.26) into (1.23) yields

$$F = -J \sum_{(ij) \in B} m_i m_j - h \sum_i m_i$$
$$+ T \sum_i \left(\frac{1 + m_i}{2} \log \frac{1 + m_i}{2} + \frac{1 - m_i}{2} \log \frac{1 - m_i}{2} \right). \tag{1.27}$$

Variation of this expression with respect to m_i leads to

$$m_i = \tanh \beta \left(J \sum_{j \in I} m_j + h \right) \tag{1.28}$$

which is identical to (1.10) for uniform magnetization ($m_i = m$, $\forall i$). We have again rederived the previous result of mean-field theory.

Bibliographical note

A compact exposition of the theory of phase transitions including mean-field theory is found in Yeomans (1992). For a full account of the theory of phase transitions and critical phenomena, see Stanley (1987). In Opper and Saad (2001), one finds an extensive coverage of recent developments in applications of mean-field theory to interdisciplinary fields as well as a detailed elucidation of various aspects of mean-field theory.

2

MEAN-FIELD THEORY OF SPIN GLASSES

We next discuss the problem of spin glasses. If the interactions between spins are not uniform in space, the analysis of the previous chapter does not apply in the naïve form. In particular, when the interactions are ferromagnetic for some bonds and antiferromagnetic for others, then the spin orientation cannot be uniform in space, unlike the ferromagnetic system, even at low temperatures. Under such a circumstance it sometimes happens that spins become randomly frozen—random in space but frozen in time. This is the intuitive picture of the spin glass phase. In the present chapter we investigate the condition for the existence of the spin glass phase by extending the mean-field theory so that it is applicable to the problem of disordered systems with random interactions. In particular we elucidate the properties of the so-called replica-symmetric solution. The replica method introduced here serves as a very powerful tool of analysis throughout this book.

2.1 Spin glass and the Edwards–Anderson model

Atoms are located on lattice points at regular intervals in a crystal. This is not the case in glasses where the positions of atoms are random in space. An important point is that in glasses the apparently random locations of atoms do not change in a day or two into another set of random locations. A state with spatial randomness apparently does not change with time. The term *spin glass* implies that the spin orientation has a similarity to this type of location of atom in glasses: spins are randomly frozen in spin glasses. The goal of the theory of spin glasses is to clarify the conditions for the existence of spin glass states.[1]

It is established within mean-field theory that the spin glass phase exists at low temperatures when random interactions of certain types exist between spins. The present and the next chapters are devoted to the mean-field theory of spin glasses. We first introduce a model of random systems and explain the replica method, a general method of analysis of random systems. Then the replica-symmetric solution is presented.

---

[1]More rigorously, the spin glass state is considered stable for an infinitely long time at least within the mean-field theory, whereas ordinary glasses will transform to crystals without randomness after a very long period.

11

2.1.1   *Edwards–Anderson model*

Let us suppose that the interaction J_{ij} between a spin pair (ij) changes from one pair to another. The Hamiltonian in the absence of an external field is then expressed as

$$H = - \sum_{(ij) \in B} J_{ij} S_i S_j. \tag{2.1}$$

The spin variables are assumed to be of the Ising type ($S_i = \pm 1$) here. Each J_{ij} is supposed to be distributed independently according to a probability distribution $P(J_{ij})$. One often uses the *Gaussian model* and the $\pm J$ *model* as typical examples of the distribution of $P(J_{ij})$. Their explicit forms are

$$P(J_{ij}) = \frac{1}{\sqrt{2\pi J^2}} \exp\left\{ -\frac{(J_{ij} - J_0)^2}{2J^2} \right\} \tag{2.2}$$

$$P(J_{ij}) = p\delta(J_{ij} - J) + (1 - p)\delta(J_{ij} + J), \tag{2.3}$$

respectively. Equation (2.2) is a Gaussian distribution with mean J_0 and variance J^2 while in (2.3) J_{ij} is either J (> 0) (with probability p) or $-J$ (with probability $1 - p$).

Randomness in J_{ij} has various types of origin depending upon the specific problem. For example, in some spin glass materials, the positions of atoms carrying spins are randomly distributed, resulting in randomness in interactions. It is impossible in such a case to identify the location of each atom precisely and therefore it is essential in theoretical treatments to introduce a probability distribution for J_{ij}. In such a situation (2.1) is called the *Edwards–Anderson model* (Edwards and Anderson 1975). The randomness in site positions (site randomness) is considered less relevant to the macroscopic properties of spin glasses compared to the randomness in interactions (bond randomness). Thus J_{ij} is supposed to be distributed randomly and independently at each bond (ij) according to a probability like (2.2) and (2.3). The Hopfield model of neural networks treated in Chapter 7 also has the form of (2.1). The type of randomness of J_{ij} in the Hopfield model is different from that of the Edwards–Anderson model. The randomness in J_{ij} of the Hopfield model has its origin in the randomness of memorized patterns. We focus our attention on the spin glass problem in Chapters 2 to 4.

2.1.2   *Quenched system and configurational average*

Evaluation of a physical quantity using the Hamiltonian (2.1) starts from the trace operation over the spin variables $\boldsymbol{S} = \{S_i\}$ for a given fixed (*quenched*) set of J_{ij} generated by the probability distribution $P(J_{ij})$. For instance, the free energy is calculated as

$$F = -T \log \text{Tr} \, e^{-\beta H}, \tag{2.4}$$

which is a function of $\boldsymbol{J} \equiv \{J_{ij}\}$. The next step is to average (2.4) over the distribution of $\boldsymbol{J}$ to obtain the final expression of the free energy. The latter procedure of averaging is called the *configurational average* and will be denoted by brackets $[\cdots]$ in this book,

$$[F] = -T[\log Z] = -T \int \prod_{(ij)} \mathrm{d}J_{ij}\, P(J_{ij}) \log Z. \tag{2.5}$$

Differentiation of this averaged free energy $[F]$ by the external field h or the temperature T leads to the magnetization or the internal energy, respectively. The reason to trace out first S for a given fixed J is that the positions of atoms carrying spins are random in space but fixed in the time scale of rapid thermal motions of spins. It is thus appropriate to evaluate the trace over S first with the interactions J fixed.

It happens that the free energy per degree of freedom $f(J) = F(J)/N$ has vanishingly small deviations from its mean value $[f]$ in the thermodynamic limit $N \to \infty$. The free energy f for a given J thus agrees with the mean $[f]$ with probability 1, which is called the *self-averaging property* of the free energy. Since the raw value f for a given J agrees with its configurational average $[f]$ with probability 1 in the thermodynamic limit, we may choose either of these quantities in actual calculations. The mean $[f]$ is easier to handle because it has no explicit dependence on J even for finite-size systems. We shall treat the average free energy in most of the cases hereafter.

2.1.3 *Replica method*

The dependence of $\log Z$ on J is very complicated and it is not easy to calculate the configurational average $[\log Z]$. The manipulations are greatly facilitated by the relation

$$[\log Z] = \lim_{n \to 0} \frac{[Z^n] - 1}{n}. \tag{2.6}$$

One prepares n replicas of the original system, evaluates the configurational average of the product of their partition functions Z^n, and then takes the limit $n \to 0$. This technique, the *replica method*, is useful because it is easier to evaluate $[Z^n]$ than $[\log Z]$.

Equation (2.6) is an identity and is always correct. A problem in actual replica calculations is that one often evaluates $[Z^n]$ with positive integer n in mind and then extrapolates the result to $n \to 0$. We therefore should be careful to discuss the significance of the results of replica calculations.

2.2 Sherrington–Kirkpatrick model

The mean-field theory of spin glasses is usually developed for the *Sherrington–Kirkpatrick* (SK) *model*, the infinite-range version of the Edwards–Anderson model (Sherrington and Kirkpatrick 1975). In this section we introduce the SK model and explain the basic methods of calculations using the replica method.

2.2.1 SK model

The infinite-range model of spin glasses is expected to play the role of mean-field theory analogously to the case of the ferromagnetic Ising model. We therefore start from the Hamiltonian

$$H = -\sum_{i<j} J_{ij} S_i S_j - h \sum_i S_i. \tag{2.7}$$

The first sum on the right hand side runs over all distinct pairs of spins, $N(N-1)/2$ of them. The interaction J_{ij} is a quenched variable with the Gaussian distribution function

$$P(J_{ij}) = \frac{1}{J}\sqrt{\frac{N}{2\pi}} \exp\left\{-\frac{N}{2J^2}\left(J_{ij} - \frac{J_0}{N}\right)^2\right\}. \tag{2.8}$$

The mean and variance of this distribution are both proportional to $1/N$:

$$[J_{ij}] = \frac{J_0}{N}, \quad [(\Delta J_{ij})^2] = \frac{J^2}{N}. \tag{2.9}$$

The reason for such a normalization is that extensive quantities (e.g. the energy and specific heat) are found to be proportional to N if one takes the above normalization of interactions, as we shall see shortly.

2.2.2 Replica average of the partition function

According to the prescription of the replica method, one first has to take the configurational average of the nth power of the partition function

$$[Z^n] = \int \left(\prod_{i<j} dJ_{ij} P(J_{ij})\right) \operatorname{Tr} \exp\left(\beta \sum_{i<j} J_{ij} \sum_{\alpha=1}^n S_i^\alpha S_j^\alpha + \beta h \sum_{i=1}^N \sum_{\alpha=1}^n S_i^\alpha\right), \tag{2.10}$$

where α is the replica index. The integral over J_{ij} can be carried out independently for each (ij) using (2.8). The result, up to a trivial constant, is

$$\operatorname{Tr} \exp\left\{\frac{1}{N}\sum_{i<j}\left(\frac{1}{2}\beta^2 J^2 \sum_{\alpha,\beta} S_i^\alpha S_j^\alpha S_i^\beta S_j^\beta + \beta J_0 \sum_\alpha S_i^\alpha S_j^\alpha\right) + \beta h \sum_i \sum_\alpha S_i^\alpha\right\}. \tag{2.11}$$

By rewriting the sums over $i < j$ and α, β in the above exponent, we find the following form, for sufficiently large N:

$$[Z^n] = \exp\left(\frac{N\beta^2 J^2 n}{4}\right) \operatorname{Tr} \exp\left\{\frac{\beta^2 J^2}{2N} \sum_{\alpha<\beta}\left(\sum_i S_i^\alpha S_i^\beta\right)^2\right.$$

$$\left. + \frac{\beta J_0}{2N} \sum_\alpha \left(\sum_i S_i^\alpha\right)^2 + \beta h \sum_i \sum_\alpha S_i^\alpha\right\}. \tag{2.12}$$

2.2.3 Reduction by Gaussian integral

We could carry out the trace over S_i^α independently at each site i in (2.12) if the quantities in the exponent were linear in the spin variables. It is therefore useful to linearize those squared quantities by Gaussian integrals with the integration variables $q_{\alpha\beta}$ for the term $(\sum_i S_i^\alpha S_i^\beta)^2$ and m_α for $(\sum_i S_i^\alpha)^2$:

$$[Z^n] = \exp\left(\frac{N\beta^2 J^2 n}{4}\right) \int \prod_{\alpha<\beta} dq_{\alpha\beta} \int \prod_\alpha dm_\alpha$$

$$\cdot \exp\left(-\frac{N\beta^2 J^2}{2}\sum_{\alpha<\beta} q_{\alpha\beta}^2 - \frac{N\beta J_0}{2}\sum_\alpha m_\alpha^2\right)$$

$$\cdot \mathrm{Tr}\exp\left(\beta^2 J^2 \sum_{\alpha<\beta} q_{\alpha\beta}\sum_i S_i^\alpha S_i^\beta + \beta\sum_\alpha (J_0 m_\alpha + h)\sum_i S_i^\alpha\right). \quad (2.13)$$

If we represent the sum over the variable at a single site $(\sum_{S_i^\alpha})$ also by the symbol Tr, the third line of the above equation is

$$\left\{\mathrm{Tr}\exp\left(\beta^2 J^2 \sum_{\alpha<\beta} q_{\alpha\beta}S^\alpha S^\beta + \beta\sum_\alpha (J_0 m_\alpha + h)S^\alpha\right)\right\}^N \equiv \exp(N\log\mathrm{Tr}\,e^L),$$
$$(2.14)$$

where

$$L = \beta^2 J^2 \sum_{\alpha<\beta} q_{\alpha\beta}S^\alpha S^\beta + \beta\sum_\alpha (J_0 m_\alpha + h)S^\alpha. \quad (2.15)$$

We thus have

$$[Z^n] = \exp\left(\frac{N\beta^2 J^2 n}{4}\right) \int \prod_{\alpha<\beta} dq_{\alpha\beta} \int \prod_\alpha dm_\alpha$$

$$\cdot \exp\left(-\frac{N\beta^2 J^2}{2}\sum_{\alpha<\beta} q_{\alpha\beta}^2 - \frac{N\beta J_0}{2}\sum_\alpha m_\alpha^2 + N\log\mathrm{Tr}\,e^L\right). \quad (2.16)$$

2.2.4 Steepest descent

The exponent of the above integrand is proportional to N, so that it is possible to evaluate the integral by steepest descent. We then find in the thermodynamic limit $N \to \infty$

$$[Z^n] \approx \exp\left(-\frac{N\beta^2 J^2}{2}\sum_{\alpha<\beta} q_{\alpha\beta}^2 - \frac{N\beta J_0}{2}\sum_\alpha m_\alpha^2 + N\log\mathrm{Tr}\,e^L + \frac{N}{4}\beta^2 J^2 n\right)$$

$$\approx 1 + Nn \left\{ -\frac{\beta^2 J^2}{4n} \sum_{\alpha \neq \beta} q_{\alpha\beta}^2 - \frac{\beta J_0}{2n} \sum_{\alpha} m_{\alpha}^2 + \frac{1}{n} \log \operatorname{Tr} e^L + \frac{1}{4}\beta^2 J^2 \right\}.$$

In deriving this last expression, the limit $n \to 0$ has been taken with N kept very large but finite. The values of $q_{\alpha\beta}$ and m_{α} in the above expression should be chosen to extremize (maximize or minimize) the quantity in the braces $\{\ \}$. The replica method therefore gives the free energy as

$$-\beta[f] = \lim_{n \to 0} \frac{[Z^n] - 1}{nN} = \lim_{n \to 0} \left\{ -\frac{\beta^2 J^2}{4n} \sum_{\alpha \neq \beta} q_{\alpha\beta}^2 \right.$$

$$\left. -\frac{\beta J_0}{2n} \sum_{\alpha} m_{\alpha}^2 + \frac{1}{4}\beta^2 J^2 + \frac{1}{n} \log \operatorname{Tr} e^L \right\}. \tag{2.17}$$

The saddle-point condition that the free energy is extremized with respect to the variable $q_{\alpha\beta}$ $(\alpha \neq \beta)$ is

$$q_{\alpha\beta} = \frac{1}{\beta^2 J^2} \frac{\partial}{\partial q_{\alpha\beta}} \log \operatorname{Tr} e^L = \frac{\operatorname{Tr} S^\alpha S^\beta e^L}{\operatorname{Tr} e^L} = \langle S^\alpha S^\beta \rangle_L, \tag{2.18}$$

where $\langle \cdots \rangle_L$ is the average by the weight e^L. The saddle-point condition for m_{α} is, similarly,

$$m_{\alpha} = \frac{1}{\beta J_0} \frac{\partial}{\partial m_{\alpha}} \log \operatorname{Tr} e^L = \frac{\operatorname{Tr} S^\alpha e^L}{\operatorname{Tr} e^L} = \langle S^\alpha \rangle_L. \tag{2.19}$$

2.2.5   Order parameters

The variables $q_{\alpha\beta}$ and m_{α} have been introduced for technical convenience in Gaussian integrals. However, these variables turn out to represent order parameters in a similar manner to the ferromagnetic model explained in §1.4. To confirm this fact, we first note that (2.18) can be written in the following form:

$$q_{\alpha\beta} = [\langle S_i^\alpha S_i^\beta \rangle] = \left[\frac{\operatorname{Tr} S_i^\alpha S_i^\beta \exp(-\beta \sum_\gamma H_\gamma)}{\operatorname{Tr} \exp(-\beta \sum_\gamma H_\gamma)} \right], \tag{2.20}$$

where H_γ denotes the γth replica Hamiltonian

$$H_\gamma = -\sum_{i<j} J_{ij} S_i^\gamma S_j^\gamma - h \sum_i S_i^\gamma. \tag{2.21}$$

It is possible to check, by almost the same calculations as in the previous sections, that (2.18) and (2.20) represent the same quantity. First of all the denominator of (2.20) is Z^n, which approaches one as $n \to 0$ so that it is unnecessary to evaluate them explicitly. The numerator corresponds to the quantity in the calculation of $[Z^n]$ with $S_i^\alpha S_i^\beta$ inserted after the Tr symbol. With these points in mind, one can follow the calculations in §2.2.2 and afterwards to find the following quantity instead of (2.14):

$$(\mathrm{Tr}\,\mathrm{e}^L)^{N-1} \cdot \mathrm{Tr}\,(S^\alpha S^\beta \mathrm{e}^L). \tag{2.22}$$

The quantity $\log \mathrm{Tr}\,\mathrm{e}^L$ is proportional to n as is seen from (2.17) and thus $\mathrm{Tr}\,\mathrm{e}^L$ approaches one as $n \to 0$. Hence (2.22) reduces to $\mathrm{Tr}\,(S^\alpha S^\beta \mathrm{e}^L)$ in the limit $n \to 0$. One can then check that (2.22) agrees with (2.18) from the fact that the denominator of (2.18) approaches one. We have thus established that (2.20) and (2.18) represent the same quantity. Similarly we find

$$m_\alpha = [\langle S_i^\alpha \rangle]. \tag{2.23}$$

The parameter m is the ordinary ferromagnetic order parameter according to (2.23), and is the value of m_α when the latter is independent of α. The other parameter $q_{\alpha\beta}$ is the *spin glass order parameter*. This may be understood by remembering that traces over all replicas other than α and β cancel out in the denominator and numerator in (2.20). One then finds

$$q_{\alpha\beta} = \left[\frac{\mathrm{Tr}\,S_i^\alpha \mathrm{e}^{-\beta H_\alpha}}{\mathrm{Tr}\,\mathrm{e}^{-\beta H_\alpha}} \frac{\mathrm{Tr}\,S_i^\beta \mathrm{e}^{-\beta H_\beta}}{\mathrm{Tr}\,\mathrm{e}^{-\beta H_\beta}} \right] = [\langle S_i^\alpha \rangle \langle S_i^\beta \rangle] = [\langle S_i \rangle^2] \equiv q \tag{2.24}$$

if we cannot distinguish one replica from another. In the paramagnetic phase at high temperatures, $\langle S_i \rangle$ (which is $\langle S_i^\alpha \rangle$ for any single α) vanishes at each site i and therefore $m = q = 0$. The ferromagnetic phase has ordering almost uniform in space, and if we choose that orientation of ordering as the positive direction, then $\langle S_i \rangle > 0$ at most sites. This implies $m > 0$ and $q > 0$.

If the spin glass phase characteristic of the Edwards–Anderson model or the SK model exists, the spins in that phase should be randomly frozen. In the spin glass phase $\langle S_i \rangle$ is not vanishing at any site because the spin does not fluctuate significantly in time. However, the sign of this expectation value would change from site to site, and such an apparently random spin pattern does not change in time. The spin configuration frozen in time is replaced by another frozen spin configuration for a different set of interactions $\boldsymbol{J}$ since the environment of a spin changes drastically. Thus the configurational average of $\langle S_i \rangle$ over the distribution of $\boldsymbol{J}$ corresponds to the average over various environments at a given spin, which would yield both $\langle S_i \rangle > 0$ and $\langle S_i \rangle < 0$, suggesting the possibility of $m = [\langle S_i \rangle] = 0$. The spin glass order parameter q is not vanishing in the same situation because it is the average of a positive quantity $\langle S_i \rangle^2$. Thus there could exist a phase with $m = 0$ and $q > 0$, which is the spin glass phase with q as the spin glass order parameter. It will indeed be shown that the equations of state for the SK model have a solution with $m = 0$ and $q > 0$.

2.3   Replica-symmetric solution

2.3.1   *Equations of state*

It is necessary to have the explicit dependence of $q_{\alpha\beta}$ and m_α on replica indices α and β in order to calculate the free energy and order parameters from (2.17) to (2.19). Naïvely, the dependence on these replica indices should not affect the physics of the system because replicas have been introduced artificially for the convenience of the configurational average. It therefore seems natural to assume *replica symmetry* (RS), $q_{\alpha\beta} = q$ and $m_\alpha = m$ (which we used in the previous section), and derive the replica-symmetric solution.

The free energy (2.17) is, before taking the limit $n \to 0$,

$$-\beta[f] = \frac{\beta^2 J^2}{4n} \left\{-n(n-1)q^2\right\} - \frac{\beta J_0}{2n} nm^2 + \frac{1}{n} \log \mathrm{Tr}\, e^L + \frac{1}{4}\beta^2 J^2. \qquad (2.25)$$

The third term on the right hand side can be evaluated using the definition of L, (2.15), and a Gaussian integral as

$$\log \mathrm{Tr}\, e^L = \log \mathrm{Tr}\, \sqrt{\frac{\beta^2 J^2 q}{2\pi}} \int dz$$

$$\cdot \exp \left(-\frac{\beta^2 J^2 q}{2} z^2 + \beta^2 J^2 q z \sum_\alpha S^\alpha - \frac{n}{2}\beta^2 J^2 q + \beta(J_0 m + h) \sum_\alpha S^\alpha\right)$$

$$= \log \int Dz \exp \left(n \log 2 \cosh(\beta J \sqrt{q} z + \beta J_0 m + \beta h) - \frac{n}{2}\beta^2 J^2 q\right)$$

$$= \log \left(1 + n \int Dz \log 2 \cosh \beta \tilde{H}(z) - \frac{n}{2}\beta^2 J^2 q + \mathcal{O}(n^2)\right). \qquad (2.26)$$

Here $Dz = dz \exp(-z^2/2)/\sqrt{2\pi}$ is the Gaussian measure and $\tilde{H}(z) = J\sqrt{q}z + J_0 m + h$. Inserting (2.26) into (2.25) and taking the limit $n \to 0$, we have

$$-\beta[f] = \frac{\beta^2 J^2}{4}(1-q)^2 - \frac{1}{2}\beta J_0 m^2 + \int Dz \log 2 \cosh \beta \tilde{H}(z). \qquad (2.27)$$

The extremization condition of the free energy (2.27) with respect to m is

$$m = \int Dz \tanh \beta \tilde{H}(z). \qquad (2.28)$$

This is the equation of state of the ferromagnetic order parameter m and corresponds to (2.19) with the trace operation being carried out explicitly. This operation is performed by inserting $q_{\alpha\beta} = q$ and $m_\alpha = m$ into (2.15) and taking the trace in the numerator of (2.18). The denominator reduces to one as $n \to 0$. It is convenient to decompose the double sum over α and β by a Gaussian integral.

Comparison of (2.28) with the equation of state for a single spin in a field $m = \tanh \beta h$ (obtained from (1.10) by setting $J = 0$) suggests the interpretation that the internal field has a Gaussian distribution due to randomness.

The extremization condition with respect to q is

$$\frac{\beta^2 J^2}{2}(q - 1) + \int Dz (\tanh \beta \tilde{H}(z)) \cdot \frac{\beta J}{2\sqrt{q}} z = 0, \qquad (2.29)$$

partial integration of which yields

$$q = 1 - \int Dz \operatorname{sech}^2 \beta \tilde{H}(z) = \int Dz \tanh^2 \beta \tilde{H}(z). \tag{2.30}$$

2.3.2 Phase diagram

The behaviour of the solution of the equations of state (2.28) and (2.30) is determined by the parameters β and J_0. For simplicity let us restrict ourselves to the case without external field $h = 0$ for the rest of this chapter. If the distribution of J_{ij} is symmetric ($J_0 = 0$), we have $\tilde{H}(z) = J\sqrt{q}z$ so that $\tanh \beta \tilde{H}(z)$ is an odd function. Then the magnetization vanishes ($m = 0$) and there is no ferromagnetic phase. The free energy is

$$-\beta[f] = \frac{1}{4}\beta^2 J^2 (1 - q)^2 + \int Dz \log 2 \cosh(\beta J \sqrt{q}z). \tag{2.31}$$

To investigate the properties of the system near the critical point where the spin glass order parameter q is small, it is convenient to expand the right hand side of (2.31) as

$$\beta[f] = -\frac{1}{4}\beta^2 J^2 - \log 2 - \frac{\beta^2 J^2}{4}(1 - \beta^2 J^2)q^2 + \mathcal{O}(q^3). \tag{2.32}$$

The Landau theory tells us that the critical point is determined by the condition of vanishing coefficient of the second-order term q^2 as we saw in (1.12). Hence the spin glass transition is concluded to exist at $T = J \equiv T_f$.

It should be noted that the coefficient of q^2 in (2.32) is negative if $T > T_f$. This means that the paramagnetic solution ($q = 0$) at high temperatures *maximizes* the free energy. Similarly the spin glass solution $q > 0$ for $T < T_f$ *maximizes* the free energy in the low-temperature phase. This pathological behaviour is a consequence of the replica method in the following sense. As one can see from (2.25), the coefficient of q^2, which represents the number of replica pairs, changes the sign at $n = 1$ and we have a negative number of pairs of replicas in the limit $n \to 0$, which causes maximization, instead of minimization, of the free energy. By contrast the coefficient of m does not change as in (2.25) and the free energy can be minimized with respect to this order parameter as is usually the case in statistical mechanics.

A ferromagnetic solution ($m > 0$) may exist if the distribution of J_{ij} is not symmetric around zero ($J_0 > 0$). Expanding the right hand side of (2.30) and keeping only the lowest order terms in q and m, we have

$$q = \beta^2 J^2 q + \beta^2 J_0^2 m^2. \tag{2.33}$$

If $J_0 = 0$, the critical point is identified as the temperature where the coefficient $\beta^2 J^2$ becomes one by the same argument as in §1.3.2. This result agrees with the conclusion already derived from the expansion of the free energy, $T_f = J$.

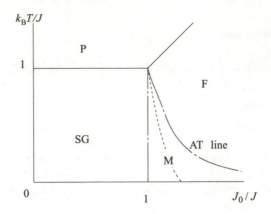

FIG. 2.1. Phase diagram of the SK model. The dashed line is the boundary between the ferromagnetic (F) and spin glass (SG) phases and exists only under the ansatz of replica symmetry. The dash–dotted lines will be explained in detail in the next chapter: the replica-symmetric solution is unstable below the AT line, and a mixed phase (M) emerges between the spin glass and ferromagnetic phases. The system is in the paramagnetic phase (P) in the high-temperature region.

If $J_0 > 0$ and $m > 0$, (2.33) implies $q = \mathcal{O}(m^2)$. We then expand the right hand side of the equation of state (2.28) bearing this in mind and keep only the lowest order term to obtain

$$m = \beta J_0 m + \mathcal{O}(q). \tag{2.34}$$

It has thus been shown that the ferromagnetic critical point, where m starts to assume a non-vanishing value, is $\beta J_0 = 1$ or $T_c = J_0$.

We have so far derived the boundaries between the paramagnetic and spin glass phases and between the paramagnetic and ferromagnetic phases. The boundary between the spin glass and ferromagnetic phases is given only by numerically solving (2.28) and (2.30). Figure 2.1 is the phase diagram thus obtained. The spin glass phase ($q > 0, m = 0$) exists as long as J_0 is smaller than J. This spin glass phase extends below the ferromagnetic phase in the range $J_0 > J$, which is called the *re-entrant transition*. The dashed line in Fig. 2.1 is the phase boundary between the spin glass and ferromagnetic phases representing the re-entrant transition. As explained in the next chapter, this dashed line actually disappears if we take into account the effects of replica symmetry breaking. Instead, the vertical line (separating the spin glass and mixed phases, shown dash–dotted) and the curve marked 'AT line' (dash–dotted) emerge under replica symmetry breaking. The properties of the mixed phase will be explained in the next chapter.

2.3.3 Negative entropy

Failure of the assumption of replica symmetry at low temperatures manifests itself in the negative value $-1/2\pi$ of the ground-state entropy for $J_0 = 0$. This is a clear inconsistency for the Ising model with discrete degrees of freedom. To verify this result from (2.31), we first derive the low-temperature form of the spin glass order parameter q. According to (2.30), q tends to one as $T \to 0$. We thus assume $q = 1 - aT$ $(a > 0)$ for T very small and check the consistency of this linear form. The q in $\text{sech}^2 \tilde{H}(z)$ of (2.30) can be approximated by one to leading order. Then we have, for $\beta \to \infty$,

$$
\int Dz\, \text{sech}^2 \beta Jz = \frac{1}{\beta J} \int Dz \frac{d}{dz} \tanh \beta Jz
$$

$$
\longrightarrow \frac{1}{\beta J} \int Dz\, \{2\delta(z)\} = \sqrt{\frac{2}{\pi}\frac{T}{J}}. \tag{2.35}
$$

This result confirms the consistency of the assumption $q = 1 - aT$ with $a = \sqrt{2/\pi}/J$.

To obtain the ground-state entropy, it is necessary to investigate the behaviour of the first term on the right hand side of (2.31) in the limit $T \to 0$. Substitution of $q = 1 - aT$ into this term readily leads to the contribution $-T/2\pi$ to the free energy. The second term, the integral of $\log 2 \cosh \tilde{H}(z)$, is evaluated by separating the integration range into positive and negative parts. These two parts actually give the same value, and it is sufficient to calculate one of them and multiply the result by the factor 2. The integral for large β is then

$$
2 \int_0^\infty Dz \left\{ \beta J \sqrt{q} z + \log(1 + e^{-2\beta J \sqrt{q} z}) \right\} \approx \frac{2\beta J(1 - aT/2)}{\sqrt{2\pi}} + 2 \int_0^\infty Dz\, e^{-2\beta J \sqrt{q} z}. \tag{2.36}
$$

The second term can be shown to be of $\mathcal{O}(T^2)$, and we may neglect it in our evaluation of the ground-state entropy. The first term contributes $-\sqrt{2/\pi}J + T/\pi$ to the free energy. The free energy in the low-temperature limit therefore behaves as

$$
[f] \approx -\sqrt{\frac{2}{\pi}}J + \frac{T}{2\pi}, \tag{2.37}
$$

from which we conclude that the ground-state entropy is $-1/2\pi$ and the ground-state energy is $-\sqrt{2/\pi}J$.

It was suspected at an early stage of research that this negative entropy might have been caused by the inappropriate exchange of limits $n \to 0$ and $N \to \infty$ in deriving (2.17). The correct order is $N \to \infty$ after $n \to 0$, but we took the limit $N \to \infty$ first so that the method of steepest descent is applicable. However, it has now been established that the assumption of replica symmetry $q_{\alpha\beta} = q \ (\forall(\alpha\beta); \alpha \neq \beta)$ is the real source of the trouble. We shall study this problem in the next chapter.

Bibliographical note

Developments following the pioneering contributions of Edwards and Anderson (1975) and Sherrington and Kirkpatrick (1975) up to the mid 1980s are summarized in Mézard *et al.* (1987), Binder and Young (1986), Fischer and Hertz (1991), and van Hemmen and Morgenstern (1987). See also the arguments and references in the next chapter.

3

REPLICA SYMMETRY BREAKING

Let us continue our analysis of the SK model. The free energy of the SK model derived under the ansatz of replica symmetry has the problem of negative entropy at low temperatures. It is therefore natural to investigate the possibility that the order parameter $q_{\alpha\beta}$ may assume various values depending upon the replica indices α and β and possibly the α-dependence of m_α. The theory of replica symmetry breaking started in this way as a mathematical effort to avoid unphysical conclusions of the replica-symmetric solution. It turned out, however, that the scheme of replica symmetry breaking has a very rich physical implication, namely the existence of a vast variety of stable states with ultrametric structure in the phase space. The present chapter is devoted to the elucidation of this story.

3.1 Stability of replica-symmetric solution

It was shown in the previous chapter that the replica-symmetric solution of the SK model has a spin glass phase with negative entropy at low temperatures. We now test the appropriateness of the assumption of replica symmetry from a different point of view.

A necessary condition for the replica-symmetric solution to be reliable is that the free energy is stable for infinitesimal deviations from that solution. To check such a stability, we expand the exponent appearing in the calculation of the partition function (2.16) to second order in $(q_{\alpha\beta} - q)$ and $(m_\alpha - m)$, deviations from the replica-symmetric solution, as

$$\int \prod_\alpha \mathrm{d}m_\alpha \prod_{\alpha<\beta} \mathrm{d}q_{\alpha\beta} \exp\left[-\beta N\{f_{\mathrm{RS}} + (\text{quadratic in } (q_{\alpha\beta} - q) \text{ and } (m_\alpha - m))\}\right],$$

(3.1)

where f_{RS} is the replica-symmetric free energy. This integral should not diverge in the limit $N \to \infty$ and thus the quadratic form must be positive definite (or at least positive semi-definite). We show in the present section that this stability condition of the replica-symmetric solution is not satisfied in the region below a line, called the *de Almeida–Thouless (AT) line*, in the phase diagram (de Almeida and Thouless 1978). The explicit form of the solution with replica symmetry breaking below the AT line and its physical significance will be discussed in subsequent sections.

3.1.1   *Hessian*

We restrict ourselves to the case $h = 0$ unless stated otherwise. It is convenient to rescale the variables as

$$\beta J \, q_{\alpha\beta} = y^{\alpha\beta}, \quad \sqrt{\beta J_0} \, m_\alpha = x^\alpha. \tag{3.2}$$

Then the free energy is, from (2.17),

$$[f] = -\frac{\beta J^2}{4} - \lim_{n \to 0} \frac{1}{\beta n} \left\{ -\sum_{\alpha < \beta} \frac{1}{2}(y^{\alpha\beta})^2 - \sum_\alpha \frac{1}{2}(x^\alpha)^2 \right.$$
$$\left. + \log \mathrm{Tr} \exp \left(\beta J \sum_{\alpha < \beta} y^{\alpha\beta} S^\alpha S^\beta + \sqrt{\beta J_0} \sum_\alpha x^\alpha S^\alpha \right) \right\}. \tag{3.3}$$

Let us expand $[f]$ to second order in small deviations around the replica-symmetric solution to check the stability,

$$x^\alpha = x + \epsilon^\alpha, \quad y^{\alpha\beta} = y + \eta^{\alpha\beta}. \tag{3.4}$$

The final term of (3.3) is expanded to second order in ϵ^α and $\eta^{\alpha\beta}$ as, with the notation $L_0 = \beta J y \sum_{\alpha < \beta} S^\alpha S^\beta + \sqrt{\beta J_0} x \sum_\alpha S_\alpha$,

$$\log \mathrm{Tr} \exp \left(L_0 + \beta J \sum_{\alpha < \beta} \eta^{\alpha\beta} S^\alpha S^\beta + \sqrt{\beta J_0} \sum_\alpha \epsilon^\alpha S^\alpha \right)$$
$$\approx \log \mathrm{Tr} \, e^{L_0} + \frac{\beta J_0}{2} \sum_{\alpha\beta} \epsilon^\alpha \epsilon^\beta \langle S^\alpha S^\beta \rangle_{L_0} + \frac{\beta^2 J^2}{2} \sum_{\alpha < \beta} \sum_{\gamma < \delta} \eta^{\alpha\beta} \eta^{\gamma\delta} \langle S^\alpha S^\beta S^\gamma S^\delta \rangle_{L_0}$$
$$- \frac{\beta J_0}{2} \sum_{\alpha\beta} \epsilon^\alpha \epsilon^\beta \langle S^\alpha \rangle_{L_0} \langle S^\beta \rangle_{L_0} - \frac{\beta^2 J^2}{2} \sum_{\alpha < \beta} \sum_{\gamma < \delta} \eta^{\alpha\beta} \eta^{\gamma\delta} \langle S^\alpha S^\beta \rangle_{L_0} \langle S^\gamma S^\delta \rangle_{L_0}$$
$$- \beta J \sqrt{\beta J_0} \sum_\delta \sum_{\alpha < \beta} \epsilon^\delta \eta^{\alpha\beta} \langle S^\delta \rangle_{L_0} \langle S^\alpha S^\beta \rangle_{L_0}$$
$$+ \beta J \sqrt{\beta J_0} \sum_\delta \sum_{\alpha < \beta} \epsilon^\delta \eta^{\alpha\beta} \langle S^\delta S^\alpha S^\beta \rangle_{L_0}. \tag{3.5}$$

Here $\langle \cdots \rangle_{L_0}$ denotes the average by the replica-symmetric weight e^{L_0}. We have used the facts that the replica-symmetric solution extremizes (3.3) (so that the terms linear in ϵ^α and $\eta^{\alpha\beta}$ vanish) and that $\mathrm{Tr} \, e^{L_0} \to 1$ as $n \to 0$ as explained in §2.3.1. We see that the second-order term of $[f]$ with respect to ϵ^α and $\eta^{\alpha\beta}$ is, taking the first and second terms in the braces $\{\cdots\}$ in (3.3) into account,

$$\Delta \equiv \frac{1}{2} \sum_{\alpha\beta} \{ \delta_{\alpha\beta} - \beta J_0 (\langle S^\alpha S^\beta \rangle_{L_0} - \langle S^\alpha \rangle_{L_0} \langle S^\beta \rangle_{L_0}) \} \epsilon^\alpha \epsilon^\beta$$

$$+\beta J\sqrt{\beta J_0}\sum_{\delta}\sum_{\alpha<\beta}(\langle S^{\delta}\rangle_{L_0}\langle S^{\alpha}S^{\beta}\rangle_{L_0}-\langle S^{\alpha}S^{\beta}S^{\delta}\rangle_{L_0})\epsilon^{\delta}\eta^{\alpha\beta}$$

$$+\frac{1}{2}\sum_{\alpha<\beta}\sum_{\gamma<\delta}\{\delta_{(\alpha\beta)(\delta\gamma)}-\beta^2 J^2(\langle S^{\alpha}S^{\beta}S^{\gamma}S^{\delta}\rangle_{L_0}$$

$$-\langle S^{\alpha}S^{\beta}\rangle_{L_0}\langle S^{\gamma}S^{\delta}\rangle_{L_0})\}\eta^{\alpha\beta}\eta^{\gamma\delta} \tag{3.6}$$

up to the trivial factor of βn (which is irrelevant to the sign). We denote the matrix of coefficients of this quadratic form in ϵ^{α} and $\eta^{\alpha\beta}$ by G which is called the *Hessian* matrix. Stability of the replica-symmetric solution requires that the eigenvalues of G all be positive.

To derive the eigenvalues, let us list the matrix elements of G. Since $\langle\cdots\rangle_{L_0}$ represents the average by weight of the replica-symmetric solution, the coefficient of the second-order terms in ϵ has only two types of values. To simplify the notation we omit the suffix L_0 in the present section.

$$G_{\alpha\alpha}=1-\beta J_0(1-\langle S^{\alpha}\rangle^2)\equiv A \tag{3.7}$$

$$G_{\alpha\beta}=-\beta J_0(\langle S^{\alpha}S^{\beta}\rangle-\langle S^{\alpha}\rangle^2)\equiv B. \tag{3.8}$$

The coefficients of the second-order term in η have three different values, the diagonal and two types of off-diagonal elements. One of the off-diagonal elements has a matched single replica index and the other has all indices different:

$$G_{(\alpha\beta)(\alpha\beta)}=1-\beta^2 J^2(1-\langle S^{\alpha}S^{\beta}\rangle^2)\equiv P \tag{3.9}$$

$$G_{(\alpha\beta)(\alpha\gamma)}=-\beta^2 J^2(\langle S^{\beta}S^{\gamma}\rangle-\langle S^{\alpha}S^{\beta}\rangle^2)\equiv Q \tag{3.10}$$

$$G_{(\alpha\beta)(\gamma\delta)}=-\beta^2 J^2(\langle S^{\alpha}S^{\beta}S^{\gamma}S^{\delta}\rangle-\langle S^{\alpha}S^{\beta}\rangle^2)\equiv R. \tag{3.11}$$

Finally there are two kinds of cross-terms in ϵ and η:

$$G_{\alpha(\alpha\beta)}=\beta J\sqrt{\beta J_0}(\langle S^{\alpha}\rangle\langle S^{\alpha}S^{\beta}\rangle-\langle S^{\beta}\rangle)\equiv C \tag{3.12}$$

$$G_{\gamma(\alpha\beta)}=\beta J\sqrt{\beta J_0}(\langle S^{\gamma}\rangle\langle S^{\alpha}S^{\beta}\rangle-\langle S^{\alpha}S^{\beta}S^{\gamma}\rangle)\equiv D. \tag{3.13}$$

These complete the elements of G.

The expectation values appearing in (3.7) to (3.13) can be evaluated from the replica-symmetric solution. The elements of G are written in terms of $\langle S^{\alpha}\rangle=m$ and $\langle S^{\alpha}S^{\beta}\rangle=q$ satisfying (2.28) and (2.30) as well as

$$\langle S^{\alpha}S^{\beta}S^{\gamma}\rangle\equiv t=\int Dz\tanh^3\beta\tilde{H}(z) \tag{3.14}$$

$$\langle S^{\alpha}S^{\beta}S^{\gamma}S^{\delta}\rangle\equiv r=\int Dz\tanh^4\beta\tilde{H}(z). \tag{3.15}$$

The integrals on the right of (3.14) and (3.15) can be derived by the method in §2.3.1.

3.1.2   *Eigenvalues of the Hessian and the AT line*

We start the analysis of stability by the simplest case of paramagnetic solution. All order parameters m, q, r, and t vanish in the paramagnetic phase. Hence B, Q, R, C, and D (the off-diagonal elements of G) are all zero. The stability condition for infinitesimal deviations of the ferromagnetic order parameter ϵ^α is $A > 0$, which is equivalent to $1 - \beta J_0 > 0$ or $T > J_0$ from (3.7). Similarly the stability for spin-glass-like infinitesimal deviations $\eta^{\alpha\beta}$ is $P > 0$ or $T > J$. These two conditions precisely agree with the region of existence of the paramagnetic phase derived in §2.3.2 (see Fig. 2.1). Therefore the replica-symmetric solution is stable in the paramagnetic phase.

It is a more elaborate task to investigate the stability condition of the ordered phases. It is necessary to calculate all eigenvalues of the Hessian. Details are given in Appendix A, and we just mention the results here.

Let us write the eigenvalue equation in the form

$$G\boldsymbol{\mu} = \lambda\boldsymbol{\mu}, \quad \boldsymbol{\mu} = \begin{pmatrix} \{\epsilon^\alpha\} \\ \{\eta^{\alpha\beta}\} \end{pmatrix}. \tag{3.16}$$

The symbol $\{\epsilon^\alpha\}$ denotes a column from ϵ^1 at the top to ϵ^n at the bottom, and $\{\eta^{\alpha\beta}\}$ is for η^{12} to $\eta^{n-1,n}$.

The first eigenvector $\boldsymbol{\mu}_1$ has $\epsilon^\alpha = a$ and $\eta^{\alpha\beta} = b$, uniform in both parts. Its eigenvalue is, in the limit $n \to 0$,

$$\lambda_1 = \frac{1}{2}\left\{ A - B + P - 4Q + 3R \pm \sqrt{(A - B - P + 4Q - 3R)^2 - 8(C - D)^2} \right\}. \tag{3.17}$$

The second eigenvector $\boldsymbol{\mu}_2$ has $\epsilon^\theta = a$ for a specific replica θ and $\epsilon^\alpha = b$ otherwise, and $\eta^{\alpha\beta} = c$ when α or β is equal to θ and $\eta^{\alpha\beta} = d$ otherwise. The eigenvalue of this eigenvector becomes degenerate with λ_1 in the limit $n \to 0$. The third and final eigenvector $\boldsymbol{\mu}_3$ has $\epsilon^\theta = a, \epsilon^\nu = a$ for two specific replicas θ, ν and $\epsilon^\alpha = b$ otherwise, and $\eta^{\theta\nu} = c, \eta^{\theta\alpha} = \eta^{\nu\alpha} = d$ and $\eta^{\alpha\beta} = e$ otherwise. Its eigenvalue is

$$\lambda_3 = P - 2Q + R. \tag{3.18}$$

A sufficient condition for $\lambda_1, \lambda_2 > 0$ is, from (3.17),

$$A - B = 1 - \beta J_0(1 - q) > 0, \quad P - 4Q + 3R = 1 - \beta^2 J^2(1 - 4q + 3r) > 0. \tag{3.19}$$

These two conditions are seen to be equivalent to the saddle-point condition of the replica-symmetric free energy (2.27) with respect to m and q as can be verified by the second-order derivatives:

$$A - B = \frac{1}{J_0}\left.\frac{\partial^2[f]}{\partial m^2}\right|_{\text{RS}} > 0, \quad P - 4Q + 3R = -\frac{2}{\beta J^2}\left.\frac{\partial^2[f]}{\partial q^2}\right|_{\text{RS}} > 0. \tag{3.20}$$

These inequalities always hold as has been mentioned in §2.3.2.

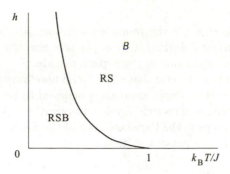

FIG. 3.1. Stability limit of the replica-symmetric (RS) solution in the h–T phase diagram (the AT line) below which replica symmetry breaking (RSB) occurs.

The condition for positive λ_3 is

$$P - 2Q + R = 1 - \beta^2 J^2 (1 - 2q + r) > 0 \qquad (3.21)$$

or more explicitly

$$\left(\frac{T}{J}\right)^2 > \int Dz \, \text{sech}^4(\beta J \sqrt{q} z + \beta J_0 m). \qquad (3.22)$$

By numerically solving the equations of state of the replica-symmetric order parameters (2.28) and (2.30), one finds that the stability condition (3.22) is not satisfied in the spin glass and mixed phases in Fig. 2.1. The line of the limit of stability within the ferromagnetic phase (i.e. the boundary between the ferromagnetic and mixed phases) is the AT line. The mixed phase has finite ferromagnetic order but replica symmetry is broken there. More elaborate analysis is required in the mixed phase as shown in the next section.

The stability of replica symmetry in the case of finite h with symmetric distribution $J_0 = 0$ can be studied similarly. Let us just mention the conclusion that the stability condition in such a case is given simply by replacing $J_0 m$ by h in (3.22). The phase diagram thus obtained is depicted in Fig. 3.1. A phase with broken replica symmetry extends into the low-temperature region. This phase is also often called the spin glass phase. The limit of stability in the present case is also termed the AT line.

3.2   Replica symmetry breaking

The third eigenvector $\boldsymbol{\mu}_3$, which causes replica symmetry breaking, is called the *replicon mode*. There is no replica symmetry breaking in m_α since the replicon mode has $a = b$ for ϵ^θ and ϵ^ν in the limit $n \to 0$, as in the relation (A.19) or (A.21) in Appendix A. Only $q_{\alpha\beta}$ shows dependence on α and β. It is necessary to clarify how $q_{\alpha\beta}$ depends on α and β, but unfortunately we are not aware of any first-principle argument which can lead to the exact solution. One thus proceeds

by trial and error to check if the tentative solution satisfies various necessary
conditions for the correct solution, such as positive entropy at low temperatures
and the non-negative eigenvalue of the replicon mode.

The only solution found so far that satisfies all necessary conditions is the one
by Parisi (1979, 1980). The *Parisi solution* is believed to be the exact solution of
the SK model also because of its rich physical implications. The replica symmetry
is broken in multiple steps in the Parisi solution of the SK model. We shall explain
mainly its first step in the present section.

3.2.1 *Parisi solution*

Let us regard $q_{\alpha\beta}$ ($\alpha \neq \beta$) of the replica-symmetric solution of the SK model
as an element of an $n \times n$ matrix. Then all the elements except those along the
diagonal have the common value q, and we may write

$$\{q_{\alpha\beta}\} = \begin{pmatrix} 0 & & & & \\ & 0 & & q & \\ & & 0 & & \\ & & & 0 & \\ & q & & & 0 \\ & & & & & 0 \end{pmatrix}. \tag{3.23}$$

In the first step of replica symmetry breaking (1RSB), one introduces a positive integer m_1 ($\leq n$) and divides the replicas into n/m_1 blocks. Off-diagonal
blocks have q_0 as their elements and diagonal blocks are assigned q_1. All diagonal
elements are kept 0. The following example is for the case of $n = 6, m_1 = 3$.

$$\begin{pmatrix} \begin{array}{ccc|} 0 & q_1 & q_1 \\ q_1 & 0 & q_1 \\ q_1 & q_1 & 0 \\ \hline \end{array} & q_0 \\ q_0 & \begin{array}{|ccc} 0 & q_1 & q_1 \\ q_1 & 0 & q_1 \\ q_1 & q_1 & 0 \end{array} \end{pmatrix}. \tag{3.24}$$

In the second step, the off-diagonal blocks are left untouched and the diagonal
blocks are further divided into m_1/m_2 blocks. The elements of the innermost
blocks are assumed to be q_2 and all the other elements of the larger diagonal
blocks are kept as q_1. For example, if we have $n = 12, m_1 = 6, m_2 = 3$,

$$
\left(
\begin{array}{ccc}
\begin{array}{ccc} 0 & q_2 & q_2 \\ q_2 & 0 & q_2 \\ q_2 & q_2 & 0 \end{array} & & \\
& q_1 & \\
& \begin{array}{ccc} 0 & q_2 & q_2 \\ q_2 & 0 & q_2 \\ q_2 & q_2 & 0 \end{array} & \\
q_1 & & \\
& & q_0 \\
& & q_0 \\
& \begin{array}{ccc} 0 & q_2 & q_2 \\ q_2 & 0 & q_2 \\ q_2 & q_2 & 0 \end{array} & \\
& & q_1 \\
q_0 & & \begin{array}{ccc} 0 & q_2 & q_2 \\ q_2 & 0 & q_2 \\ q_2 & q_2 & 0 \end{array} \\
& q_1 & \\
& & \begin{array}{ccc} 0 & q_2 & q_2 \\ q_2 & 0 & q_2 \\ q_2 & q_2 & 0 \end{array}
\end{array}
\right). \tag{3.25}
$$

The numbers $n, m_1, m_2, \ldots$ are integers by definition and are ordered as $n \geq m_1 \geq m_2 \geq \cdots \geq 1$.

Now we define the function $q(x)$ as

$$
q(x) = q_i \quad (m_{i+1} < x \leq m_i) \tag{3.26}
$$

and take the limit $n \to 0$ following the prescription of the replica method. We somewhat arbitrarily reverse the above inequalities

$$
0 \leq m_1 \leq \cdots \leq 1 \quad (0 \leq x \leq 1) \tag{3.27}
$$

and suppose that $q(x)$ becomes a continuous function defined between 0 and 1. This is the basic idea of the Parisi solution.

3.2.2 First-step RSB

We derive expressions of the physical quantities by the first-step RSB (1RSB) represented in (3.24). The first term on the right hand side of the single-body effective Hamiltonian (2.15) reduces to

$$
\sum_{\alpha<\beta} q_{\alpha\beta} S^\alpha S^\beta = \frac{1}{2} \left\{ q_0 \left(\sum_\alpha^n S^\alpha \right)^2 + (q_1 - q_0) \sum_{\text{block}}^{n/m_1} \left(\sum_{\alpha \in \text{block}}^{m_1} S^\alpha \right)^2 - nq_1 \right\}. \tag{3.28}
$$

The first term on the right hand side here fills all elements of the matrix $\{q_{\alpha\beta}\}$ with q_0 but the block-diagonal part is replaced with q_1 by the second term. The last term forces the diagonal elements to zero. Similarly the quadratic term of $q_{\alpha\beta}$ in the free energy (2.17) is

$$
\lim_{n\to 0} \frac{1}{n} \sum_{\alpha\neq\beta} q_{\alpha\beta}^2 = \lim_{n\to 0} \frac{1}{n} \left\{ n^2 q_0^2 + \frac{n}{m_1} m_1^2 (q_1^2 - q_0^2) - nq_1^2 \right\} = (m_1 - 1)q_1^2 - m_1 q_0^2. \tag{3.29}
$$

We insert (3.28) and (3.29) into (2.17) and linearize $(\sum_\alpha S^\alpha)^2$ in (3.28) by a Gaussian integral in a similar manner as in the replica-symmetric calculations.

It is necessary to introduce $1 + n/m_1$ Gaussian variables corresponding to the number of terms of the form $(\sum_\alpha S^\alpha)^2$ in (3.28). Finally we take the limit $n \to 0$ to find the free energy with 1RSB as

$$\beta f_{1RSB} = \frac{\beta^2 J^2}{4} \left\{ (m_1 - 1)q_1^2 - m_1 q_0^2 + 2q_1 - 1 \right\} + \frac{\beta J_0}{2} m^2 - \log 2$$
$$- \frac{1}{m_1} \int Du \log \int Dv \cosh^{m_1} \Xi \tag{3.30}$$
$$\Xi = \beta(J\sqrt{q_0}\, u + J\sqrt{q_1 - q_0}\, v + J_0 m + h). \tag{3.31}$$

Here we have used the replica symmetry of magnetization $m = m_\alpha$.

The variational parameters q_0, q_1, m, and m_1 all fall in the range between 0 and 1. The variational (extremization) conditions of (3.30) with respect to m, q_0, and q_1 lead to the equations of state:

$$m = \int Du \, \frac{\int Dv \cosh^{m_1} \Xi \tanh \Xi}{\int Dv \cosh^{m_1} \Xi} \tag{3.32}$$

$$q_0 = \int Du \left(\frac{\int Dv \cosh^{m_1} \Xi \tanh \Xi}{\int Dv \cosh^{m_1} \Xi} \right)^2 \tag{3.33}$$

$$q_1 = \int Du \, \frac{\int Dv \cosh^{m_1} \Xi \tanh^2 \Xi}{\int Dv \cosh^{m_1} \Xi}. \tag{3.34}$$

Comparison of these equations of state for the order parameters with those for the replica-symmetric solution (2.28) and (2.30) suggests the following interpretation. In (3.32) for the magnetization, the integrand after Du represents magnetization within a block of the 1RSB matrix (3.24), which is averaged over all blocks with the Gaussian weight. Analogously, (3.34) for q_1 is the spin glass order parameter within a diagonal block averaged over all blocks. In (3.33) for q_0, on the other hand, one first calculates the magnetization within a block and takes its product between blocks, an interblock spin glass order parameter. Indeed, if one carries out the trace operation in the definition of $q_{\alpha\beta}$, (2.18), by taking α and β within a single block and assuming 1RSB, one obtains (3.34), whereas (3.33) results if α and β belong to different blocks. The Schwarz inequality assures $q_1 \geq q_0$.

We omit the explicit form of the extremization condition of the free energy (3.31) by the parameter m_1 since the form is a little complicated and is not used later.

When $J_0 = h = 0$, Ξ is odd in u, v, and thus $m = 0$ is the only solution of (3.32). The order parameter q_1 can be positive for $T < T_f = J$ because the first term in the expansion of the right hand side of (3.34) for small q_0 and q_1 is $\beta^2 J^2 q_1$. Therefore the RS and 1RSB give the same transition point. The parameter m_1 is one at T_f and decreases with temperature.

3.2.3 *Stability of the first-step RSB*

The stability of 1RSB can be investigated by a direct generalization of the argument in §3.1. We mention only a few main points for the case $J_0 = h = 0$. It is sufficient to treat two cases: one with all indices $\alpha, \beta, \gamma, \delta$ of the Hessian elements within the same block and the other with indices in two different blocks.

If α and β in $q_{\alpha\beta}$ belong to the same block, the stability condition of the replicon mode for infinitesimal deviations from 1RSB is expressed as

$$\lambda_3 = P - 2Q + R = 1 - \beta^2 J^2 \int Du \, \frac{\int Dv \cosh^{m_1-4} \Xi}{\int Dv \cosh^{m_1} \Xi} > 0. \tag{3.35}$$

For the replicon mode between two different blocks, the stability condition reads

$$\lambda_3 = P - 2Q + R = 1 - \beta^2 J^2 \int Du \left(\frac{\int Dv \cosh^{m_1-1} \Xi}{\int Dv \cosh^{m_1} \Xi} \right)^4 > 0. \tag{3.36}$$

According to the Schwarz inequality, the right hand side of (3.35) is less than or equal to that of (3.36), and therefore the former is sufficient. Equation (3.35) is not satisfied in the spin glass phase similar to the case of the RSB solution. However, the absolute value of the eigenvalue is confirmed by numerical evaluation to be smaller than that of the RS solution although λ_3 is still negative. This suggests an improvement towards a stable solution. The entropy per spin at $J_0 = 0, T = 0$ reduces from $-0.16 (= -1/2\pi)$ for the RS solution to -0.01 for the 1RSB. Thus we may expect to obtain still better results if we go further into replica symmetry breaking.

3.3   Full RSB solution

Let us proceed with the calculation of the free energy (2.17) by a multiple-step RSB. We restrict ourselves to the case $J_0 = 0$ for simplicity.

3.3.1 *Physical quantities*

The sum involving $q_{\alpha\beta}^2$ in the free energy (2.17) can be expressed at the Kth step of RSB (K-RSB) as follows by counting the number of elements in a similar way to the 1RSB case (3.29):

$$\sum_{\alpha \neq \beta} q_{\alpha\beta}^l$$
$$= q_0^l n^2 + (q_1^l - q_0^l)m_1^2 \cdot \frac{n}{m_1} + (q_2^l - q_1^l)m_2^2 \cdot \frac{m_1}{m_2} \cdot \frac{n}{m_1} + \cdots \quad q_K^l \quad n$$
$$= n \sum_{j=0}^{K} (m_j - m_{j+1})q_j^l, \tag{3.37}$$

where l is an arbitrary integer and $m_0 = n, m_{K+1} = 1$. In the limit $n \to 0$, we may use the replacement $m_j - m_{j+1} \to -dx$ to find

$$\frac{1}{n}\sum_{\alpha\neq\beta}q_{\alpha\beta}^l \rightarrow -\int_0^1 q^l(x)\mathrm{d}x. \tag{3.38}$$

The internal energy for $J_0 = 0, h = 0$ is given by differentiation of the free energy (2.17) by β as[2]

$$E = -\frac{\beta J^2}{2}\left(1+\frac{2}{n}\sum_{\alpha<\beta}q_{\alpha\beta}^2\right) \rightarrow -\frac{\beta J^2}{2}\left(1-\int_0^1 q^2(x)\mathrm{d}x\right). \tag{3.39}$$

The magnetic susceptibility can be written down from the second derivative of (2.17) by h as

$$\chi = \beta\left(1+\frac{1}{n}\sum_{\alpha\neq\beta}q_{\alpha\beta}\right) \rightarrow \beta\left(1-\int_0^1 q(x)\mathrm{d}x\right). \tag{3.40}$$

It needs some calculations to derive the free energy in the full RSB scheme. Details are given in Appendix B. The final expression of the free energy (2.17) is

$$\beta f = -\frac{\beta^2 J^2}{4}\left\{1+\int_0^1 q(x)^2\mathrm{d}x - 2q(1)\right\} - \int \mathrm{D}u\, f_0(0,\sqrt{q(0)}u). \tag{3.41}$$

Here f_0 satisfies the *Parisi equation*

$$\frac{\partial f_0(x,h)}{\partial x} = -\frac{J^2}{2}\frac{\mathrm{d}q}{\mathrm{d}x}\left\{\frac{\partial^2 f_0}{\partial h^2} + x\left(\frac{\partial f_0}{\partial h}\right)^2\right\} \tag{3.42}$$

to be solved under the initial condition $f_0(1,h) = \log 2\cosh\beta h$.

3.3.2 *Order parameter near the critical point*

It is in general very difficult to find a solution to the extremization condition of the free energy (3.41) with respect to the order function $q(x)$. It is nevertheless possible to derive some explicit results by the Landau expansion when the temperature is close to the critical point and consequently $q(x)$ is small. Let us briefly explain the essence of this procedure.

When $J_0 = h = 0$, the expansion of the free energy (2.17) to fourth order in $q_{\alpha\beta}$ turns out to be

$$\beta f = \lim_{n\to 0}\frac{1}{n}\left\{\frac{1}{4}\left(\frac{T^2}{T_\mathrm{f}^2}-1\right)\mathrm{Tr}\, Q^2 - \frac{1}{6}\mathrm{Tr}\, Q^3\right.$$

$$\left. -\frac{1}{8}\mathrm{Tr}\, Q^4 + \frac{1}{4}\sum_{\alpha\neq\beta\neq\gamma}Q_{\alpha\beta}^2 Q_{\alpha\gamma}^2 - \frac{1}{12}\sum_{\alpha\neq\beta}Q_{\alpha\beta}^4\right\}, \tag{3.43}$$

where we have dropped q-independent terms. The operator Tr here denotes the diagonal sum in the replica space. We have introduced the notation $Q_{\alpha\beta} = $

---

[2]The symbol of configurational average $[\cdots]$ will be omitted in the present chapter as long as it does not lead to confusion.

$(\beta J)^2 q_{\alpha\beta}$. Only the last term is relevant to the RSB. It can indeed be verified that the eigenvalue of the replicon mode that determines stability of the RS solution is, by setting the coefficient of the last term to $-y$ (which is actually $-1/12$),

$$\lambda_3 = -16 y \theta^2, \qquad (3.44)$$

where $\theta = (T_{\mathrm{f}} - T)/T_{\mathrm{f}}$. We may thus neglect all fourth-order terms except $Q_{\alpha\beta}^4$ to discuss the essential features of the RSB and let $n \to 0$ to get

$$\beta f = \frac{1}{2} \int_0^1 \mathrm{d}x \left\{ |\theta| q^2(x) - \frac{1}{3} x q^3(x) - q(x) \int_0^x q^2(y) \mathrm{d}y + \frac{1}{6} q^4(x) \right\}. \qquad (3.45)$$

The extremization condition with respect to $q(x)$ is written explicitly as

$$2|\theta| q(x) - x q^2(x) - \int_0^x q^2(y) \mathrm{d}y - 2q(x) \int_x^1 q(y) \mathrm{d}y + \frac{2}{3} q^3(x) = 0. \qquad (3.46)$$

Differentiation of this formula gives

$$|\theta| - x q(x) - \int_x^1 q(y) \mathrm{d}y + q^2(x) = 0 \quad \text{or} \quad q'(x) = 0. \qquad (3.47)$$

Still further differentiation leads to

$$q(x) = \frac{x}{2} \quad \text{or} \quad q'(x) = 0. \qquad (3.48)$$

The RS solution corresponds to a constant $q(x)$. This constant is equal to $|\theta|$ according to (3.46). There also exists an x-dependent solution

$$q(x) = \frac{x}{2} \quad (0 \le x \le x_1 = 2q(1)) \qquad (3.49)$$

$$q(x) = q(1) \quad (x_1 \le x \le 1). \qquad (3.50)$$

By inserting this solution in the variational condition (3.46), we obtain

$$q(1) = |\theta| + \mathcal{O}(\theta^2). \qquad (3.51)$$

Figure 3.2 shows the resulting behaviour of $q(x)$ near the critical point where θ is close to zero.

3.3.3   Vertical phase boundary

The susceptibility is a constant $\chi = 1/J$ near the critical point T_{f} because the integral in (3.40) is $1 - T/T_{\mathrm{f}}$ according to (3.49)–(3.51). It turns out in fact that this result remains valid not only near T_{f} but over the whole temperature range below the critical point. We use this fact to show that the phase boundary between the spin glass and ferromagnetic phases is a vertical line at $J_0 = J$ as in Fig. 2.1 (Toulouse 1980).

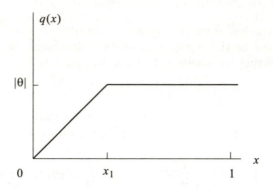

FIG. 3.2. $q(x)$ near the critical point

The Hamiltonian of the SK model (2.7) suggests that a change of the centre of distribution of J_{ij} from 0 to J_0/N shifts the energy per spin by $-J_0 m^2/2$. Thus the free energy $f(T, m, J_0)$ as a function of T and m satisfies

$$f(T, m, J_0) = f(T, m, 0) - \frac{1}{2} J_0 m^2. \tag{3.52}$$

From the thermodynamic relation

$$\frac{\partial f(T, m, 0)}{\partial m} = h \tag{3.53}$$

and the fact that $m = 0$ when $J_0 = 0$ and $h = 0$, we obtain

$$\chi^{-1} = \left(\frac{\partial m}{\partial h} \right]_{h \to 0} \right)^{-1} = \frac{\partial^2 f(T, m, 0)}{\partial m^2} \bigg]_{m \to 0}. \tag{3.54}$$

Thus, for sufficiently small m, we have

$$f(T, m, 0) = f_0(T) + \frac{1}{2} \chi^{-1} m^2. \tag{3.55}$$

Combining (3.52) and (3.55) gives

$$f(T, m, J_0) = f_0(T) + \frac{1}{2} (\chi^{-1} - J_0) m^2. \tag{3.56}$$

This formula shows that the coefficient of m^2 in $f(T, m, J_0)$ vanishes when $\chi = 1/J_0$ and therefore there is a phase transition between the ferromagnetic and non-ferromagnetic phases according to the Landau theory. Since $\chi = 1/J$ in the whole range $T < T_f$, we conclude that the boundary between the ferromagnetic and spin glass phases exists at $J_0 = J$.

Stability analysis of the Parisi solution has revealed that the eigenvalue of the replicon mode is zero, implying marginal stability of the Parisi RSB solution. No other solutions have been found with a non-negative replicon eigenvalue of the Hessian.

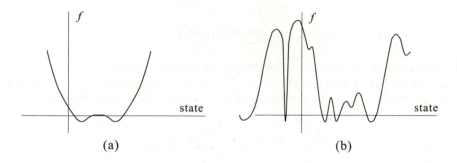

FIG. 3.3. Simple free energy (a) and multivalley structure (b)

3.4 Physical significance of RSB

The RSB of the Parisi type has been introduced as a mathematical tool to resolve controversies in the RS solution. It has, however, been discovered that the solution has a profound physical significance. The main results are sketched here (Mézard *et al.* 1987; Binder and Young 1986).

3.4.1 *Multivalley structure*

In a ferromagnet the free energy as a function of the state of the system has a simple structure as depicted in Fig. 3.3(a). The free energy of the spin glass state, on the other hand, is considered to have many minima as in Fig. 3.3(b), and the barriers between them are expected to grow indefinitely as the system size increases. It is possible to give a clear interpretation of the RSB solution if we accept this physical picture.

Suppose that the system size is large but not infinite. Then the system is trapped in the valley around one of the minima of the free energy for quite a long time. However, after a very long time, the system climbs the barriers and reaches all valleys eventually. Hence, within some limited time scale, the physical properties of a system are determined by one of the valleys. But, after an extremely long time, one would observe behaviour reflecting the properties of all the valleys. This latter situation is the one assumed in the conventional formulation of equilibrium statistical mechanics.

We now label free energy valleys by the index a and write $m_i^a = \langle S_i \rangle_a$ for the magnetization calculated by restricting the system to a specific valley a. This is analogous to the restriction of states to those with $m > 0$ (neglecting $m < 0$) in a simple ferromagnet.

3.4.2 q_{EA} *and* $\bar{q}$

To understand the spin ordering in a single valley, it is necessary to take the thermodynamic limit to separate the valley from the others by increasing the barriers indefinitely. Then we may ignore transitions between valleys and observe the long-time behaviour of the system in a valley. It therefore makes sense to define the order parameter q_{EA} for a single valley as

$$q_{\text{EA}} = \lim_{t \to \infty} \lim_{N \to \infty} \left[\langle S_i(t_0) S_i(t_0 + t) \rangle \right]. \tag{3.57}$$

This quantity measures the similarity (or overlap) of a spin state at site i after a long time to the initial condition at t_0. The physical significance of this quantity suggests its equivalence to the average of the squared local magnetization $(m_i^a)^2$:

$$q_{\text{EA}} = \left[\sum_a P_a \left(m_i^a \right)^2 \right] = \left[\sum_a P_a \frac{1}{N} \sum_i (m_i^a)^2 \right]. \tag{3.58}$$

Here P_a is the probability that the system is located in a valley (*a pure state*) a, that is $P_a = \text{e}^{-\beta F_a}/Z$. In the second equality of (3.58), we have assumed that the averaged squared local magnetization does not depend on the location.

We may also define another order parameter $\bar{q}$ that represents the average over all valleys corresponding to the long-time observation (the usual statistical-mechanical average). This order parameter can be expressed explicitly as

$$\bar{q} = \left[\left(\sum_a P_a m_i^a \right)^2 \right] = \left[\sum_{ab} P_a P_b m_i^a m_i^b \right] = \frac{1}{N} \left[\sum_{ab} P_a P_b \sum_i m_i^a m_i^b \right], \tag{3.59}$$

which is rewritten using $m_i = \sum_a P_a m_i^a$ as

$$\bar{q} = [m_i^2] = [\langle S_i \rangle^2]. \tag{3.60}$$

As one can see from (3.59), $\bar{q}$ is the average with overlaps between valleys taken into account and is an appropriate quantity for time scales longer than transition times between valleys.

If there exists only a single valley (and its totally reflected state), the relation $q_{\text{EA}} = \bar{q}$ should hold, but in general we have $q_{\text{EA}} > \bar{q}$. The difference of these two order parameters $q_{\text{EA}} - \bar{q}$ is a measure of the existence of a *multivalley structure*. We generally expect a continuous spectrum of order parameters between $\bar{q}$ and q_{EA} corresponding to the variety of degrees of transitions between valleys. This would correspond to the continuous function $q(x)$ of the Parisi RSB solution.

3.4.3   *Distribution of overlaps*

Similarity between two valleys a and b is measured by the overlap q_{ab} defined by

$$q_{ab} = \frac{1}{N} \sum_i m_i^a m_i^b. \tag{3.61}$$

This q_{ab} takes its maximum when the two valleys a and b coincide and is zero when they are completely uncorrelated. Let us define the distribution of q_{ab} for a given random interaction $\boldsymbol{J}$ as

$$P_J(q) = \langle \delta(q - q_{ab}) \rangle = \sum_{ab} P_a P_b \delta(q - q_{ab}), \tag{3.62}$$

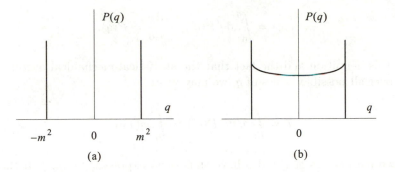

FIG. 3.4. Distribution function $P(q)$ of a simple system (a) and that of a system with multivalley structure (b)

and write $P(q)$ for the configurational average of $P_J(q)$:

$$P(q) = [P_J(q)]. \tag{3.63}$$

In a simple system like a ferromagnet, there are only two different valleys connected by overall spin reversal and q_{ab} assumes only $\pm m^2$. Then $P(q)$ is constituted only by two delta functions at $q = \pm m^2$, Fig. 3.4(a). If there is a multivalley structure with continuously different states, on the other hand, q_{ab} assumes various values and $P(q)$ has a continuous part as in Fig. 3.4(b).

3.4.4   Replica representation of the order parameter

Let us further investigate the relationship between the RSB and the continuous part of the distribution function $P(q)$. The quantity $q_{\alpha\beta}$ in the replica formalism is the overlap between two replicas α and β at a specific site

$$q_{\alpha\beta} = \langle S_i^\alpha S_i^\beta \rangle. \tag{3.64}$$

In the RSB this quantity has different values from one pair of replicas $\alpha\beta$ to another pair. The genuine statistical-mechanical average should be the mean of all possible values of $q_{\alpha\beta}$ and is identified with $\bar{q}$ defined in (3.59),

$$\bar{q} = \lim_{n \to 0} \frac{1}{n(n-1)} \sum_{\alpha \neq \beta} q_{\alpha\beta}. \tag{3.65}$$

The spin glass order parameter for a single valley, on the other hand, does not reflect the difference between valleys caused by transitions between them and therefore is expected to be larger than any other possible values of the order parameter. We may then identify q_{EA} with the largest value of $q_{\alpha\beta}$ in the replica method:

$$q_{EA} = \max_{(\alpha\beta)} q_{\alpha\beta} = \max_x q(x). \tag{3.66}$$

Let us define $x(q)$ as the accumulated distribution of $P(q)$:

$$x(q) = \int_0^q \mathrm{d}q' P(q'), \quad \frac{\mathrm{d}x}{\mathrm{d}q} = P(q). \tag{3.67}$$

Using this definition and the fact that the statistical-mechanical average is the mean over all possible values of q, we may write

$$\bar{q} = \int_0^1 q' \mathrm{d}q' P(q') = \int_0^1 q(x) \mathrm{d}x. \tag{3.68}$$

The two parameters q_{EA} and $\bar{q}$ have thus been expressed by $q(x)$. If there are many valleys, $q_{\alpha\beta}$ takes various values, and $P(q)$ cannot be expressed simply in terms of two delta functions. The order function $q(x)$ under such a circumstance has a non-trivial structure as one can see from (3.67), which corresponds to the RSB of Parisi type. The functional form of $q(x)$ mentioned in §3.3.2 reflects the multivalley structure of the space of states of the spin glass phase.

3.4.5 Ultrametricity

The Parisi RSB solution shows a remarkable feature of *ultrametricity*. The configurational average of the distribution function between three different states

$$P_J(q_1, q_2, q_3) = \sum_{abc} P_a P_b P_c \delta(q_1 - q_{ab})\delta(q_2 - q_{bc})\delta(q_3 - q_{ca}) \tag{3.69}$$

can be evaluated by the RSB method to yield

$$[P_J(q_1, q_2, q_3)] = \frac{1}{2}P(q_1)x(q_1)\delta(q_1 - q_2)\delta(q_1 - q_3)$$

$$+\frac{1}{2}\left\{P(q_1)P(q_2)\Theta(q_1 - q_2)\delta(q_2 - q_3) + (\text{two terms with } 1, 2, 3 \text{ permuted})\right\}.$$

Here $x(q)$ has been defined in (3.67), and $\Theta(q_1 - q_2)$ is the step function equal to 1 for $q_1 > q_2$ and 0 for $q_1 < q_2$. The first term on the right hand side is non-vanishing only if the three overlaps are equal to each other, and the second term requires that the overlaps be the edges of an isosceles triangle ($q_1 > q_2, q_2 = q_3$). This means that the distances between three states should form either an equilateral or an isosceles triangle. We may interpret this result as a tree-like (or equivalently, nested) structure of the space of states as in Fig. 3.5. A metric space where the distances between three points satisfy this condition is called an ultrametric space.

3.5 TAP equation

A different point of view on spin glasses is provided by the equation of state due to Thouless, Anderson, and Palmer (TAP) which concerns the local magnetization in spin glasses (Thouless *et al.* 1977).

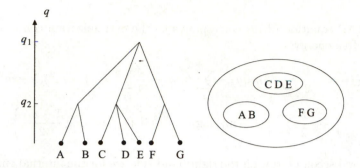

FIG. 3.5. Tree-like and nested structures in an ultrametric space. The distance between C and D is equal to that between C and E and to that between D and E, which is smaller than that between A and C and that between C and F.

3.5.1 TAP equation

The local magnetization of the SK model satisfies the following *TAP equation*, given the random interactions $\boldsymbol{J} = \{J_{ij}\}$:

$$m_i = \tanh\beta\left\{\sum_j J_{ij}m_j + h_i - \beta\sum_j J_{ij}^2(1 - m_j^2)m_i\right\}. \tag{3.70}$$

The first term on the right hand side represents the usual internal field, a generalization of (1.19). The third term is called the *reaction field of Onsager* and is added to remove the effects of self-response in the following sense. The magnetization m_i affects site j through the internal field $J_{ij}m_i$ that changes the magnetization of site j by the amount $\chi_{jj}J_{ij}m_i$. Here

$$\chi_{jj} = \left.\frac{\partial m_j}{\partial h_j}\right|_{h_j \to 0} = \beta(1 - m_j^2). \tag{3.71}$$

Then the internal field at site i would increase by

$$J_{ij}\chi_{jj}J_{ij}m_i = \beta J_{ij}^2(1 - m_j^2)m_i. \tag{3.72}$$

The internal field at site i should not include such a rebound of itself. The third term on the right hand side of (3.70) removes this effect. In a usual ferromagnet with infinite-range interactions, the interaction scales as $J_{ij} = J/N$ and the third term is negligible since it is of $\mathcal{O}(1/N)$. In the SK model, however, we have $J_{ij}^2 = \mathcal{O}(1/N)$ and the third term is of the same order as the first and second terms and cannot be neglected. The TAP equation gives a basis to treat the spin glass problem without taking the configurational average over the distribution of $\boldsymbol{J}$.

The TAP equation (3.70) corresponds to the extremization condition of the following free energy:

$$
f_{\text{TAP}} = -\frac{1}{2} \sum_{i \neq j} J_{ij} m_i m_j - \sum_i h_i m_i - \frac{\beta}{4} \sum_{i \neq j} J_{ij}^2 (1 - m_i^2)(1 - m_j^2)
$$
$$
+ \frac{T}{2} \sum_i \left\{ (1 + m_i) \log \frac{1 + m_i}{2} + (1 - m_i) \log \frac{1 - m_i}{2} \right\}. \qquad (3.73)
$$

The first and second terms on the right hand side are for the internal energy, and the final term denotes the entropy. The third term corresponds to the reaction field.

This free energy can be derived from an expansion of the free energy with magnetization specified

$$
-\beta \tilde{f}(\alpha, \beta, \boldsymbol{m}) = \log \operatorname{Tr} e^{-\beta H(\alpha)} - \beta \sum_i h_i m_i, \qquad (3.74)
$$

where $H(\alpha) = \alpha H_0 - \sum_i h_i S_i$ and $\boldsymbol{m} = \{m_i\}$. The Hamiltonian H_0 denotes the usual SK model, and $h_i(\alpha, \beta, \boldsymbol{m})$ is the Lagrange multiplier to enforce the constraint $m_i = \langle S_i \rangle_\alpha$, where $\langle \cdot \rangle_\alpha$ is the thermal average with $H(\alpha)$. Expanding $\tilde{f}$ to second order in α around $\alpha = 0$ and setting α equal to one, we can derive (3.73). This is called the *Plefka expansion* (Plefka 1982).

To see it, let us first carry out the differentiations

$$
\frac{\partial \tilde{f}}{\partial \alpha} = \langle H_0 \rangle_\alpha \qquad (3.75)
$$

$$
\frac{\partial^2 \tilde{f}}{\partial \alpha^2} = -\beta \left\langle H_0 \left(H_0 - \langle H_0 \rangle_\alpha - \sum_i \frac{\partial h_i}{\partial \alpha} (S_i - m_i) \right) \right\rangle_\alpha. \qquad (3.76)
$$

The first two terms of the expansion $\tilde{f}(1) \approx \tilde{f}(0) + \tilde{f}'(0)$ give (3.73) except the Onsager term with J_{ij}^2 since

$$
\tilde{f}(0) = T \sum_i \left(\frac{1 + m_i}{2} \log \frac{1 + m_i}{2} + \frac{1 - m_i}{2} \log \frac{1 - m_i}{2} \right) \qquad (3.77)
$$

$$
\tilde{f}'(0) = -\frac{1}{2} \sum_{i \neq j} J_{ij} m_i m_j. \qquad (3.78)
$$

The second derivative (3.76) can be evaluated at $\alpha = 0$ from the relation

$$
\left. \frac{\partial h_i}{\partial \alpha} \right]_{\alpha=0} = \left. \frac{\partial}{\partial \alpha} \frac{\partial \tilde{f}}{\partial m_i} \right]_{\alpha=0} = -\sum_{j(\neq i)} J_{ij} m_j. \qquad (3.79)
$$

Inserting this relation into (3.76) we find

$$\tilde{f}''(0) = -\frac{1}{2}\beta \sum_{i \neq j} J_{ij}^2 (1 - m_i^2)(1 - m_j^2) \tag{3.80}$$

which gives the Onsager term in (3.73). It has hence been shown that $f_{\text{TAP}} = \tilde{f}(0) + \tilde{f}'(0) + \tilde{f}''(0)/2$. It can also be shown that the convergence condition of the above expansion for $\alpha \geq 1$ is equivalent to the stability condition of the free energy that the eigenvalues of the Hessian $\{\partial^2 f_{\text{TAP}}/\partial m_i \partial m_j\}$ be non-negative. All higher order terms in the expansion vanish in the thermodynamic limit as long as the stability condition is satisfied.

3.5.2 Cavity method

The *cavity method* is useful to derive the TAP equation from a different perspective (Mézard *et al.* 1986, 1987). It also attracts attention in relation to practical algorithms to solve information processing problems (Opper and Saad 2001). The argument in the present section is restricted to the case of the SK model for simplicity (Opper and Winther 2001).

Let us consider the local magnetization $m_i = \langle S_i \rangle$, where the thermal average is taken within a single valley. The goal is to show that this local magnetization satisfies the TAP equation. For this purpose it suffices to derive the distribution function of local spin $P_i(S_i)$, with which the above thermal average is carried out. It is assumed for simplicity that there is no external field $h_i = 0$. The local magnetization is determined by the local field $\tilde{h}_i = \sum_j J_{ij}S_j$. Hence the joint distribution of S_i and $\tilde{h}_i$ can be written as

$$P(S_i, \tilde{h}_i) \propto e^{\beta \tilde{h}_i S_i} P(\tilde{h}_i \setminus S_i), \tag{3.81}$$

where $P(\tilde{h}_i \setminus S_i)$ is the distribution of the local field when the spin S_i is removed from the system (i.e. when we set $J_{ij} = 0$ for all j) (*cavity field*). More explicitly,

$$P(\tilde{h}_i \setminus S_i) \equiv \text{Tr}_{\boldsymbol{S} \setminus S_i} \delta(\tilde{h}_i - \sum_j J_{ij}S_j) P(\boldsymbol{S} \setminus S_i), \tag{3.82}$$

where $P(\boldsymbol{S} \setminus S_i)$ is the probability distribution of the whole system without S_i ($J_{ij} = 0$ for all j). The distribution

$$P_i(S_i) \propto \int d\tilde{h}_i \, e^{\beta \tilde{h}_i S_i} P(\tilde{h}_i \setminus S_i) \tag{3.83}$$

is thus determined once $P(\tilde{h}_i \setminus S_i)$ is known.

In the SK model the range of interaction is unlimited and the number of terms appearing in the sum $\sum_j J_{ij}S_j$ is $N-1$. If all these terms are independent and identically distributed, then the central limit theorem assures that the cavity field $\tilde{h}_i$ is Gaussian distributed. This is certainly the case on the Bethe lattice (Fig. 3.6) that breaks up into independent trees as soon as a site is removed. Let

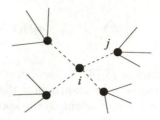

FIG. 3.6. There are no loops of bonds on the Bethe lattice, so that removal of any single site breaks the system into independent trees.

us assume that this is also true for the SK model in which the correlations of different sites are weak. We then have

$$P(\tilde{h}_i \setminus S_i) = \frac{1}{\sqrt{2\pi V_i^2}} \exp\left\{-\frac{(\tilde{h}_i - \langle \tilde{h}_i \rangle_{\setminus i})^2}{2V_i}\right\}.$$ (3.84)

Combination of this Gaussian form and (3.83) yields

$$m_i = \tanh \beta \langle \tilde{h}_i \rangle_{\setminus i}.$$ (3.85)

It is therefore necessary to evaluate the average $\langle \tilde{h}_i \rangle_{\setminus i}$.

The standard average of the local field (without the cavity),

$$\langle \tilde{h}_i \rangle = \mathrm{Tr}_{S_i} \int \mathrm{d}\tilde{h}_i \, \tilde{h}_i P(S_i, \tilde{h}_i),$$ (3.86)

together with the Gaussian cavity field (3.84) leads to the relation

$$\langle \tilde{h}_i \rangle = \langle \tilde{h}_i \rangle_{\setminus i} + V_i \langle S_i \rangle$$ (3.87)

or, in terms of m_i,

$$\langle \tilde{h}_i \rangle_{\setminus i} = \sum_j J_{ij} m_j - V_i m_i.$$ (3.88)

We then have to evaluate the variance of the local field

$$V_i = \sum_{j,k} J_{ij} J_{ik} (\langle S_j S_k \rangle_{\setminus i} - \langle S_j \rangle_{\setminus i} \langle S_k \rangle_{\setminus i}).$$ (3.89)

Only the diagonal terms ($j = k$) survive in the above sum because of the clustering property in a single valley

$$\frac{1}{N^2} \sum_{j,k} (\langle S_k S_j \rangle - \langle S_k \rangle \langle S_j \rangle)^2 \to 0$$ (3.90)

as $N \to \infty$ as well as the independence of J_{ij} and J_{ik} ($j \neq k$). We then have

$$V_i \approx \sum_j J_{ij}^2 (1 - \langle S_j \rangle_{\setminus i}^2) \approx \sum_j J_{ij}^2 (1 - \langle S_j \rangle^2) = \sum_j J_{ij}^2 (1 - m_j^2).$$ (3.91)

From this, (3.85), and (3.88), we arrive at the TAP equation (3.70).

The equations of state within the RS ansatz (2.28) and (2.30) can be derived from the TAP equation with the cavity method taken into account. We first separate the interaction J_{ij} into the ferromagnetic and random terms,

$$J_{ij} = \frac{J_0}{N} + \frac{J}{\sqrt{N}} z_{ij}, \tag{3.92}$$

where z_{ij} is a Gaussian random variable with vanishing mean and unit variance. Then (3.88) is

$$\langle \tilde{h}_i \rangle_{\backslash i} = \frac{J_0}{N} \sum_j m_j + \frac{J}{\sqrt{N}} \sum_j z_{ij} m_j - V_i m_i. \tag{3.93}$$

The first term on the right hand side is identified with $J_0 m$, where m is the ferromagnetic order parameter. The effects of the third term, the cavity correction, can be taken into account by treating only the second term under the assumption that each term $z_{ij} m_j$ is an independent quenched random variable; the whole expression of (3.93) is the thermal average of the cavity field and therefore the contribution from one site j would not interfere with that from another site j. Then the second term is Gaussian distributed according to the central limit theorem. The mean vanishes and the variance is

$$\sum_j \sum_k [z_{ij} z_{ik}] m_j m_k = \sum_j m_j^2 = Nq. \tag{3.94}$$

Thus the second term (with the third term taken into account) is expressed as $\sqrt{Nq}\, z$ with z being a Gaussian quenched random variable with vanishing mean and variance unity. Hence, averaging (3.85) over the distribution of z, we find

$$m = \int \mathrm{D}z \tanh \beta(J_0 m + \sqrt{Jq}\, z), \tag{3.95}$$

which is the RS equation of state (2.28). Averaging the square of (3.85) gives (2.30). One should remember that the amount of information in the TAP equation is larger than that of the RS equations of state because the former is a set of equations for N variables whereas the latter is for the macroscopic order parameters m and q. It is also possible to derive the results of RSB calculations with more elaborate arguments (Mézard et al. 1986).

3.5.3   Properties of the solution

To investigate the behaviour of the solution of the TAP equation (3.70) around the spin glass transition point, we assume that both the m_i and h_i are small and expand the right hand side to first order to obtain

$$m_i = \beta \sum_j J_{ij} m_j + \beta h_i - \beta^2 J^2 m_i. \tag{3.96}$$

This linear equation (3.96) can be solved by the eigenvalues and eigenvectors of the symmetric matrix $\boldsymbol{J}$. Let us for this purpose expand J_{ij} by its eigenvectors as

$$J_{ij} = \sum_\lambda J_\lambda \langle i|\lambda\rangle \langle \lambda|j\rangle. \tag{3.97}$$

We define the λ-magnetization and λ-field by

$$m_\lambda = \sum_i \langle \lambda|i\rangle m_i, \quad h_\lambda = \sum_i \langle \lambda|i\rangle h_i \tag{3.98}$$

and rewrite (3.96) as

$$m_\lambda = \beta m_\lambda J_\lambda + \beta h_\lambda - \beta^2 J^2 m_\lambda. \tag{3.99}$$

Thus the λ-susceptibility acquires the expression

$$\chi_\lambda = \frac{\partial m_\lambda}{\partial h_\lambda} = \frac{\beta}{1 - \beta J_\lambda + (\beta J)^2}. \tag{3.100}$$

The eigenvalues of the random matrix $\boldsymbol{J}$ are known to be distributed between $-2J$ and $2J$ with the density

$$\rho(J_\lambda) = \frac{\sqrt{4J^2 - J_\lambda^2}}{2\pi J^2}. \tag{3.101}$$

It is thus clear from (3.100) that the susceptibility corresponding to the largest eigenvalue $J_\lambda = 2J$ diverges at $T_f = J$, implying a phase transition. This transition point $T_f = J$ agrees with the replica result.

In a uniform ferromagnet, the uniform magnetization corresponding to the conventional susceptibility (which diverges at the transition point) develops below the transition point to form an ordered phase. Susceptibilities to all other external fields (such as a field with random sign at each site) do not diverge at any temperature. In the SK model, by contrast, there is a continuous spectrum of J_λ and therefore, according to (3.100), various modes continue to diverge one after another below the transition point T_f where the mode with the largest eigenvalue shows a divergent susceptibility. In this sense there exist continuous phase transitions below T_f. This fact corresponds to the marginal stability of the Parisi solution with zero eigenvalue of the Hessian and is characteristic of the spin glass phase of the SK model.

The local magnetization m_i^a which appeared in the argument of the multivalley structure in the previous section is considered to be the solution of the TAP equation that minimizes the free energy. Numerical analysis indicates that solutions of the TAP equation at low temperatures lie on the border of the stability condition, reminiscent of the marginal stability of the Parisi solution (Nemoto and Takayama 1985). General solutions of the TAP equation may, however, correspond to local minima, not the global minima of the free energy (3.73). It is indeed expected that the solutions satisfying the minimization condition of the free energy occupy only a fraction of the whole set of solutions of the TAP equation that has very many solutions of $\mathcal{O}(e^{aN})$ $(a > 0)$.

Bibliographical note

Extensive accounts of the scheme of the RSB can be found in Mézard *et al.* (1987), Binder and Young (1986), and Fischer and Hertz (1991). All of these volumes and van Hemmen and Morgenstern (1987) cover most of the developments related to the mean-field theory until the mid 1980s including dynamics and experiments not discussed in the present book. One of the major topics of high current activity in spin glass theory is slow dynamics. The goal is to clarify the mechanism of anomalous long relaxations in glassy systems at the mean-field level as well as in realistic finite-dimensional systems. Reviews on this problem are found in Young (1997) and Miyako *et al.* (2000). Another important issue is the existence and properties of the spin glass state in three (and other finite) dimensions. The mean-field theory predicts a very complicated structure of the spin glass state as represented by the full RSB scheme for the SK model. Whether or not this picture applies to three-dimensional systems is not a trivial problem, and many theoretical and experimental investigations are still going on. The reader will find summaries of recent activities in the same volumes as above (Young 1997; Miyako *et al.* 2000). See also Dotsenko (2001) for the renormalization group analyses of finite-dimensional systems.

4

GAUGE THEORY OF SPIN GLASSES

We introduced the mean-field theory of spin glasses in the previous chapters and saw that a rich structure of the phase space emerges from the replica symmetry breaking. The next important problem would be to study how reliable the predictions of mean-field theory are in realistic finite-dimensional systems. It is in general very difficult to investigate two- and three-dimensional systems by analytical methods, and current studies in this field are predominantly by numerical methods. It is not the purpose of this book, however, to review the status of numerical calculations; we instead introduce a different type of argument, the gauge theory, which uses the symmetry of the system to derive a number of rigorous/exact results. The gauge theory does not directly answer the problem of the existence of the spin glass phase in finite dimensions. Nevertheless it places strong constraints on the possible structure of the phase diagram. Also, the gauge theory will be found to be closely related to the Bayesian method frequently encountered in information processing problems to be discussed in subsequent chapters.

4.1 Phase diagram of finite-dimensional systems

The SK model may be regarded as the Edwards–Anderson model in the limit of infinite spatial dimension. The phase diagram of the finite-dimensional $\pm J$ Ising model (2.3) is expected to have a structure like Fig. 4.1. The case of $p = 1$ is the pure ferromagnetic Ising model with a ferromagnetic phase for $T < T_c$ and paramagnetic phase for $T > T_c$. As p decreases, antiferromagnetic interactions gradually destabilize the ferromagnetic phase, resulting in a decreased transition temperature. The ferromagnetic phase eventually disappears completely for p below a threshold p_c. Numerical evidence shows that the spin glass phase exists adjacent to the ferromagnetic phase if the spatial dimensionality is three or larger. There might be a mixed phase with the RSB at low temperatures within the ferromagnetic phase. The Gaussian model (2.2) is expected to have an analogous phase diagram.

It is very difficult to determine the structure of this phase diagram accurately, and active investigations, mainly numerical, are still going on. The gauge theory does not give a direct answer to the existence problem of the spin glass phase in finite dimensions, but it provides a number of powerful tools to restrict possibilities. It also gives the exact solution for the energy under certain conditions (Nishimori 1980, 1981; Morita and Horiguchi 1980; Horiguchi 1981).

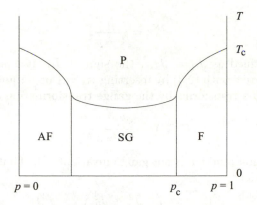

FIG. 4.1. Phase diagram of the $\pm J$ model

4.2 Gauge transformation

Let us consider the symmetry of the Edwards–Anderson model

$$H = -\sum_{\langle ij \rangle} J_{ij} S_i S_j \tag{4.1}$$

to show that a simple transformation of variables using the symmetry leads to a number of non-trivial conclusions. We do not restrict the range of the sum $\langle ij \rangle$ in (4.1) here; it could be over nearest neighbours or it may include farther pairs. We first discuss the $\pm J$ model and add comments on the Gaussian and other models later.

We define the *gauge transformation* of spins and interactions as follows:

$$S_i \to S_i \sigma_i, \quad J_{ij} \to J_{ij} \sigma_i \sigma_j. \tag{4.2}$$

Here σ_i is an Ising spin variable at site i fixed to either 1 or -1 arbitrarily independently of S_i. This transformation is performed at all sites. Then the Hamiltonian (4.1) is transformed as

$$H \to -\sum_{\langle ij \rangle} J_{ij} \sigma_i \sigma_j \cdot S_i \sigma_i \cdot S_j \sigma_j = H, \tag{4.3}$$

which shows that the Hamiltonian is *gauge invariant*.

To see how the probability distribution (2.3) of the $\pm J$ model changes under the gauge transformation, it is convenient to rewrite the expression (2.3) as

$$P(J_{ij}) = \frac{e^{K_p \tau_{ij}}}{2 \cosh K_p}, \tag{4.4}$$

where K_p is a function of the probability p,

$$e^{2K_p} = \frac{p}{1-p}. \tag{4.5}$$

In (4.4), τ_{ij} is defined as $J_{ij} = J\tau_{ij}$, the sign of J_{ij}. It is straightforward to check that (4.4) agrees with (2.3) by inserting $\tau_{ij} = 1$ or -1 and using (4.5). The distribution function transforms by the gauge transformation as

$$P(J_{ij}) \rightarrow \frac{e^{K_p \tau_{ij} \sigma_i \sigma_j}}{2 \cosh K_p}. \tag{4.6}$$

Thus the distribution function is not gauge invariant. The Gaussian distribution function transforms as

$$P(J_{ij}) \rightarrow \frac{1}{\sqrt{2\pi J^2}} \exp\left(-\frac{J_{ij}^2 + J_0^2}{2J^2}\right) \exp\left(\frac{J_0}{J^2} J_{ij} \sigma_i \sigma_j\right). \tag{4.7}$$

It should be noted here that the arguments developed below in the present chapter apply to a more generic model with a distribution function of the form

$$P(J_{ij}) = P_0(|J_{ij}|) e^{aJ_{ij}}. \tag{4.8}$$

The Gaussian model clearly has this form with $a = J_0/J^2$, and the same is true for the $\pm J$ model since its distribution can be expressed as

$$P(J_{ij}) = p\delta(\tau_{ij}-1) + (1-p)\delta(\tau_{ij}+1) = \frac{e^{K_p \tau_{ij}}}{2 \cosh K_p}\{\delta(\tau_{ij}-1) + \delta(\tau_{ij}+1)\} \tag{4.9}$$

for which we choose $a = K_p/J$. The probability distribution (4.8) transforms as

$$P(J_{ij}) \rightarrow P_0(|J_{ij}|) e^{aJ_{ij}\sigma_i\sigma_j}. \tag{4.10}$$

4.3   Exact solution for the internal energy

An appropriate application of gauge transformation allows us to calculate the exact value of the internal energy of the Edwards–Anderson model under a certain condition. We mainly explain the case of the $\pm J$ model and state only the results for the Gaussian model. Other models in the class (4.8) can be treated analogously.

4.3.1   Application of gauge transformation

The internal energy is the configurational average of the statistical-mechanical average of the Hamiltonian:

$$[E] = \left[\frac{\mathrm{Tr}_{\boldsymbol{S}}\, H e^{-\beta H}}{\mathrm{Tr}_{\boldsymbol{S}}\, e^{-\beta H}}\right] = \sum_{\tau} \frac{\exp(K_p \sum_{\langle ij \rangle} \tau_{ij})}{(2 \cosh K_p)^{N_B}}$$
$$\cdot \frac{\mathrm{Tr}_{\boldsymbol{S}}\left(-J \sum_{\langle ij \rangle} \tau_{ij} S_i S_j\right) \exp(K \sum_{\langle ij \rangle} \tau_{ij} S_i S_j)}{\mathrm{Tr}_{\boldsymbol{S}}\, \exp(K \sum_{\langle ij \rangle} \tau_{ij} S_i S_j)}. \tag{4.11}$$

Here $\mathrm{Tr}_{\boldsymbol{S}}$ denotes the sum over $\boldsymbol{S} = \{S_i = \pm 1\}$, $K = \beta J$, and N_B is the number of terms in the sum $\sum_{\langle ij \rangle}$ or the total number of interaction bonds, $N_B = |B|$.

We write Tr for the sum over spin variables at sites and reserve the symbol $\sum_\tau$ for variables $\{\tau_{ij}\}$ defined on bonds.

Let us now perform gauge transformation. The gauge transformation (4.2) just changes the order of sums in Tr_S appearing in (4.11) and those in $\sum_\tau$. For example, the sum over $S_i = \pm 1$ in the order '+1 first and then -1' is changed by the gauge transformation to the other order '-1 first and then 1', if $\sigma_i = -1$. Thus the value of the internal energy is independent of the gauge transformation. We then have

$$[E] = \sum_\tau \frac{\exp(K_p \sum \tau_{ij}\sigma_i\sigma_j)}{(2\cosh K_p)^{N_B}} \cdot \frac{\text{Tr}_S \left(-J\sum \tau_{ij}S_iS_j\right)\exp(K\sum \tau_{ij}S_iS_j)}{\text{Tr}_S \exp(K\sum \tau_{ij}S_iS_j)}, \quad (4.12)$$

where gauge invariance of the Hamiltonian has been used. It should further be noted that the value of the above formula does not depend on the choice of the gauge variables $\sigma \equiv \{\sigma_i\}$ (of the Ising type). This implies that the result remains unchanged if we sum the above equation over all possible values of σ and divide the result by 2^N, the number of possible configurations of gauge variables:

$$[E] = \frac{1}{2^N(2\cosh K_p)^{N_B}} \sum_\tau \text{Tr}_\sigma \exp(K_p \sum \tau_{ij}\sigma_i\sigma_j)$$

$$\cdot \frac{\text{Tr}_S \left(-J\sum \tau_{ij}S_iS_j\right)\exp(K\sum \tau_{ij}S_iS_j)}{\text{Tr}_S \exp(K\sum \tau_{ij}S_iS_j)}. \quad (4.13)$$

4.3.2 Exact internal energy

One observes in (4.13) that, if $K = K_p$, the sum over S in the denominator (the partition function) cancels out the sum over σ that has been obtained by gauge transformation from the probability distribution $P(J_{ij})$. Then the internal energy becomes

$$[E] = \frac{1}{2^N(2\cosh K)^{N_B}} \sum_\tau \text{Tr}_S \left(-J\sum_{\langle ij\rangle} \tau_{ij}S_iS_j\right) \exp(K\sum \tau_{ij}S_iS_j). \quad (4.14)$$

The sums over τ and S in (4.14) can be carried out as follows:

$$[E] = -\frac{J}{2^N(2\cosh K)^{N_B}} \sum_\tau \text{Tr}_S \frac{\partial}{\partial K} \exp(K\sum \tau_{ij}S_iS_j)$$

$$= -\frac{J}{2^N(2\cosh K)^{N_B}} \frac{\partial}{\partial K} \text{Tr}_S \prod_{\langle ij\rangle} \left(\sum_{\tau_{ij}=\pm 1} \exp(K\tau_{ij}S_iS_j)\right)$$

$$= -N_B J \tanh K. \quad (4.15)$$

This is the exact solution for the internal energy under the condition $K = K_p$. The above calculations hold for any lattice. Special features of each lattice are reflected only through N_B, the total number of bonds.

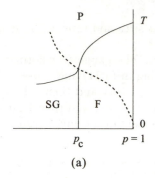 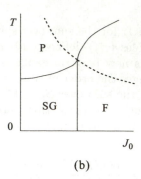

(a)   (b)

FIG. 4.2. Nishimori line (dashed) in the (a) $\pm J$ and (b) Gaussian models

4.3.3 *Relation with the phase diagram*

The condition $K = K_p$ relates the temperature $T\,(= J/K)$ and the probability $p\,(= (\tanh K_p + 1)/2)$, which defines a curve in the T–p phase diagram. The curve $K = K_p$ is called the *Nishimori line* and connects $(T = 0, p = 1)$ and $(T = \infty, p = 1/2)$ in the phase diagram of the $\pm J$ model (Fig. 4.2(a)).

The exact internal energy (4.15) on the Nishimori line has no singularity as a function of the temperature. The Nishimori line, on the other hand, extends from the ferromagnetic ground state at $(T = 0, p = 1)$ to the high-temperature limit $(T = \infty, p = 1/2)$ as shown in Fig. 4.2(a); it inevitably goes across a phase boundary. It might seem strange that the internal energy is non-singular when the line crosses a phase boundary at a transition point. We should, however, accept those two apparently contradicting results since (4.15) is, after all, the exact solution.[3] One possibility is that the singular part of the internal energy happens to vanish on the Nishimori line. This is probably a feature only of the internal energy, and the other physical quantities (e.g. the free energy, specific heat, and magnetic susceptibility) should have singularities at the crossing point. In almost all cases investigated so far, this transition point on the Nishimori line is a multicritical point where paramagnetic, ferromagnetic, and spin glass phases merge.

Similar arguments apply to the Gaussian model. The Nishimori line in this case is $J_0/J^2 = \beta$, from the cancellation condition of numerator and denominator as in the previous subsection. It is shown as the dashed line in Fig. 4.2(b). The energy for $J_0/J^2 = \beta$ is

$$[E] = -N_B J_0. \tag{4.16}$$

It is possible to confirm (4.16) for the infinite-range version of the Gaussian model, the SK model, with $h = 0$. One can easily verify that $m = q$ from the RS solution (2.28) and (2.30) under the condition $\beta J^2 = J_0$. The internal energy is, from the free energy (2.27),

---

[3]The existence of a finite region of ferromagnetic phase in two and higher dimensions has been proved (Horiguchi and Morita 1982*a, b*).

$$[E] = -\frac{N}{2} \left\{ J_0 m^2 + \beta J^2 (1 - q^2) \right\}. \tag{4.17}$$

Insertion of $m = q$ and $\beta J^2 = J_0$ into the above formula gives $[E] = -J_0 N/2$, which agrees in the limit $N \to \infty$ with (4.16) with $N_B = N(N-1)/2$ and $J_0 \to J_0/N$. Therefore the RS solution of the SK model is exact on the Nishimori line at least as far as the internal energy is concerned. The AT line lies below the Nishimori line and the RS solution is stable. It will indeed be proved in §4.6.3 that the structure of the phase space is always simple on the Nishimori line.

4.3.4 Distribution of the local energy

We can calculate the expectation value of the distribution function of the energy of a single bond $J_{ij} S_i S_j$

$$P(E) = [\langle \delta(E - J_{ij} S_i S_j) \rangle] \tag{4.18}$$

by the same method as above (Nishimori 1986a). Since $\delta(E - J_{ij} S_i S_j)$ is gauge invariant, arguments in §4.3 apply and the following relation corresponding to (4.14) is derived when $K = K_p$:

$$P(E) = \frac{1}{2^N (2 \cosh K)^{N_B}} \sum_{\tau} \text{Tr}_{\boldsymbol{S}} \, \delta(E - J_{ij} S_i S_j) \exp(K \sum \tau_{lm} S_l S_m). \tag{4.19}$$

Summing over bond variables other than the specific one (ij), which we are treating, can be carried out. The result cancels out with the corresponding factors in the denominator. The problem then reduces to the sum over the three variables $\tau_{ij}, S_i,$ and S_j, which is easily performed to yield

$$P(E) = p\delta(E - J) + (1 - p)\delta(E + J). \tag{4.20}$$

It is also possible to show that the simultaneous distribution of two different bonds is decoupled to the product of distributions of single bonds when $K = K_p$:

$$P_2(E_1, E_2) = [\langle \delta(E_1 - J_{ij} S_i S_j) \delta(E_2 - J_{kl} S_k S_l) \rangle] = P(E_1) P(E_2). \tag{4.21}$$

The same holds for distributions of more than two bonds. According to (4.20) and (4.21), when $K = K_p$, the local energy of a bond is determined independently of the other bonds or spin variables on average but depends only on the original distribution function (2.3). The same is true for the Gaussian model.

4.3.5 Distribution of the local field

The distribution function of the local field to site i

$$P(h) = [\langle \delta(h - \sum_j J_{ij} S_j) \rangle] \tag{4.22}$$

can be evaluated exactly if $K = K_p$ by the same method (Nishimori 1986a). Since the Hamiltonian (4.1) is invariant under the overall spin flip $S_i \to -S_i \, (\forall i)$, (4.22) is equal to

$$P(h) = \frac{1}{2}[\langle \delta(h - \sum_j J_{ij}S_j) \rangle + \langle \delta(h + \sum_j J_{ij}S_j) \rangle] \tag{4.23}$$

which is manifestly gauge invariant. We can therefore evaluate it as before under the condition $K = K_p$. After gauge transformation and cancellation of denominator and numerator, one is left with the variables related to site i to find

$$P(h) = \frac{1}{(2\cosh K)^z} \sum_\tau \frac{\delta(h - J\sum_j \tau_{ij}) + \delta(h + J\sum_j \tau_{ij})}{2} \exp(K \sum_j \tau_{ij})$$

$$= \frac{1}{(2\cosh K)^z} \sum_\tau \delta(h - J\sum_j \tau_{ij}) \cosh \beta h, \tag{4.24}$$

where z is the coordination number, and the sum over τ runs over the bonds connected to i. This result shows again that each bond connected to i behaves independently of other bonds and spins with the appropriate probability weight $p = e^{K_p}/2\cosh K_p$ or $1 - p = e^{-K_p}/2\cosh K_p$. The same argument for the Gaussian model leads to the distribution

$$P(h) = \frac{1}{2\sqrt{2\pi z}\,J} \left\{ \exp\left(-\frac{(h - z\beta J)^2}{2zJ^2}\right) + \exp\left(-\frac{(h + z\beta J)^2}{2zJ^2}\right) \right\} \tag{4.25}$$

when $\beta J^2 = J_0$.

4.4   Bound on the specific heat

It is not possible to derive the exact solution of the specific heat. We can nevertheless estimate its upper bound. The specific heat is the temperature derivative of the internal energy:

$$T^2[C] = -\frac{\partial[E]}{\partial\beta} = \left[\frac{\text{Tr}_S H^2 e^{-\beta H}}{\text{Tr}_S e^{-\beta H}} - \left(\frac{\text{Tr}_S H e^{-\beta H}}{\text{Tr}_S e^{-\beta H}} \right)^2 \right]. \tag{4.26}$$

For the $\pm J$ model, the first term of the above expression ($\equiv C_1$) can be calculated in the same manner as before:

$$C_1 = \sum_\tau \frac{\exp(K_p \sum \tau_{ij})}{(2\cosh K_p)^{N_B}} \cdot \frac{\text{Tr}_S (-J\sum \tau_{ij}S_iS_j)^2 \exp(K\sum \tau_{ij}S_iS_j)}{\text{Tr}_S \exp(K\sum \tau_{ij}S_iS_j)}$$

$$= \frac{J^2}{2^N(2\cosh K_p)^{N_B}} \sum_\tau \text{Tr}_\sigma \exp(K_p \sum \tau_{ij}\sigma_i\sigma_j)$$

$$\cdot \frac{(\partial^2/\partial K^2)\text{Tr}_S \exp(K\sum \tau_{ij}S_iS_j)}{\text{Tr}_S \exp(K\sum \tau_{ij}S_iS_j)}. \tag{4.27}$$

Cancellation of the denominator and numerator is observed when $K = K_p$ to give

$$C_1 = \frac{J^2}{2^N(2\cosh K)^{N_B}} \frac{\partial^2}{\partial K^2} \text{Tr}_S \exp(K \sum \tau_{ij}S_iS_j)$$

$$= \frac{J^2}{2^N (2\cosh K)^{N_B}} 2^N \frac{\partial^2}{\partial K^2} (2\cosh K)^{N_B}$$
$$= J^2 (N_B^2 \tanh^2 K + N_B \mathrm{sech}^2 K). \tag{4.28}$$

The second term on the right hand side of (4.26), C_2, cannot be evaluated directly but its lower bound is obtained by the Schwarz inequality:

$$C_2 = [E^2] \geq [E]^2 = J^2 N_B^2 \tanh^2 K. \tag{4.29}$$

From (4.28) and (4.29),

$$T^2[C] \leq J^2 N_B \mathrm{sech}^2 K. \tag{4.30}$$

Hence the specific heat on the Nishimori line does not diverge although the line crosses a phase boundary and thus the specific heat would be singular at the transition point.

For the Gaussian distribution, the upper bound on the specific heat is

$$T^2[C] \leq J^2 N_B. \tag{4.31}$$

4.5 Bound on the free energy and internal energy

We can derive an interesting inequality involving the free energy and, simultaneously, rederive the internal energy and the bound on the specific heat from an inequality on the Kullback–Leibler divergence (Iba 1999). Let us suppose that $P(x)$ and $Q(x)$ are probability distribution functions of a stochastic variable x. These functions satisfy the normalization condition $\sum_x P(x) = \sum_x Q(x) = 1$. The following quantity is called the *Kullback–Leibler divergence* of $P(x)$ and $Q(x)$:

$$G = \sum_x P(x) \log \frac{P(x)}{Q(x)}. \tag{4.32}$$

Since G vanishes when $P(x) = Q(x)$ $(\forall x)$, it measures the similarity of the two distributions. The Kullback–Leibler divergence is also called the *relative entropy*.

The Kullback–Leibler divergence is positive semi-definite:

$$G = \sum_x P(x) \left\{ \log \frac{P(x)}{Q(x)} + \frac{Q(x)}{P(x)} - 1 \right\} \geq 0. \tag{4.33}$$

Here we have used the inequality $-\log y + y - 1 \geq 0$ for positive y. The inequality (4.33) leads to an inequality on the free energy. We restrict ourselves to the $\pm J$ model for simplicity.

Let us choose the set of signs $\tau = \{\tau_{ij}\}$ of $J_{ij} \equiv J\tau_{ij}$ as the stochastic variable x and define $P(x)$ and $Q(x)$ as

$$P(\tau) = \frac{\mathrm{Tr}_{\boldsymbol{\sigma}} \exp(K_p \sum_{\langle ij \rangle} \tau_{ij}\sigma_i\sigma_j)}{2^N (2\cosh K_p)^{N_B}}, \quad Q(\tau) = \frac{\mathrm{Tr}_{\boldsymbol{\sigma}} \exp(K \sum_{\langle ij \rangle} \tau_{ij}\sigma_i\sigma_j)}{2^N (2\cosh K)^{N_B}}. \tag{4.34}$$

It is easy to check that these functions satisfy the normalization condition. Then (4.32) has the following expression:

$$G = \sum_\tau \frac{\text{Tr}_\sigma \, \exp(K_p \sum \tau_{ij}\sigma_i\sigma_j)}{2^N (2\cosh K_p)^{N_B}}$$
$$\cdot \left\{ \log \text{Tr}_\sigma \, \exp(K_p \sum \tau_{ij}\sigma_i\sigma_j) - \log \text{Tr}_\sigma \, \exp(K \sum \tau_{ij}\sigma_i\sigma_j) \right\}$$
$$- N_B \log 2\cosh K_p + N_B \log 2\cosh K. \tag{4.35}$$

This equation can be shown to be equivalent to the following relation by using gauge transformation:

$$G = \sum_\tau \frac{\exp(K_p \sum \tau_{ij})}{(2\cosh K_p)^{N_B}}$$
$$\cdot \left\{ \log \text{Tr}_\sigma \, \exp(K_p \sum \tau_{ij}\sigma_i\sigma_j) - \log \text{Tr}_\sigma \, \exp(K \sum \tau_{ij}\sigma_i\sigma_j) \right\}$$
$$- N_B \log 2\cosh K_p + N_B \log 2\cosh K. \tag{4.36}$$

The second term on the right hand side of (4.36) is nothing more than the logarithm of the partition function of the $\pm J$ model, the configurational average of which is the free energy $F(K, p)$ divided by temperature. The first term on the right hand side is the same quantity on the Nishimori line ($K = K_p$). Hence the inequality $G \geq 0$ is rewritten as

$$\beta F(K, p) + N_B \log 2\cosh K \geq \beta_p F(K_p, p) + N_B \log 2\cosh K_p. \tag{4.37}$$

The function $\beta F_0(K) = -N_B \log 2\cosh K$ is equal to the free energy of the one-dimensional $\pm J$ model with N_B bonds and free boundary condition. Then (4.37) is written as

$$\beta\{F(K, p) - F_0(K)\} \geq \beta_p\{F(K_p, p) - F_0(K_p)\}. \tag{4.38}$$

This inequality suggests that the system becomes closest to the one-dimensional model on the Nishimori line as far as the free energy is concerned.

Let us write the left hand side of (4.38) as $g(K, p)$. Minimization of $g(K, p)$ at $K = K_p$ gives the following relations:

$$\left. \frac{\partial g(K, p)}{\partial K} \right]_{K=K_p} = 0, \quad \left. \frac{\partial^2 g(K, p)}{\partial K^2} \right]_{K=K_p} \geq 0. \tag{4.39}$$

The equality in the second relation holds when $g(K, p)$ is flat at $K = K_p$. This happens, for instance, for the one-dimensional model where $g(K, p) = 0$ identically. However, such a case is exceptional and the strict inequality holds in most systems. By noting that the derivative of βF by β is the internal energy, we can confirm that the first equation of (4.39) agrees with the exact solution for the internal energy (4.15). The second formula of (4.39) is seen to be equivalent to the upper bound of the specific heat (4.30).

When the strict inequality holds in the second relation of (4.39), the following inequality follows for K close to K_p:

$$J \frac{\partial g(K,p)}{\partial K} = [E] + N_B J \tanh K \begin{cases} < 0 & (K < K_p) \\ > 0 & (K > K_p). \end{cases} \tag{4.40}$$

The energy thus satisfies

$$[E(T,p)] \begin{cases} < -N_B J \tanh K & (T > T_p = J/K_p) \\ > -N_B J \tanh K & (T < T_p) \end{cases} \tag{4.41}$$

when T is not far away from T_p. The internal energy for generic T and p is smaller (larger) than the one-dimensional value when T is slightly larger (smaller) than T_p corresponding to the point on the Nishimori line for the given p.

4.6 Correlation functions

One can apply the gauge theory to correlation functions to derive an upper bound that strongly restricts the possible structure of the phase diagram (Nishimori 1981; Horiguchi and Morita 1981). For simplicity, the formulae below are written only in terms of two-point correlation functions (the expectation value of the product of two spin variables) although the same arguments apply to any other many-point correlation functions. It will also be shown that the distribution function of the spin glass order parameter has a simple structure on the Nishimori line (Nishimori and Sherrington 2001; Gillin *et al.* 2001) and that the spin configuration is a non-monotonic function of the temperature (Nishimori 1993).

4.6.1 *Identities*

Let us consider the $\pm J$ model. The two-point correlation function is defined by

$$[\langle S_0 S_r \rangle_K] = \left[\frac{\text{Tr}_S \, S_0 S_r e^{-\beta H}}{\text{Tr}_S \, e^{-\beta H}} \right]$$

$$= \sum_\tau \frac{\exp(K_p \sum \tau_{ij})}{(2 \cosh K_p)^{N_B}} \cdot \frac{\text{Tr}_S \, S_0 S_r \exp(K \sum \tau_{ij} S_i S_j)}{\text{Tr}_S \, \exp(K \sum \tau_{ij} S_i S_j)}. \tag{4.42}$$

A gauge transformation changes the above expression to

$$[\langle S_0 S_r \rangle_K] = \frac{1}{2^N (2 \cosh K_p)^{N_B}} \sum_\tau \text{Tr}_\sigma \, \sigma_0 \sigma_r \exp(K_p \sum \tau_{ij} \sigma_i \sigma_j)$$

$$\cdot \frac{\text{Tr}_S \, S_0 S_r \exp(K \sum \tau_{ij} S_i S_j)}{\text{Tr}_S \, \exp(K \sum \tau_{ij} S_i S_j)}. \tag{4.43}$$

We do not observe a cancellation of the numerator and denominator here even when $K = K_p$ because of the factor $\sigma_0 \sigma_r$ caused by the gauge transformation of

$S_0 S_r$. However, an interesting identity results if one inserts the partition function to the numerator and denominator:

$$[\langle S_0 S_r \rangle_K] = \frac{1}{2^N (2 \cosh K_p)^{N_B}} \sum_\tau \left\{ \mathrm{Tr}_{\boldsymbol{\sigma}} \, \exp(K_p \sum \tau_{ij} \sigma_i \sigma_j) \right\}$$

$$\cdot \left(\frac{\mathrm{Tr}_{\boldsymbol{\sigma}} \, \sigma_0 \sigma_r \exp(K_p \sum \tau_{ij} \sigma_i \sigma_j)}{\mathrm{Tr}_{\boldsymbol{\sigma}} \, \exp(K_p \sum \tau_{ij} \sigma_i \sigma_j)} \right) \cdot \left(\frac{\mathrm{Tr}_{\boldsymbol{S}} \, S_0 S_r \exp(K \sum \tau_{ij} S_i S_j)}{\mathrm{Tr}_{\boldsymbol{S}} \, \exp(K \sum \tau_{ij} S_i S_j)} \right). \quad (4.44)$$

The last two factors here represent correlation functions $\langle \sigma_0 \sigma_r \rangle_{K_p}$ and $\langle S_0 S_r \rangle_K$ for interaction strengths K_p and K, respectively. The above expression turns out to be equal to the configurational average of the product of these two correlation functions

$$[\langle S_0 S_r \rangle_K] = [\langle \sigma_0 \sigma_r \rangle_{K_p} \langle S_0 S_r \rangle_K]. \quad (4.45)$$

To see this, we write the definition of the right hand side as

$$[\langle \sigma_0 \sigma_r \rangle_{K_p} \langle S_0 S_r \rangle_K] = \frac{1}{(2 \cosh K_p)^{N_B}} \sum_\tau \exp(K_p \sum \tau_{ij})$$

$$\cdot \frac{\mathrm{Tr}_{\boldsymbol{S}} \, S_0 S_r \exp(K_p \sum \tau_{ij} S_i S_j)}{\mathrm{Tr}_{\boldsymbol{S}} \, \exp(K_p \sum \tau_{ij} S_i S_j)} \cdot \frac{\mathrm{Tr}_{\boldsymbol{S}} \, S_0 S_r \exp(K \sum \tau_{ij} S_i S_j)}{\mathrm{Tr}_{\boldsymbol{S}} \, \exp(K \sum \tau_{ij} S_i S_j)}. \quad (4.46)$$

Here we have used the variable $\boldsymbol{S}$ instead of $\boldsymbol{\sigma}$ in writing $\langle \sigma_0 \sigma_r \rangle_{K_p}$. The result is independent of such a choice because, after all, we sum it over ± 1. The product of the two correlation functions in (4.46) is clearly gauge invariant. Thus, after gauge transformation, (4.46) is seen to be equal to (4.44), and (4.45) has been proved.

By taking the limit $r \to \infty$ in (4.45) with $K = K_p$, site 0 should become independent of site r so that the left hand side would approach $[\langle S_0 \rangle_K][\langle S_r \rangle_K]$, the square of the ferromagnetic order parameter m. The right hand side, on the other hand, approaches $[\langle \sigma_0 \rangle_K \langle S_0 \rangle_K][\langle \sigma_r \rangle_K \langle S_r \rangle_K]$, the square of the spin glass order parameter q. We therefore have $m = q$ on the Nishimori line. Since the spin glass phase has $m = 0$ and $q > 0$ by definition, we conclude that the Nishimori line never enters the spin glass phase (if any). In other words, we have obtained a restriction on the possible location of the spin glass phase. The result $m = q$ will be confirmed from a different argument in §4.6.3.

Another interesting relation on the correlation function can be derived as follows. Let us consider the configurational average of the inverse of the correlation function $[\langle S_0 S_r \rangle_K^{-1}]$. The same manipulation as above leads to the following relation:

$$\left[\frac{1}{\langle S_0 S_r \rangle_K} \right] = \left[\frac{\langle \sigma_0 \sigma_r \rangle_{K_p}}{\langle S_0 S_r \rangle_K} \right]. \quad (4.47)$$

The right hand side is unity if $K = K_p$. Therefore the expectation value of the inverse of an arbitrary correlation function is one on the Nishimori line. This is not as unnatural as it might seem; the inverse correlation $\langle S_0 S_r \rangle_K^{-1}$ is either greater than 1 or less than -1 depending upon its sign. The former contribution dominates on the Nishimori line (a part of which lies within the ferromagnetic phase), resulting in the positive constant.

A more general relation holds if we multiply the numerator of the above equation (4.47) by an arbitrary gauge-invariant quantity Q:

$$\left[\frac{Q}{\langle S_0 S_r\rangle_K}\right] = \left[\frac{Q\langle\sigma_0\sigma_r\rangle_{K_p}}{\langle S_0 S_r\rangle_K}\right]. \tag{4.48}$$

Here Q may be, for instance, the energy at an arbitrary temperature $\langle H\rangle_{K'}$ or the absolute value of an arbitrary correlation $|\langle S_i S_j\rangle_{K'}|$. This identity (4.48) shows that the gauge-invariant quantity Q is completely uncorrelated with the inverse of the correlation function on the Nishimori line because the configurational average decouples:

$$\left[\frac{Q}{\langle S_0 S_r\rangle_{K_p}}\right] = [Q]\left(= [Q]\left[\frac{1}{\langle S_0 S_r\rangle_{K_p}}\right]\right). \tag{4.49}$$

It is somewhat counter-intuitive that any gauge-invariant quantity Q takes a value independent of an arbitrarily chosen inverse correlation function. More work should be done to clarify the significance of this result.

4.6.2 Restrictions on the phase diagram

A useful inequality is derived from the correlation identity (4.45). By taking the absolute values of both sides of (4.45), we find

$$|[\langle S_0 S_r\rangle_K]| = \left|[\langle\sigma_0\sigma_r\rangle_{K_p}\langle S_0 S_r\rangle_K]\right| \le [|\langle\sigma_0\sigma_r\rangle_{K_p}|\cdot|\langle S_0 S_r\rangle_K|] \le [|\langle\sigma_0\sigma_r\rangle_{K_p}|]. \tag{4.50}$$

It has been used here that an upper bound is obtained by taking the absolute value before the expectation value and that the correlation function does not exceed unity.

The right hand side of (4.50) represents a correlation function on the Nishimori line $K = K_p$. This is a correlation between site 0 and site r, ignoring the sign of the usual correlation function $\langle\sigma_0\sigma_r\rangle_{K_p}$. It does not decay with increasing r if spins are frozen at each site as in the spin glass and ferromagnetic phases. The left hand side, on the other hand, reduces to the square of the usual ferromagnetic order parameter in the limit $r \to \infty$ and therefore approaches zero in the spin glass and paramagnetic phases. The right hand side of (4.50) vanishes as $r \to \infty$ if the point on the Nishimori line corresponding to a given p lies within the paramagnetic phase. Then the left hand side vanishes irrespective of K, implying the absence of ferromagnetic ordering. This fact can be interpreted as follows.

Let us define a point A as the crossing of the vertical (constant p) line L (dash–dotted in Fig. 4.3) and the Nishimori line (shown dashed). If A lies in the paramagnetic phase as in Fig. 4.3, no point on L is in the ferromagnetic phase by the above argument. We therefore conclude that the phase boundary

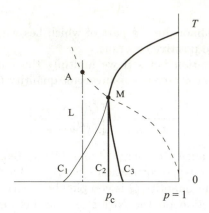

FIG. 4.3. Structure of the phase diagram compatible with the correlation inequality: C_2 and C_3. The shape C_1 is not allowed.

between the ferromagnetic and spin glass (or paramagnetic, in the absence of the
spin glass phase) phases does not extend below the spin glass (or paramagnetic)
phase like C_1. The boundary is either vertical as C_2 or re-entrant as C_3 where
the spin glass (or paramagnetic) phase lies below the ferromagnetic phase. It
can also be seen that there is no ferromagnetic phase to the left of the crossing
point M of the Nishimori line and the boundary between the ferromagnetic and
non-ferromagnetic phases. It is then natural to expect that M is a multicritical
point (where paramagnetic, ferromagnetic, and spin glass phases merge) as has
been confirmed by the renormalization-group and numerical calculations cited
at the end of this chapter.

4.6.3   Distribution of order parameters

The remarkable simplicity of the exact energy (4.15) and independence of energy
distribution (4.21) suggest that the state of the system would be a simple one on
the Nishimori line. This observation is reinforced by the relation $q = m$, which is
interpreted as the absence of the spin glass phase. We show in the present subsection that a general relation between the order parameter distribution functions
confirms this conclusion. In particular, we prove that the distribution functions
of q and m coincide, $P_q(x) = P_m(x)$, a generalization of the relation $q = m$. Since
the magnetization shows no RSB, the structure of $P_m(x)$ is simple (i.e. composed
of at most two delta functions). Then the relation $P_q(x) = P_m(x)$ implies that
$P_q(x)$ is also simple, leading to the absence of a complicated structure of the
phase space on the Nishimori line.

The distribution function of the spin glass order parameter for a generic
finite-dimensional system is defined using two replicas of the system with spins
σ and S:

$$P_q(x) = \left[\frac{\mathrm{Tr}_{\sigma}\,\mathrm{Tr}_{S}\,\delta(x - \frac{1}{N}\sum_i \sigma_i S_i)\,e^{-\beta H(\sigma) - \beta H(S)}}{\mathrm{Tr}_{\sigma}\,\mathrm{Tr}_{S}\,e^{-\beta H(\sigma) - \beta H(S)}} \right], \qquad (4.51)$$

where the Hamiltonians are

$$H(\sigma) = -\sum J_{ij}\sigma_i \sigma_j, \quad H(S) = -\sum J_{ij} S_i S_j. \qquad (4.52)$$

The two replicas share the same set of bonds J. There is no interaction between the two replicas. If the system has a complicated phase space, the spins take various different states so that the overlap of two independent spin configurations $\sum_i \sigma_i S_i/N$ is expected to have more than two values ($\pm m^2$) for a finite-step RSB or a continuous spectrum for full RSB as in the SK model.

It is convenient to define a generalization of $P_q(x)$ that compares spin configurations at two different (inverse) temperatures β_1 and β_2,

$$P_q(x; \beta_1 J, \beta_2 J) = \left[\frac{\mathrm{Tr}_{\sigma}\,\mathrm{Tr}_{S}\,\delta(x - \frac{1}{N}\sum_i \sigma_i S_i)\,e^{-\beta_1 H(\sigma) - \beta_2 H(S)}}{\mathrm{Tr}_{\sigma}\,\mathrm{Tr}_{S}\,e^{-\beta_1 H(\sigma) - \beta_2 H(S)}} \right]. \qquad (4.53)$$

The distribution of magnetization is defined similarly but without replicas:

$$P_m(x) = \left[\frac{\mathrm{Tr}_{S}\,\delta(x - \frac{1}{N}\sum_i S_i)\,e^{-\beta H(S)}}{\mathrm{Tr}_{S}\,e^{-\beta H(S)}} \right]. \qquad (4.54)$$

By applying the same procedure as in §4.6.1 to the right hand side of (4.54), we find

$$P_m(x) = \frac{1}{2^N (2\cosh K_p)^{N_B}} \sum_{\tau} \mathrm{Tr}_{\sigma}\,\exp(K_p \sum \tau_{ij}\sigma_i \sigma_j)$$

$$\cdot \frac{\mathrm{Tr}_{\sigma}\,\mathrm{Tr}_{S}\,\delta(x - \frac{1}{N}\sum_i \sigma_i S_i)\exp(K_p \sum \tau_{ij}\sigma_i \sigma_j)\exp(K \sum \tau_{ij} S_i S_j)}{\mathrm{Tr}_{\sigma}\,\mathrm{Tr}_{S}\,\exp(K_p \sum \tau_{ij}\sigma_i \sigma_j)\exp(K \sum \tau_{ij} S_i S_j)}. \qquad (4.55)$$

Similarly, gauge transformation of the right hand side of (4.53) yields the same expression as above when $\beta_1 J = K_p$ and $\beta_2 J = K$. We therefore have

$$P_m(x; K) = P_q(x; K_p, K). \qquad (4.56)$$

Here the K-dependence of the left hand side has been written out explicitly. By setting $K = K_p$, we obtain $P_q(x) = P_m(x)$. This relation shows that the phase space is simple on the Nishimori line as mentioned above. Comparison of the first moments of $P_q(x)$ and $P_m(x)$ proves that $q = m$. Since $q = m$ means that the ordered state on the Nishimori line should be a ferromagnetic phase, the absence of complicated phase space implies the absence of a mixed phase (ferromagnetic phase with complicated phase space).

Equation (4.56) may be interpreted that the spin state at K projected to the spin state at K_p (the right hand side) always has a simple distribution function

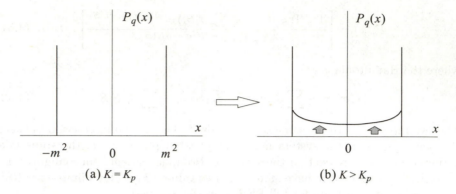

FIG. 4.4. The distribution function of the spin glass order parameter on the Nishimori line is simple (a). If there exists a phase with full RSB immediately below the Nishimori line, the derivative of $P_q(x)$ with respect to K should be positive in a finite range of x (b).

(the left hand side). Physically, this implies that the spin state at K_p, on the Nishimori line, is much like a perfect ferromagnetic state since the left hand side of the above equation represents the distribution of the spin state relative to the perfect ferromagnetic state, $\sum_i (S_i \cdot 1)$. This observation will be confirmed from a different point of view in §4.6.4.

More information can be extracted from (4.56) on the possible location of the AT line. Differentiation of both sides of (4.56) by K at $K = K_p$ yields

$$\frac{\partial}{\partial K} P_m(x; K)\bigg]_{K=K_p} = \frac{\partial}{\partial K} P_q(x; K_p, K)\bigg]_{K=K_p} = \frac{1}{2}\frac{\partial}{\partial K} P_q(x; K, K)\bigg]_{K=K_p}. \tag{4.57}$$

The left hand side is

$$\frac{\partial}{\partial K} P_m(x; K) = -\frac{1}{2}\delta'(x - m(K))m'(K) + \frac{1}{2}\delta'(x + m(K))m'(K) \tag{4.58}$$

which vanishes at almost all $x\,(\neq \pm m(K))$. The right hand side thus vanishes at $x \neq \pm m(K)$. It follows that there does not exist a phase with full RSB immediately below the Nishimori line $K = K_p$; otherwise the derivative of $P_q(x; K, K)$ with respect to the inverse temperature K should be positive in a finite range of x to lead to a continuous spectrum of $P_q(x; K, K)$ at a point slightly below the Nishimori line, see Fig. 4.4. Clearly the same argument applies to any step of the RSB because the right hand side of (4.57) would have non-vanishing (divergent) values at some $x \neq \pm m(K)$ if a finite-step RSB occurs just below the Nishimori line. Therefore we conclude that the Nishimori line does not coincide with the AT line marking the onset of RSB if any. Note that the present argument does not exclude the anomalous possibility of the RSB just below the Nishimori line with infinitesimally slow emergence of the non-trivial structure like $P_q(x) \propto f(x)\,e^{-1/(K-K_p)}$.

It is possible to develop the same argument for the Gaussian model.

4.6.4 Non-monotonicity of spin configurations

Let us next investigate how spin orientations are aligned with each other when we neglect the reduction of spin magnitude by thermal fluctuations in the $\pm J$ model. We only look at the sign of correlation functions:

$$\left[\frac{\langle S_0 S_r \rangle_K}{|\langle S_0 S_r \rangle_K|}\right] = \frac{1}{(2\cosh K_p)^{N_B}} \sum_\tau \exp(K_p \sum \tau_{ij}) \frac{\mathrm{Tr}_S \, S_0 S_r \exp(K \sum \tau_{ij} S_i S_j)}{|\mathrm{Tr}_S \, S_0 S_r \exp(K \sum \tau_{ij} S_i S_j)|}.$$
(4.59)

After gauge transformation, we find

$$\left[\frac{\langle S_0 S_r \rangle_K}{|\langle S_0 S_r \rangle_K|}\right]$$

$$= \frac{1}{2^N (2\cosh K_p)^{N_B}} \sum_\tau \mathrm{Tr}_\sigma \, \exp(K_p \sum \tau_{ij}\sigma_i\sigma_j) \langle \sigma_0 \sigma_r \rangle_{K_p} \frac{\langle S_0 S_r \rangle_K}{|\langle S_0 S_r \rangle_K|}$$

$$\leq \frac{1}{2^N (2\cosh K_p)^{N_B}} \sum_\tau \mathrm{Tr}_\sigma \, \exp(K_p \sum \tau_{ij}\sigma_i\sigma_j) |\langle \sigma_0 \sigma_r \rangle_{K_p}|.$$
(4.60)

We have taken the absolute value and replaced the sign of $\langle S_0 S_r \rangle_K$ by its upper bound 1. The right hand side is equivalent to

$$\left[\frac{\langle \sigma_0 \sigma_r \rangle_{K_p}}{|\langle \sigma_0 \sigma_r \rangle_{K_p}|}\right]$$
(4.61)

because, by rewriting (4.61) using the gauge transformation, we have

$$\left[\frac{\langle \sigma_0 \sigma_r \rangle_{K_p}}{|\langle \sigma_0 \sigma_r \rangle_{K_p}|}\right]$$

$$= \frac{1}{2^N (2\cosh K_p)^{N_B}} \sum_\tau \frac{\{\mathrm{Tr}_\sigma \, \sigma_0 \sigma_r \exp(K_p \sum \tau_{ij}\sigma_i\sigma_j)\}^2}{|\mathrm{Tr}_\sigma \, \sigma_0 \sigma_r \exp(K_p \sum \tau_{ij}\sigma_i\sigma_j)|}$$

$$= \frac{1}{2^N (2\cosh K_p)^{N_B}} \sum_\tau \left|\mathrm{Tr}_\sigma \, \sigma_0 \sigma_r \exp(K_p \sum \tau_{ij}\sigma_i\sigma_j)\right|$$

$$= \frac{1}{2^N (2\cosh K_p)^{N_B}} \sum_\tau \mathrm{Tr}_\sigma \, \exp(K_p \sum \tau_{ij}\sigma_i\sigma_j) |\langle \sigma_0 \sigma_r \rangle_{K_p}|.$$
(4.62)

Thus the following relation has been proved:

$$[\mathrm{sgn}\langle \sigma_0 \sigma_r \rangle_K] \leq [\mathrm{sgn}\langle S_0 S_r \rangle_{K_p}].$$
(4.63)

This inequality shows that the expectation value of the relative orientation of two arbitrarily chosen spins is a maximum at $K = K_p$ as a function of K with p fixed. Spins become best aligned with each other on the Nishimori line

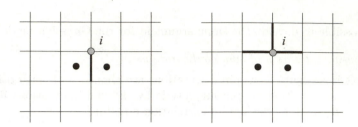

FIG. 4.5. Different bond configurations with the same distribution of frustration. Bold lines denote antiferromagnetic interactions. Black dots indicate
the frustrated plaquettes ($f_c = -1$). One of the two configurations changes
to the other by the gauge transformation with $\sigma_i = -1$.

when the temperature is decreased from a high value at a fixed p, and then the
relative orientation decreases as the temperature is further lowered, implying
non-monotonic behaviour of spin alignment. Note that the correlation function
itself $[\langle S_0 S_r \rangle_K]$ is not expected to be a maximum on the Nishimori line.

4.7   Entropy of frustration

We further develop an argument for the $\pm J$ model that the phase boundary
below the Nishimori line is expected to be vertical like C_2 of Fig. 4.3 (Nishimori
1986b). Starting from the definition of the configurational average of the free
energy

$$-\beta[F] = \sum_\tau \frac{\exp(K_p \sum \tau_{ij})}{(2\cosh K_p)^{N_B}} \cdot \log \mathrm{Tr}_S \exp(K \sum \tau_{ij} S_i S_j), \qquad (4.64)$$

we can derive the following expression by gauge transformation under the condition $K = K_p$:

$$-\beta[F] = \frac{1}{2^N (2\cosh K)^{N_B}} \sum_\tau \mathrm{Tr}_\sigma \exp(K \sum \tau_{ij}\sigma_i\sigma_j) \cdot \log \mathrm{Tr}_S \exp(K \sum \tau_{ij} S_i S_j)$$

$$\equiv \frac{1}{2^N (2\cosh K)^{N_B}} \sum_\tau Z(K) \log Z(K). \qquad (4.65)$$

Let us recall here that $Z(K)$ in front of $\log Z(K)$ was obtained from the gauge
transformation of the distribution function $P(J_{ij})$ and the sum over gauge variables. Since gauge transformation does not change the product of bonds $f_c =
\prod_c J_{ij}$ over an arbitrary closed loop c, this f_c is a gauge-invariant quantity,
called *frustration* (see Fig. 4.5).[4] The sum of all bond configurations with the

---

[4] A more accurate statement is that the loop c is said to be frustrated when $f_c < 0$. One often
talks about frustration of the smallest possible loop, a *plaquette* (the basic square composed of
four bonds in the case of the square lattice, for example).

same distribution of frustration $\{f_c\}$ gives $Z(K_p)$ (up to a normalization factor). This $Z(K_p)$ is therefore identified with the probability of the distribution of frustration. Then, (4.65) may be regarded as the average of the logarithm of the probability of frustration distribution on the Nishimori line, which is nothing but the entropy of frustration distribution. We are therefore able to interpret the free energy on the Nishimori line as the entropy of frustration distribution.

It should be noted here that the distribution of frustration is determined only by the bond configuration $\boldsymbol{J}$ and is independent of temperature. Also, it is expected that the free energy is singular at the point M in Fig. 4.3 where the Nishimori line crosses the boundary between the ferromagnetic and non-ferromagnetic phases, leading to a singularity in the frustration distribution. These observations indicate that the singularity in the free energy at M is caused by a sudden change of the frustration distribution, which is of geometrical nature. In other words, the frustration distribution is singular at the same $p(= p_c)$ as the point M as one changes p with temperature fixed. This singularity should be reflected in singularities in physical quantities at $p = p_c$. Our conclusion is that there is a vertical phase boundary at the same p as M. It should be remembered that singularities at higher temperatures than the point M are actually erased by large thermal fluctuations. This argument is not a rigorous proof for a vertical boundary, but existing numerical results are compatible with this conclusion (see the bibliographical note at the end of the chapter).

Singularities in the distribution of frustration are purely of a geometrical nature independent of spin variables. It is therefore expected that the location of a vertical boundary is universal, shared by, for instance, the XY model on the same lattice if the distribution of J_{ij} is the same (Nishimori 1992).

4.8   Modified $\pm J$ model

The existence of a vertical phase boundary discussed in §4.7 can be confirmed also from the following argument (Kitatani 1992). The probability distribution function of interactions of the $\pm J$ model is given as in (4.4) for each bond. It is instructive to modify this distribution and introduce the *modified $\pm J$ model* with the following distribution:

$$P_M(K_p, a, \boldsymbol{\tau}) = \frac{\exp\{(K_p + a)\sum_{\langle ij \rangle} \tau_{ij}\}Z(K_p, \boldsymbol{\tau})}{(2\cosh K_p)^{N_B} Z(K_p + a, \boldsymbol{\tau})}, \tag{4.66}$$

where a is a real parameter. Equation (4.66) reduces to the usual $\pm J$ model when $a = 0$. It is straightforward to show that (4.66) satisfies the normalization condition by summing it over $\boldsymbol{\tau}$ and then using gauge transformation.

4.8.1   *Expectation value of physical quantities*

The expectation value of a gauge-invariant quantity in the modified $\pm J$ model coincides with that of the conventional $\pm J$ model. We denote by $\{\cdots\}_{K_p}^a$ the configurational average by the probability (4.66) and $[\cdots]_{K_p}$ for the configurational average in the conventional $\pm J$ model. To prove the coincidence, we first write the definition of the configurational average of a gauge-invariant quantity Q in the modified $\pm J$ model and apply gauge transformation to it. Then we sum it over the gauge variables $\boldsymbol{\sigma}$ to find that $Z(K_p + a, \boldsymbol{\tau})$ appearing in both the numerator and denominator cancel to give

$$\{Q\}_{K_p}^a = \frac{1}{2^N (2 \cosh K_p)^{N_B}} \sum_{\boldsymbol{\tau}} Z(K_p, \boldsymbol{\tau}) Q = [Q]_{K_p}. \qquad (4.67)$$

The final equality can be derived by applying gauge transformation to the definition of $[Q]_{K_p}$ and summing the result over gauge variables. Equation (4.67) shows that the configurational average of a gauge-invariant quantity is independent of a.

Let us next derive a few relations for correlation functions by the same method as in the previous sections. If we take the limit $r \to \infty$ in (4.45) (which holds for the conventional $\pm J$ model), the left hand side reduces to the squared magnetization $m(K, K_p)^2$. Similarly, when $K = K_p$, the right hand side approaches the square of the usual spin glass order parameter $q(K_p, K_p)^2$. We thus have

$$m(K_p, K_p) = q(K_p, K_p). \qquad (4.68)$$

The corresponding relation for the modified $\pm J$ model is

$$m_{\mathrm{M}}(K_p + a, K_p) = q_{\mathrm{M}}(K_p + a, K_p), \qquad (4.69)$$

where the subscript M denotes that the quantities are for the modified $\pm J$ model. Another useful relation is, for general K,

$$q(K, K_p) = q_{\mathrm{M}}(K, K_p), \qquad (4.70)$$

which is valid according to (4.67) because the spin glass order parameter q is gauge invariant. It is also not difficult to prove that

$$m(K_p + a, K_p) = m_{\mathrm{M}}(K_p, K_p) \qquad (4.71)$$

from gauge transformation.

4.8.2 *Phase diagram*

Various formulae derived in the previous subsection are useful to show the close relationship between the phase diagrams of the modified and conventional $\pm J$ models. First of all, we note that the spin glass phase exists in the same region in both models according to (4.70). In the conventional $\pm J$ model, (4.68) implies that $q > 0$ if $m > 0$ on the Nishimori line $K = K_p$. Thus there does not exist a spin glass phase ($q > 0, m = 0$) when $K = K_p$, and the ordered phase at low temperatures on the Nishimori line (the part with $p > p_c$ in Fig. 4.6(a)) should be the ferromagnetic phase. Accordingly, the ordered phase to the upper right side of the Nishimori line cannot be the spin glass phase but is the ferromagnetic phase.

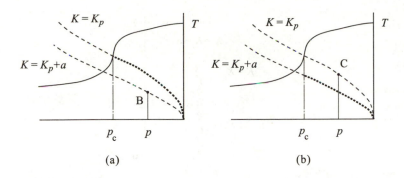

FIG. 4.6. Phase diagram of the conventional (a) and modified (b) $\pm J$ models

In the modified $\pm J$ model, on the other hand, (4.69) holds when $K = K_p + a$, so that the lower part of the curve $K = K_p + a$ is in the ferromagnetic phase (the part with $p > p_c$ in Fig. 4.6(b)). It is very plausible that the ordered phase to the upper right of the line $K = K_p + a$ is the ferromagnetic phase, similar to the case of the conventional $\pm J$ model.

We next notice that the magnetization at the point B on the line $K = K_p + a$ in the conventional $\pm J$ model (Fig. 4.6(a)) is equal to that at C on $K = K_p$ in the modified $\pm J$ model (Fig. 4.6(b)) according to (4.71). Using the above argument that the ferromagnetic phase exists to the upper right of $K = K_p + a$ in the modified $\pm J$ model, we see that $m_M(K_p, K_p) > 0$ at C and thus $m(K_p + a, K_p) > 0$ at B. If we vary $a\,(> 0)$ with p fixed, B moves along the vertical line below $K = K_p$. Therefore, if $m(K_p, K_p) > 0$ at a point on the line $K = K_p$, we are sure to have $m(K, K_p) > 0$ at all points below it. It is therefore concluded that $m > 0$ below the line $K = K_p$ for all p in the range $p > p_c$. We have already proved in §4.6 that there is no ferromagnetic phase in the range $p < p_c$, and hence a vertical boundary at $p = p_c$ is concluded to exist in the conventional $\pm J$ model.

We have assumed in the above argument that the modified $\pm J$ model has the ferromagnetic phase on the line $K = K_p$, which has not been proved rigorously to be true. However, this assumption is a very plausible one and it is quite reasonable to expect that the existence of a vertical boundary is valid generally.

4.8.3 *Existence of spin glass phase*

Active investigations are still being carried out concerning the existence of a spin glass phase as an equilibrium state in the Edwards–Anderson model (including the conventional $\pm J$ and Gaussian models) in finite dimensions. Numerical methods are used mainly, and it is presently believed that the Edwards–Anderson model with Ising spins has a spin glass phase in three and higher dimensions. In the modified $\pm J$ model, it is possible to prove the existence of a spin glass phase relatively easily. Let us suppose $a < 0$ in the present section.

As pointed out previously, the conventional $\pm J$ and modified $\pm J$ models

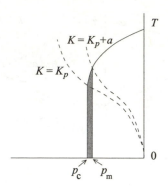

FIG. 4.7. Phase diagram of the modified $\pm J$ model with $a < 0$

share the same region with $q > 0$ (spin glass or ferromagnetic phase) in the phase diagram. In the modified $\pm J$ model, we have $m_M = q_M > 0$ in the low-temperature part of the line $K = K_p + a$ and hence this part lies in the ferromagnetic phase (Fig. 4.7). It is by the way possible to prove the following inequality similarly to (4.50):

$$|\{\langle S_0 S_r \rangle_K\}_{K_p}^a| \le \{|\langle S_0 S_r \rangle_{K_p+a}|\}_{K_p}^a. \tag{4.72}$$

If we set $r \to \infty$ in this relation, the left hand side reduces to the squared magnetization on the line $K = K_p$ in the modified $\pm J$ model and the right hand side to an order parameter on $K = K_p + a$. Thus the right hand side approaches zero in the paramagnetic phase (where $p < p_m$ in Fig. 4.7) and consequently the left hand side vanishes as well. It then follows that the shaded region in the range $p_c < p < p_m$ in Fig. 4.7 has $q_M > 0, m_M = 0$, the spin glass phase.

The only assumption in the above argument is the existence of the ferromagnetic phase in the conventional $\pm J$ model at low temperature, which has already been proved in two dimensions (Horiguchi and Morita 1982b), and it is straightforward to apply the same argument to higher dimensions. Hence it has been proved rigorously that the modified $\pm J$ model has a spin glass phase in two and higher dimensions.

We note that the bond variables τ are not distributed independently at each bond (ij) in the modified $\pm J$ model in contrast to the conventional $\pm J$ model. However, the physical properties of the modified $\pm J$ model should not be essentially different from those of the conventional $\pm J$ model since gauge-invariant quantities (the spin glass order parameter, free energy, specific heat, and so on) assume the same values. When $a > 0$, the distribution (4.66) gives larger probabilities to ferromagnetic configurations with $\tau_{ij} > 0$ than in the conventional $\pm J$ model, and the ferromagnetic phase tends to be enhanced. The case $a < 0$ has the opposite tendency, which may be the reason for the existence of the spin glass phase.

It is nevertheless remarkable that a relatively mild modifications of the $\pm J$ model leads to a model for which a spin glass phase is proved to exist.

4.9 Gauge glass

The gauge theory applies not just to the Ising models but to many other models (Nishimori 1981; Nishimori and Stephen 1983; Georges *et al.* 1985; Ozeki and Nishimori 1993; Nishimori 1994). We explain the idea using the example of the XY model with the Hamiltonian (*gauge glass*)

$$H = -J \sum_{\langle ij \rangle} \cos(\theta_i - \theta_j - \chi_{ij}), \tag{4.73}$$

where quenched randomness exists in the phase variable χ_{ij}. The case $\chi_{ij} = 0$ is the usual ferromagnetic XY model. The gauge theory works in this model if the randomly quenched phase variable follows the distribution

$$P(\chi_{ij}) = \frac{1}{2\pi I_0(K_p)} \exp(K_p \cos \chi_{ij}), \tag{4.74}$$

where $I_0(K_p)$ is the modified Bessel function for normalization. The gauge transformation in this case is

$$\theta_i \rightarrow \theta_i - \phi_i, \quad \chi_{ij} \rightarrow \chi_{ij} - \phi_i + \phi_j. \tag{4.75}$$

Here ϕ_i denotes the gauge variable arbitrarily fixed to a real value at each i. The Hamiltonian is gauge invariant. The distribution function (4.74) transforms as

$$P(\chi_{ij}) \rightarrow \frac{1}{2\pi I_0(K_p)} \exp\{K_p \cos(\phi_i - \phi_j - \chi_{ij})\}. \tag{4.76}$$

4.9.1 *Energy, specific heat, and correlation*

To evaluate the internal energy, we first write its definition

$$[E] = \frac{N_B}{(2\pi I_0(K_p))^{N_B}} \int_0^{2\pi} \prod_{\langle ij \rangle} d\chi_{ij} \, \exp(K_p \sum \cos \chi_{ij})$$

$$\cdot \frac{\int_0^{2\pi} \prod_i d\theta_i \, \{-J\cos(\theta_i - \theta_j - \chi_{ij})\} \exp\{K \sum \cos(\theta_i - \theta_j - \chi_{ij})\}}{\int_0^{2\pi} \prod_i d\theta_i \, \exp\{K \sum \cos(\theta_i - \theta_j - \chi_{ij})\}}. \tag{4.77}$$

The integration range shifts by ϕ_i after gauge transformation, which does not affect the final value because the integrand is a periodic function with period 2π. The value of the above expression therefore does not change by gauge transformation:

$$[E] = -\frac{N_B J}{(2\pi I_0(K_p))^{N_B}} \int_0^{2\pi} \prod_{\langle ij \rangle} d\chi_{ij} \, \exp\{K_p \sum \cos(\phi_i - \phi_j - \chi_{ij})\}$$

$$\cdot \frac{\int_0^{2\pi} \prod_i d\theta_i \, \cos(\theta_i - \theta_j - \chi_{ij}) \exp\{K \sum \cos(\theta_i - \theta_j - \chi_{ij})\}}{\int_0^{2\pi} \prod_i d\theta_i \, \exp\{K \sum \cos(\theta_i - \theta_j - \chi_{ij})\}}. \quad (4.78)$$

Both sides of this formula do not depend on $\{\phi_i\}$, and consequently we may integrate the right hand side over $\{\phi_i\}$ from 0 to 2π and divide the result by $(2\pi)^N$ to get the same value:

$$[E] = -\frac{N_B J}{(2\pi)^N (2\pi I_0(K_p))^{N_B}}$$

$$\cdot \int_0^{2\pi} \prod_{\langle ij \rangle} d\chi_{ij} \int_0^{2\pi} \prod_i d\phi_i \, \exp\{K_p \sum \cos(\phi_i - \phi_j - \chi_{ij})\}$$

$$\cdot \frac{\int_0^{2\pi} \prod_i d\theta_i \, \cos(\theta_i - \theta_j - \chi_{ij}) \exp\{K \sum \cos(\theta_i - \theta_j - \chi_{ij})\}}{\int_0^{2\pi} \prod_i d\theta_i \, \exp\{K \sum \cos(\theta_i - \theta_j - \chi_{ij})\}}. \quad (4.79)$$

If $K = K_p$, the denominator and numerator cancel and we find

$$[E] = -\frac{J}{(2\pi)^N (2\pi I_0(K))^{N_B}}$$

$$\cdot \int_0^{2\pi} \prod_{\langle ij \rangle} d\chi_{ij} \frac{\partial}{\partial K} \int_0^{2\pi} \prod_i d\theta_i \, \exp\{K \sum \cos(\theta_i - \theta_j - \chi_{ij})\}. \quad (4.80)$$

Integration over χ_{ij} gives $2\pi I_0(K)$ for each $\langle ij \rangle$:

$$[E] = -\frac{J}{(2\pi)^N (2\pi I_0(K))^{N_B}} (2\pi)^N \frac{\partial}{\partial K} (2\pi I_0(K))^{N_B} = -JN_B \frac{I_1(K)}{I_0(K)}. \quad (4.81)$$

Because the modified Bessel functions $I_0(K)$ and $I_1(K)$ are not singular and $I_0(K) > 0$ for positive K, we conclude, as in the case of the Ising model, that the internal energy has no singularity along the Nishimori line $K = K_p$ although it crosses the boundary between the ferromagnetic and paramagnetic phases.

It is straightforward to evaluate an upper bound on the specific heat on the Nishimori line, similar to the Ising case, and the result is

$$T^2[C] \le J^2 N_B \left\{ \frac{1}{2} + \frac{I_2(K)}{2I_0(K)} - \left(\frac{I_1(K)}{I_0(K)} \right)^2 \right\}. \quad (4.82)$$

The right hand side remains finite anywhere on the Nishimori line.

Arguments on the correlation equality and inequality work as well. The correlation function is $[\langle \cos(\theta_i - \theta_j) \rangle_K]$, or $[\langle \exp i(\theta_i - \theta_j) \rangle_K]$, and by using the latter expression and the gauge theory, we can derive the following identity:

$$[\langle \cos(\theta_i - \theta_j) \rangle_K] = [\langle \cos(\phi_i - \phi_j) \rangle_{K_p} \langle \cos(\theta_i - \theta_j) \rangle_K]. \quad (4.83)$$

By taking the absolute value and evaluating the upper bound, we have

$$|[\langle \cos(\theta_i - \theta_j) \rangle_K]| \le [|\langle \cos(\phi_i - \phi_j) \rangle_{K_p}|]. \quad (4.84)$$

This relation can be interpreted in the same way as in the Ising model: the ferromagnetic phase is not allowed to lie below the spin glass phase.

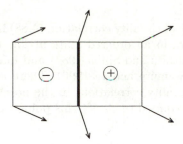

FIG. 4.8. Ground-state configuration of the six-site XY model with a single antiferromagnetic bond (bold line). The right plaquette has chirality $+$ and the left $-$.

4.9.2 Chirality

The XY model has an effective degree of freedom called *chirality*. We can show that the chirality completely loses spatial correlations when $K = K_p$.

This result is a consequence of the following relation

$$[\langle f_1(\theta_i - \theta_j - \chi_{ij}) f_2(\theta_l - \theta_m - \chi_{lm}) \rangle] = [\langle f_1(\theta_i - \theta_j - \chi_{ij}) \rangle][\langle f_2(\theta_l - \theta_m - \chi_{lm}) \rangle]$$

(4.85)

for $K = K_p$, where $\langle ij \rangle$ and $\langle lm \rangle$ are distinct bonds and f_1 and f_2 are arbitrary functions with period 2π. Equation (4.85) can be proved in the same way as we derived the exact energy (4.81) and is analogous to the decoupling of bond energy in the Ising model (4.21). Each factor on the right hand side of (4.85) is evaluated as in the previous subsection and the result is

$$[\langle f_1(\theta_i - \theta_j - \chi_{ij}) f_2(\theta_l - \theta_m - \chi_{lm}) \rangle] = \frac{\int_0^{2\pi} d\theta \, f_1(\theta) e^{K \cos \theta}}{\int_0^{2\pi} d\theta \, e^{K \cos \theta}} \cdot \frac{\int_0^{2\pi} d\theta \, f_2(\theta) e^{K \cos \theta}}{\int_0^{2\pi} d\theta \, e^{K \cos \theta}}$$

(4.86)

under the condition $K = K_p$.

Chirality has been introduced to quantify the degree of twistedness on the frustrated XY model (Villain 1977) and is defined by

$$\kappa_p = \sum \sin(\theta_i - \theta_j - \chi_{ij}),$$

(4.87)

where the sum is over a directed path (counter-clockwise, for example) around a plaquette. Frustrated plaquettes are generally neighbouring to each other as in Fig. 4.8, and such plaquettes carry the opposite signs of chirality. It is a direct consequence of the general relation (4.86) that chiralities at plaquettes without common bonds are independent. In particular, we have $[\langle \kappa_p \rangle] = 0$ at any temperature since the sine in (4.87) is an odd function and consequently

$$[\langle \kappa_{p_1} \kappa_{p_2} \rangle] = [\langle \kappa_{p_1} \rangle][\langle \kappa_{p_2} \rangle] = 0$$

(4.88)

if $K = K_p$ and the plaquettes p_1 and p_2 are not adjacent to each other sharing a bond.

The complete absence of chirality correlation (4.88) is not easy to understand intuitively. Chiralities are in an ordered state at low temperatures in regular frustrated systems (Miyashita and Shiba 1984) and in a spin glass state in the random XY model (Kawamura and Li 1996). There is no apparent reason to expect the absence of chirality correlations in the present gauge glass problem. Further investigation is required to clarify this point.

4.9.3 XY spin glass

The name 'XY spin glass' is usually used for the XY model with random interactions:

$$H = -\sum_{\langle ij \rangle} J_{ij} \cos(\theta_i - \theta_j),$$
(4.89)

where J_{ij} obeys such a distribution as the $\pm J$ or Gaussian function as in the Ising spin glass. This model is clearly different from the gauge glass (4.73). It is difficult to analyse this XY spin glass by gauge transformation because the gauge summation of the probability weight leads to a partition function of the Ising spin glass, which does not cancel out the partition function of the XY spin glass appearing in the denominator of physical quantities such as the energy. It is nevertheless possible to derive an interesting relation between the Ising and XY spin glasses using a correlation function (Nishimori 1992).

The gauge transformation of the Ising type (4.2) reveals the following relation of correlation functions:

$$[\langle \boldsymbol{S}_i \cdot \boldsymbol{S}_j \rangle_K^{XY}] = [\langle \sigma_i \sigma_j \rangle_{K_p}^{I} \langle \boldsymbol{S}_i \cdot \boldsymbol{S}_j \rangle_K^{XY}],$$
(4.90)

where $\boldsymbol{S}_i (= {}^t(\cos\theta_i, \sin\theta_i))$ denotes the XY spin and the thermal average is taken with the $\pm J\,XY$ Hamiltonian (4.89) for $\langle \cdots \rangle^{XY}$ and with the $\pm J$ Ising model for $\langle \cdots \rangle^{I}$. Taking the absolute value of both sides of (4.90) and replacing $\langle \boldsymbol{S}_i \cdot \boldsymbol{S}_j \rangle_K^{XY}$ on the right hand side by its upper bound 1, we find

$$|[\langle \boldsymbol{S}_i \cdot \boldsymbol{S}_j \rangle_K^{XY}]| \leq [|\langle \sigma_i \sigma_j \rangle_{K_p}^{I}|].$$
(4.91)

The right hand side vanishes for $p < p_c$ (see Fig. 4.3) in the limit $|i - j| \to \infty$ since the Nishimori line is in the paramagnetic phase. Thus the left hand side also vanishes in the same range of p. This proves that p_c, the lower limit of the ferromagnetic phase, for the XY model is higher than that for the Ising model

$$p_c^{XY} \geq p_c^{I}.$$
(4.92)

It is actually expected from the argument of §4.7 that the equality holds in (4.92).

The same argument can be developed for the Gaussian case as well as other models like the $\pm J$/Gaussian Heisenberg spin glass whose spin variable has three components $\boldsymbol{S} = {}^t(S_x, S_y, S_z)$ under the constraint $\boldsymbol{S}^2 = 1$.

4.10 Dynamical correlation function

It is possible to apply the gauge theory also to non-equilibrium situations (Ozeki 1995, 1997). We start our argument from the *master equation* that determines the time development of the probability $P_t(\boldsymbol{S})$ that a spin configuration $\boldsymbol{S}$ is realized at time t:

$$\frac{dP_t(\boldsymbol{S})}{dt} = \text{Tr}_{\boldsymbol{S}'} W(\boldsymbol{S}|\boldsymbol{S}') P_t(\boldsymbol{S}'). \tag{4.93}$$

Here $W(\boldsymbol{S}|\boldsymbol{S}')$ is the *transition probability* that the state changes from $\boldsymbol{S}'$ to $\boldsymbol{S}\,(\neq \boldsymbol{S}')$ in unit time. Equation (4.93) means that the probability $P_t(\boldsymbol{S})$ increases by the amount by which the state changes into $\boldsymbol{S}$. If $W < 0$, the probability decreases by the change of the state from $\boldsymbol{S}$.

An example is the *kinetic Ising model*

$$\begin{aligned} W(\boldsymbol{S}|\boldsymbol{S}') &= \delta_1(\boldsymbol{S}|\boldsymbol{S}') \frac{\exp(-\frac{\beta}{2}\Delta(\boldsymbol{S},\boldsymbol{S}'))}{2\cosh\frac{\beta}{2}\Delta(\boldsymbol{S},\boldsymbol{S}')} \\ &\quad - \delta(\boldsymbol{S},\boldsymbol{S}') \text{Tr}_{\boldsymbol{S}''} \delta_1(\boldsymbol{S}''|\boldsymbol{S}) \frac{\exp(-\frac{\beta}{2}\Delta(\boldsymbol{S}'',\boldsymbol{S}))}{2\cosh\frac{\beta}{2}\Delta(\boldsymbol{S},\boldsymbol{S}'')}, \end{aligned} \tag{4.94}$$

where $\delta_1(\boldsymbol{S}|\boldsymbol{S}')$ is a function equal to one when the difference between $\boldsymbol{S}$ and $\boldsymbol{S}'$ is just a single spin flip and is zero otherwise:

$$\delta_1(\boldsymbol{S}|\boldsymbol{S}') = \delta\{2, \sum_i (1 - S_i S_i')\}. \tag{4.95}$$

In (4.94), $\Delta(\boldsymbol{S}, \boldsymbol{S}')$ represents the energy change $H(\boldsymbol{S}) - H(\boldsymbol{S}')$. The first term on the right hand side of (4.94) is the contribution from the process where the state of the system changes from $\boldsymbol{S}'$ to $\boldsymbol{S}$ by a single spin flip with probability $e^{-\beta\Delta/2}/2\cosh(\beta\Delta/2)$. The second term is for the process where the state changes from $\boldsymbol{S}$ to another by a single spin flip. Since $\delta(\boldsymbol{S},\boldsymbol{S}'), \delta_1(\boldsymbol{S}|\boldsymbol{S}')$, and $\Delta(\boldsymbol{S},\boldsymbol{S}')$ are all gauge invariant, W is also gauge invariant: $W(\boldsymbol{S}|\boldsymbol{S}') = W(\boldsymbol{S}\boldsymbol{\sigma}|\boldsymbol{S}'\boldsymbol{\sigma})$, where $\boldsymbol{S}\boldsymbol{\sigma} = \{S_i\sigma_i\}$.

The formal solution of the master equation (4.93) is

$$P_t(\boldsymbol{S}) = \text{Tr}_{\boldsymbol{S}'} \langle \boldsymbol{S}|e^{tW}|\boldsymbol{S}'\rangle P_0(\boldsymbol{S}'). \tag{4.96}$$

We prove the following relation between the dynamical correlation function and non-equilibrium magnetization using the above formal solution:

$$[\langle S_i(0)S_i(t)\rangle_K^{K_p}] = [\langle S_i(t)\rangle_K^\Gamma]. \tag{4.97}$$

Here $\langle S_i(0)S_i(t)\rangle_K^{K_p}$ is the *autocorrelation function*, the expectation value of the spin product at site i when the system was in equilibrium at $t = 0$ with the inverse temperature K_p and was developed over time to t:

$$\langle S_i(0)S_i(t)\rangle_K^{K_p} = \text{Tr}_{\boldsymbol{S}} \text{Tr}_{\boldsymbol{S}'} S_i \langle \boldsymbol{S}|e^{tW}|\boldsymbol{S}'\rangle S_i' P_e(\boldsymbol{S}', K_p) \tag{4.98}$$

$$P_e(\boldsymbol{S}', K_p) = \frac{1}{Z(K_p)} \exp(K_p \sum \tau_{ij} S'_i S'_j), \qquad (4.99)$$

where the $K(= \beta J)$-dependence of the right hand side is in W. The expectation value $\langle S_i(t) \rangle_K^F$ is the site magnetization at time t starting from the perfect ferromagnetic state $|F\rangle$:

$$\langle S_i(t) \rangle_K^F = \mathrm{Tr}_{\boldsymbol{S}} \, S_i \langle \boldsymbol{S} | e^{tW} | F \rangle. \qquad (4.100)$$

To prove (4.97), we perform gauge transformation ($\tau_{ij} \to \tau_{ij} \sigma_i \sigma_j, S_i \to S_i \sigma_i$) to the configurational average of (4.100)

$$[\langle S_i(t) \rangle_K^F] = \frac{1}{(2 \cosh K_p)^{N_B}} \sum_\tau \exp(K_p \sum \tau_{ij}) \mathrm{Tr}_{\boldsymbol{S}} \, S_i \langle \boldsymbol{S} | e^{tW} | F \rangle. \qquad (4.101)$$

Then the term e^{tW} is transformed to $\langle \boldsymbol{S} \boldsymbol{\sigma} | e^{tW} | F \rangle$, which is equal to $\langle \boldsymbol{S} | e^{tW} | \boldsymbol{\sigma} \rangle$ by the gauge invariance of W. Hence we have

$$[\langle S_i(t) \rangle_K^F] = \frac{1}{(2 \cosh K_p)^{N_B}} \frac{1}{2^N} \sum_\tau \mathrm{Tr}_{\boldsymbol{\sigma}} \, \exp(K_p \sum \tau_{ij} \sigma_i \sigma_j) \mathrm{Tr}_{\boldsymbol{S}} \, S_i \langle \boldsymbol{S} | e^{tW} | \boldsymbol{\sigma} \rangle \sigma_i$$

$$= \frac{1}{(2 \cosh K_p)^{N_B}} \frac{1}{2^N} \sum_\tau \mathrm{Tr}_{\boldsymbol{S}} \mathrm{Tr}_{\boldsymbol{\sigma}} \, S_i \langle \boldsymbol{S} | e^{tW} | \boldsymbol{\sigma} \rangle \sigma_i Z(K_p) P_e(\boldsymbol{\sigma}, K_p). \qquad (4.102)$$

It is also possible to show that the configurational average of (4.98) is equal to the above expression, from which (4.97) follows immediately.

Equation (4.97) shows that the non-equilibrium relaxation of magnetization starting from the perfect ferromagnetic state is equal to the configurational average of the autocorrelation function under the initial condition of equilibrium state at the inverse temperature K_p. In particular, when $K = K_p$, the left hand side is the equilibrium autocorrelation function, and this identity gives a direct relation between equilibrium and non-equilibrium quantities.

A generalization of (4.97) also holds:

$$[\langle S_i(t_w) S_i(t + t_w) \rangle_K^{K_p}] = [\langle S_i(t_w) S_i(t + t_w) \rangle_K^F]. \qquad (4.103)$$

Both sides are defined as follows:

$$\langle S_i(t_w) S_i(t + t_w) \rangle_K^{K_p}$$
$$= \mathrm{Tr}_{\boldsymbol{S}_0} \mathrm{Tr}_{\boldsymbol{S}_1} \mathrm{Tr}_{\boldsymbol{S}_2} S_{2i} \langle \boldsymbol{S}_2 | e^{tW} | \boldsymbol{S}_1 \rangle S_{1i} \langle \boldsymbol{S}_1 | e^{t_w W} | \boldsymbol{S}_0 \rangle P_e(\boldsymbol{S}_0, K_p) \qquad (4.104)$$
$$\langle S_i(t_w) S_i(t + t_w) \rangle_K^F$$
$$= \mathrm{Tr}_{\boldsymbol{S}_1} \mathrm{Tr}_{\boldsymbol{S}_2} S_{2i} \langle \boldsymbol{S}_2 | e^{tW} | \boldsymbol{S}_1 \rangle S_{1i} \langle \boldsymbol{S}_1 | e^{t_w W} | F \rangle. \qquad (4.105)$$

Equation (4.104) is the autocorrelation function at time t after waiting for t_w with inverse temperature K starting at time $t = 0$ from the initial equilibrium state with the inverse temperature K_p. Equation (4.105), on the other hand, represents the correlation function with the perfect ferromagnetic state as the initial condition instead of the equilibrium state at K_p in (4.104). Equation (4.97) is a special case of (4.103) with $t_w = 0$.

To prove (4.103), we first note that the gauge transformation in (4.105) yields σ as the initial condition in this equation. Similar to the case of (4.102), we take a sum over σ and divide the result by 2^N. We next perform gauge transformation and sum it up over the gauge variables in (4.104), comparison of which with the above result leads to (4.103). It is to be noted in this calculation that (4.105) is gauge invariant.

If $K = K_p$ in (4.103),

$$[\langle S_i(0)S_i(t)\rangle_{K_p}^{\text{eq}}] = [\langle S_i(t_w)S_i(t+t_w)\rangle_{K_p}^{\text{F}}]. \tag{4.106}$$

For $K = K_p$, the left hand side of (4.103) is the equilibrium autocorrelation function and should not depend on t_w, and thus we have set $t_w = 0$. Equation (4.106) proves that the autocorrelation function does not depend on t_w *on average* if it is measured after the system is kept at the equilibrium state with inverse temperature K_p for time duration t_w and the initial condition of a perfect ferromagnetic state. The *aging* phenomenon, in which non-equilibrium quantities depend on the waiting time t_w before measurement, is considered to be an important characteristic of the spin glass phase (Young 1997; Miyako *et al.* 2000). Equation (4.106) indicates that the *configurational average* of the autocorrelation function with the perfect ferromagnetic state as the initial condition does not show aging.

Bibliographical note

The analytical framework of the gauge theory has been developed in the references cited in the text. Numerical investigations of various problems related to the gauge theory have been carried out by many authors. The main topics were the location of the multicritical point (in particular whether or not it is on the Nishimori line) and the values of the critical exponents (especially around the multicritical point). References until the early 1990s include Ozeki and Nishimori (1987), Kitatani and Oguchi (1990, 1992), Ozeki (1990), Ueno and Ozeki (1991), Singh (1991), and Le Doussal and Harris (1988, 1989). These problems have been attracting resurgent interest recently as one can see in Singh and Adler (1996), Ozeki and Ito (1998), Sørensen *et al.* (1998), Gingras and Sørensen (1998), Aarão Reis *et al.* (1999), Kawashima and Aoki (2000), Mélin and Peysson (2000), Honecker *et al.* (2000), and Hukushima (2000). Properties of the multicritical point in two dimensions have been studied also from the standpoints of the quantum Hall effect (Cho and Fisher 1997; Senthil and Fisher 2000; Read and Ludwig 2000) and supersymmetry (Gruzberg *et al.* 2001).

5

ERROR-CORRECTING CODES

Reliable transmission of information through noisy channels plays a vital role in modern society. Some aspects of this problem have close formal similarities to the theory of spin glasses. Noise in the transmission channel can be related to random interactions in spin glasses and the bit sequence representing information corresponds to the Ising spin configuration. The replica method serves as a powerful tool of analysis, and TAP-like equations can be used as a practical implementation of the algorithm to infer the original message. The gauge theory also provides an interesting point of view.

5.1 Error-correcting codes

Information theory was initiated by Shannon half a century ago. It formulates various basic notions on information transmission through noisy channels and develops a framework to manipulate those abstract objects. We first briefly review some ideas of information theory, and then restate the basic concepts, such as noise, communication, and information inference, in terms of statistical mechanics of spin glasses.

5.1.1 *Transmission of information*

Suppose that we wish to transmit a message (information) represented as a sequence of N bits from one place to another. The path for information transmission is called a (*transmission*) *channel*. A channel usually carries noise and the output from a channel is different in some bits from the input. We then ask ourselves how we can infer the original message from the noisy output.

It would be difficult to infer which bit of the output is corrupted by noise if the original message itself had been fed into the channel. It is necessary to make the message redundant before transmission by adding extra pieces of information, by use of which the noise can be removed. This process is called *channel coding* (or *encoding*), or simply coding. The encoded message is transmitted through a noisy channel. The process of information retrieval from the noisy output of a channel using redundancy is called *decoding*.

A very simple example of encoding is to repeat the same bit sequence several (for instance, three) times. If the three sequences received at the end of the channel coincide, one may infer that there was no noise. If a specific bit has different values in the three sequences, one may infer its correct value (0 or 1) by the majority rule. For example, when the original message is $(0, 0, 1, 1, 0)$ and the three output sequences from a noisy channel are $(0, 0, 1, 1, 0), (0, 1, 1, 1, 0)$,

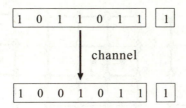

FIG. 5.1. Parity-check code

and $(0, 0, 1, 1, 0)$, the correct second bit can be inferred to be 0 whereas all the other bits coincide to be $(0, *, 1, 1, 0)$. This example shows that the redundancy is helpful to infer the original information from the noisy message.

A more sophisticated method is the *parity-check code*. For example, suppose that seven bits are grouped together and one counts the number of ones in the group. If the number is even, one adds 0 as the eighth bit (the parity bit) and adds 1 otherwise. Then there are always an even number of ones in the group of eight bits in the encoded message (*code word*). If the noise rate of the channel is not very large and at most only one bit is flipped by noise out of the eight received bits, one may infer that the output of the channel carries no noise when one finds even ones in the group of eight bits. If there are odd ones, then there should be some noise, implying *error detection* (Fig. 5.1). *Error correction* after error detection needs some further trick to be elucidated in the following sections.

5.1.2 *Similarity to spin glasses*

It is convenient to use the Ising spin ± 1 instead of 0 and 1 to treat the present problem by statistical mechanics. The basic operation on a bit sequence is the sum with modulo 2 (2=0 with mod 2), which corresponds to the product of Ising spins if one identifies 0 with $S_i = 1$ and 1 with $S_i = -1$. For example, $0 + 1 = 1$ translates into $1 \times (-1) = -1$ and $1 + 1 = 0$ into $(-1) \times (-1) = 1$. We hereafter use this identification of an Ising spin configuration with a bit sequence.

Generation of the parity bit in the parity-check code corresponds to the product of appropriate spins. By identifying such a product with the interaction between the relevant spins, we obtain a spin system very similar to the *Mattis model* of spin glasses.

In the Mattis model one allocates a randomly quenched Ising spin ξ_i to each site, and the interaction between sites i and j is chosen to be $J_{ij} = \xi_i \xi_j$. The Hamiltonian is then

$$H = -\sum_{\langle ij \rangle} \xi_i \xi_j S_i S_j. \tag{5.1}$$

The ground state is clearly identical to the configuration of the quenched spins $S_i = \xi_i$ ($\forall i$) (or its total reversal $S_i = -\xi_i$ ($\forall i$)), see Fig. 5.2(a).

Returning to the problem of error-correcting codes, we form the Mattis-type interactions $\{J^0_{i_1 \ldots i_r} = \xi_{i_1} \ldots \xi_{ir}\}$ with r an integer for appropriate combinations

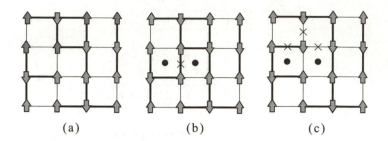

(a)   (b)   (c)

FIG. 5.2. The ground-state spin configurations of the Mattis model without noise added (a), the interaction marked by a cross flipped by noise (b), and three interactions flipped by strong noise (c). Thin lines represent ferromagnetic interactions and bold lines antiferromagnetic interactions. The plaquettes marked by dots are frustrated.

of sites $\{i_1 \ldots i_r\}$. We then feed these interactions (encoded message), instead of the original spin configuration $\boldsymbol{\xi} \equiv \{\xi_i\}$ (original message), into the noisy channel. The encoded information is redundant because the number of interactions N_B (which is the number of elements in the set $\{(i_1 \ldots i_r)\}$) is larger than the number of spins N.

For instance, the conventional $r = 2$ Mattis model on the two-dimensional square lattice has $N_B = 2N$ interactions, the number of neighbouring sites. For the original interactions without noise $\boldsymbol{J}^0 = \{\xi_i \xi_j\}$, the product of the J_{ij}^0 along an arbitrary closed loop c, $f_c = \prod J_{ij}^0 = \prod(\xi_i \xi_j)$, is always unity (positive) since all the ξ_i appear an even number of times in the product. Thus the Mattis model has no frustration. However, noise in the channel flips some elements of $\boldsymbol{J}^0$ and therefore the output of the channel, if it is regarded as interactions of the spin system, includes frustration (Fig. 5.2(b)). Nevertheless the original spin configuration is still the ground state of such a system if the noise rate is not large (i.e. only a small number of bits are flipped) as exemplified in Fig. 5.2(b). It has thus been shown that correct inference of the original message is possible even if there exists a small amount of noise in the channel, as long as an appropriate procedure is employed in encoding and decoding the message.

5.1.3 *Shannon bound*

It is necessary to introduce redundancy appropriately to transmit information accurately through a noisy channel. It can indeed be proved that such redundancy should exceed a threshold so that we are able to retrieve the original message without errors.

Let us define the *transmission rate R* of information by a channel as

$$R = \frac{N(\text{number of bits in the original message})}{N_B(\text{number of bits in the encoded message})}. \tag{5.2}$$

For smaller denominator, the redundancy is smaller and the transmission rate is larger. Strictly speaking, the numerator of (5.2) should be the 'number of informative bits in the original message'. For biased messages, where 0 and 1 appear with unequal probabilities, the amount of information measured in bits is smaller than the length of the binary message itself. This can easily be verified by the extreme case of a perfectly biased message with only ones; the message carries no information in such a case. By multiplying (5.2) by the number of bits transmitted per second, one obtains the number of information bits transmitted per second.

We consider a memoryless *binary symmetric channel* (BSC) where noise flips a bit from 1 to 0 and 0 to 1 independently at each bit with a given probability. It is known that the transmission rate should satisfy the following inequality so that error-free transmission is possible through the BSC in the limit of a very long bit sequence:

$$R < C, \tag{5.3}$$

where C is a function of the noise probability and is called the *channel capacity*. The capacity of the BSC is

$$C = 1 + p \log_2 p + (1 - p) \log_2(1 - p). \tag{5.4}$$

Here p is the probability that a bit is flipped by noise. Equation (5.3) is called the *channel coding theorem* of Shannon and implies that error-free transmission is possible if the transmission rate does not exceed the channel capacity and an appropriate procedure of encoding and decoding is employed. Similar results hold for other types of channels such as the Gaussian channel to be introduced later. A sketch of the arguments leading to the channel coding theorem is given in Appendix C.

An explicit example of a code that saturates the Shannon bound (5.3) asymptotically as an equality is the *Sourlas code* (Sourlas 1989): one takes all possible products of r spins from N sites to form Mattis-type interactions. We have mainly explained the conventional Mattis model with $r = 2$ in §5.1.2, and we discuss the general case of an arbitrary $r (= 2, 3, 4, \ldots)$ hereafter. This is nothing but the infinite-range model with r-body interactions.[5] It will be shown later that the Shannon bound (5.3) is asymptotically achieved and the error rate approaches zero in the Sourlas code if we take the limit $N \to \infty$ first and $r \to \infty$ afterwards. It should be noted, however, that the inequality (5.3) reduces to an equality with both sides approaching zero in the Sourlas code, which means that the transmission rate R is zero asymptotically. Therefore the transmission is not very efficient. A trick to improve this point was shown recently to be to take the product of a limited number of combinations, not all possible combinations of r spins, from N sites. All of these points will be elucidated in detail later in the present chapter.

---

[5]The symbol p is often used in the literature to denote the number of interacting spins. We use r instead to avoid confusion with the error probability in the BSC.

5.1.4 *Finite-temperature decoding*

Let us return to the argument of §5.1.2 and consider the problem of inference of spin configurations when the noise probability is not necessarily small. It was shown in §5.1.2 that the ground state of the Ising spin system (with the output of the channel as interactions) is the true original message (spin configuration) if the channel is not very noisy. For larger noise, the ground state is different from the original spin configuration (Fig. 5.2(c)). This suggests that the original spin configuration is one of the excited states and thus one may be able to decode more accurately by searching for states at finite temperature. It will indeed be shown that states at a specific finite temperature T_p determined by the error rate p give better results under a certain criterion. This temperature T_p turns out to coincide with the temperature at the Nishimori line discussed in Chapter 4 (Ruján 1993).

5.2 Spin glass representation

We now formulate the arguments in the previous section in a more quantitative form and proceed to explicit calculations (Sourlas 1989).

5.2.1 *Conditional probability*

Suppose that the Ising spin configuration $\boldsymbol{\xi} = \{\xi_i\}$ has been generated according to a probability distribution function $P(\boldsymbol{\xi})$. This distribution $P(\boldsymbol{\xi})$ for generating the original message is termed the *prior*. Our goal is to infer the original spin configuration from the output of a noisy channel as accurately as possible. Following the suggestion in the previous section, we form a set of products of r spins

$$J^0_{i_1 \ldots i_r} = \xi_{i_1} \ldots \xi_{i_r} (= \pm 1) \tag{5.5}$$

for appropriately chosen combinations of the ξ_i and feed the set of interactions into the channel. We first consider the BSC, and the output of the channel $J_{i_1 \ldots i_r}$ is flipped from the corresponding input $J^0_{i_1 \ldots i_r} = \xi_{i_1} \ldots \xi_{i_r}$ with probability p and is equal to $-\xi_{i_1} \ldots \xi_{i_r}$. The other possibility of correct output $\xi_{i_1} \ldots \xi_{i_r}$ has the probability $1 - p$. The output probability of a BSC can be expressed in terms of a conditional probability:

$$P(J_{i_1 \ldots i_r} | \xi_{i_1} \ldots \xi_{i_r}) = \frac{\exp(\beta_p J_{i_1 \ldots i_r} \xi_{i_1} \ldots \xi_{i_r})}{2 \cosh \beta_p}, \tag{5.6}$$

where β_p is a function of p defined as

$$e^{2\beta_p} = \frac{1 - p}{p}. \tag{5.7}$$

Equation (5.6) is equal to $1 - p$ when $J_{i_1 \ldots i_r} = \xi_{i_1} \ldots \xi_{i_r}$ according to (5.7) and is equal to p when $J_{i_1 \ldots i_r} = -\xi_{i_1} \ldots \xi_{i_r}$, implying that (5.6) is the correct conditional probability to characterize the channel. Equations (5.6) and (5.7) are analogous to the distribution function (4.4)–(4.5) in the gauge theory. The inverse of β_p defined in (5.7), denoted T_p at the end of the previous section, coincides with the temperature on the Nishimori line. One should note here that p in the present chapter is $1 - p$ in Chapter 4. The temperature T_p is sometimes called the *Nishimori temperature* in the literature of error-correcting codes.

Assume that (5.6) applies to each set $(i_1 \ldots i_r)$ independently. This is a *memoryless channel* in which each bit is affected by noise independently. The overall probability is then the product of (5.6)

$$P(\boldsymbol{J}|\boldsymbol{\xi}) = \frac{1}{(2 \cosh \beta_p)^{N_B}} \exp \left(\beta_p \sum J_{i_1 \ldots i_r} \xi_{i_1} \cdots \xi_{i_r} \right), \qquad (5.8)$$

where the sum in the exponent is taken over all sets $(i_1 \ldots i_r)$ for which the spin products are generated by (5.5). The symbol N_B is for the number of terms in this sum and is equal to the number of bits fed into the channel.

5.2.2 Bayes formula

The task is to infer the original message (spin configuration) $\boldsymbol{\xi}$ from the output $\boldsymbol{J} = \{J_{i_1 \ldots i_r}\}$. For this purpose, it is necessary to introduce the conditional probability of $\boldsymbol{\xi}$ given $\boldsymbol{J}$, which is called the *posterior*. The posterior is the conditional probability with the two entries of the left hand side of (5.8), $\boldsymbol{J}$ and $\boldsymbol{\xi}$, exchanged. The Bayes formula is useful for exchanging these two entries.

The joint probability $P(A, B)$ that two events A and B occur is expressed in terms of the product of $P(B)$ and the conditional probability $P(A|B)$ for A under the condition that B occurred. The same holds if A and B are exchanged. Thus we have

$$P(A, B) = P(A|B)P(B) = P(B|A)P(A). \qquad (5.9)$$

It follows immediately that

$$P(A|B) = \frac{P(B|A)P(A)}{P(B)} = \frac{P(B|A)P(A)}{\sum_A P(B|A)P(A)}. \qquad (5.10)$$

Equation (5.10) is the *Bayes formula*.

We can express $P(\boldsymbol{\sigma}|\boldsymbol{J})$ in terms of (5.8) using the Bayes formula:

$$P(\boldsymbol{\sigma}|\boldsymbol{J}) = \frac{P(\boldsymbol{J}|\boldsymbol{\sigma})P(\boldsymbol{\sigma})}{\text{Tr}_\sigma P(\boldsymbol{J}|\boldsymbol{\sigma})P(\boldsymbol{\sigma})}. \qquad (5.11)$$

We have written $\boldsymbol{\sigma} = \{\sigma_1, \ldots, \sigma_N\}$ for dynamical variables used for decoding. The final decoded result will be denoted by $\hat{\boldsymbol{\xi}} = \{\hat{\xi}_1, \ldots, \hat{\xi}_N\}$, and we reserve $\boldsymbol{\xi} = \{\xi_1, \ldots, \xi_n\}$ for the true original configuration. Equation (5.11) is the starting point of our argument.

The probability $P(\boldsymbol{J}|\boldsymbol{\sigma})$ represents the characteristics of a memoryless BSC and is given in (5.8). It is therefore possible to infer the original message by (5.11) if the prior $P(\boldsymbol{\sigma})$ is known. Explicit theoretical analysis is facilitated by assuming a message source that generates various messages with equal probability. This assumption is not necessarily unnatural in realistic situations where information compression before encoding usually generates a rather uniform distribution of zeros and ones. In such a case, $P(\boldsymbol{\sigma})$ can be considered a constant. The posterior is then

$$P(\boldsymbol{\sigma}|\boldsymbol{J}) = \frac{\exp\left(\beta_p \sum J_{i_1...i_r}\sigma_{i_1}\cdots\sigma_{i_r}\right)}{\mathrm{Tr}_{\boldsymbol{\sigma}}\exp\left(\beta_p \sum J_{i_1...i_r}\sigma_{i_1}\cdots\sigma_{i_r}\right)}. \tag{5.12}$$

Since $\boldsymbol{J}$ is given and fixed in the present problem, (5.12) is nothing more than the Boltzmann factor of an Ising spin glass with randomly quenched interactions $\boldsymbol{J}$. We have thus established a formal equivalence between the probabilistic inference problem of messages for a memoryless BSC and the statistical mechanics of the Ising spin glass.

5.2.3 MAP and MPM

Equation (5.12) is the probability distribution of the inferred spin configuration $\boldsymbol{\sigma}$, given the output of the channel $\boldsymbol{J}$. Then the spin configuration to maximize (5.12) is a good candidate for the decoded (inferred) spin configuration. Maximization of the Boltzmann factor is equivalent to the ground-state search of the corresponding Hamiltonian

$$H = -\sum J_{i_1...i_r}\sigma_{i_1}\cdots\sigma_{i_r}. \tag{5.13}$$

This method of decoding is called the *maximum a posteriori probability* (MAP). This is the idea already explained in §5.1.2. Maximization of the conditional probability $P(\boldsymbol{J}|\boldsymbol{\sigma})$ with respect to $\boldsymbol{\sigma}$ is equivalent to maximization of the posterior $P(\boldsymbol{\sigma}|\boldsymbol{J})$ if the prior $P(\boldsymbol{\sigma})$ is uniform. The former idea is termed the *maximum likelihood method* as $P(\boldsymbol{J}|\boldsymbol{\sigma})$ is the *likelihood function* of $\boldsymbol{\sigma}$.

The MAP maximizes the posterior of the whole bit sequence $\boldsymbol{\sigma}$. There is another strategy of decoding in which we focus our attention on a single bit i, not the whole sequence. This means we trace out all the spin variables except for a single σ_i to obtain the posterior only of σ_i. This process is called *marginalization* in statistics:

$$P(\sigma_i|\boldsymbol{J}) = \frac{\mathrm{Tr}_{\boldsymbol{\sigma}(\neq\sigma_i)}\exp\left(\beta_p \sum J_{i_1...i_r}\sigma_{i_1}\cdots\sigma_{i_r}\right)}{\mathrm{Tr}_{\boldsymbol{\sigma}}\exp\left(\beta_p \sum J_{i_1...i_r}\sigma_{i_1}\cdots\sigma_{i_r}\right)}. \tag{5.14}$$

We then compare $P(\sigma_i = 1|\boldsymbol{J})$ and $P(\sigma_i = -1|\boldsymbol{J})$ and, if the former is larger, we assign one to the decoded result of the ith bit ($\hat{\xi}_i = 1$) and assign $\hat{\xi}_i = -1$ otherwise. This process is carried out for all bits, and the set of thus decoded bits constitutes the final result. This method is called the *finite-temperature decoding* or the *maximizer of posterior marginals* (MPM) and clearly gives a different result than the MAP (Ruján 1993; Nishimori 1993; Sourlas 1994; Iba 1999).

It is instructive to consider the MPM from a different point of view. The MPM is equivalent to accepting the sign of the difference of two probabilities as the ith decoded bit

$$\hat{\xi}_i = \text{sgn} \left\{ P(\sigma_i = 1 | \boldsymbol{J}) - P(\sigma_i = -1 | \boldsymbol{J}) \right\}. \tag{5.15}$$

This may also be written as

$$\hat{\xi}_i = \text{sgn} \left(\sum_{\sigma_i = \pm 1} \sigma_i P(\sigma_i | \boldsymbol{J}) \right) = \text{sgn} \left(\frac{\text{Tr}_{\boldsymbol{\sigma}} \sigma_i P(\sigma_i | \boldsymbol{J})}{\text{Tr}_{\boldsymbol{\sigma}} P(\sigma_i | \boldsymbol{J})} \right) = \text{sgn} \langle \sigma_i \rangle_{\beta_p}. \tag{5.16}$$

Here $\langle \sigma_i \rangle_{\beta_p}$ is the local magnetization with (5.12) as the Boltzmann factor. Equation (5.16) means to calculate the local magnetization at a finite temperature $T_p = \beta_p^{-1}$ and assign its sign to the decoded bit. The MAP can be regarded as the low-temperature (large-β) limit in place of finite β_p in (5.16). The MAP was derived as the maximizer of the posterior of the whole bit sequence, which has now been shown to be equivalent to the low-temperature limit of the MPM, finite-temperature decoding. The MPM, by contrast, maximizes the posterior of a single bit. We shall study the relation between these two methods in more detail subsequently.

5.2.4   Gaussian channel

It is sometimes convenient to consider channels other than the BSC. A typical example is a *Gaussian channel*. The encoded message $\xi_{i_1} \ldots \xi_{i_r} (= \pm 1)$ is fed into the channel as a signal of some amplitude, $J_0 \xi_{i_1} \ldots \xi_{i_r}$. The output is continuously distributed around this input with the Gaussian distribution of variance J^2:

$$P(J_{i_1 \ldots i_r} | \xi_{i_1} \ldots \xi_{i_r}) = \frac{1}{\sqrt{2\pi} J} \exp \left\{ -\frac{(J_{i_1 \ldots i_r} - J_0 \xi_{i_1} \ldots \xi_{i_r})^2}{2J^2} \right\}. \tag{5.17}$$

If the prior is uniform, the posterior is written using the Bayes formula as

$$P(\boldsymbol{\sigma} | \boldsymbol{J}) = \frac{\exp \left\{ (J_0/J^2) \sum J_{i_1 \ldots i_r} \sigma_{i_1} \ldots \sigma_{i_r} \right\}}{\text{Tr}_{\boldsymbol{\sigma}} \exp \left\{ (J_0/J^2) \sum J_{i_1 \ldots i_r} \sigma_{i_1} \ldots \sigma_{i_r} \right\}}. \tag{5.18}$$

Comparison of this equation with (5.12) shows that the posterior of the Gaussian channel corresponds to that of the BSC with β_p replaced by J_0/J^2. We can therefore develop the following arguments for both of these channels almost in the same way.

5.3   Overlap

5.3.1   Measure of decoding performance

It is convenient to introduce a measure of success of decoding that represents the proximity of the decoded message to the original one. The decoded ith bit is $\hat{\xi}_i = \text{sgn} \langle \sigma_i \rangle_{\beta}$ with $\beta = \beta_p$ for the MPM and $\beta \to \infty$ for the MAP. It sometimes happens that one is not aware of the noise rate p of the channel, or equivalently β_p. Consequently it makes sense even for the MPM to develop arguments with β unspecified, which we do in the following.

The product of $\hat{\xi}_i$ and the corresponding original bit ξ_i, $\xi_i \text{sgn}\langle\sigma_i\rangle_\beta$, is 1 if these two coincide and -1 otherwise. An appropriate strategy is to increase the probability that this product is equal to one. We average this product over the output probability of the channel $P(\boldsymbol{J}|\boldsymbol{\xi})$ and the prior $P(\boldsymbol{\xi})$,

$$M(\beta) = \text{Tr}_{\boldsymbol{\xi}} \sum_{\boldsymbol{J}} P(\boldsymbol{\xi})P(\boldsymbol{J}|\boldsymbol{\xi}) \, \xi_i \, \text{sgn}\langle\sigma_i\rangle_\beta, \qquad (5.19)$$

which is the *overlap* of the original and decoded messages. We have denoted the sum over the site variables $\boldsymbol{\xi}$ by $\text{Tr}_{\boldsymbol{\xi}}$ and the one over the bond variables $\boldsymbol{J}$ by $\sum_{\boldsymbol{J}}$. A better decoding would be the one that gives a larger $M(\beta)$. For a uniform message source $P(\boldsymbol{\xi}) = 2^{-N}$, the average over $\boldsymbol{\xi}$ and $\boldsymbol{J}$ leads to the right hand side independent of i,

$$M(\beta) = \frac{1}{2^N(2\cosh\beta_p)^{N_B}} \sum_{\boldsymbol{J}} \text{Tr}_{\boldsymbol{\xi}} \, \exp(\beta_p \sum J_{i_1\dots i_r}\xi_{i_1}\dots\xi_{i_r}) \, \xi_i\text{sgn}\langle\sigma_i\rangle_\beta. \quad (5.20)$$

This expression may be regarded as the average of $\text{sgn}\langle\sigma_i\rangle_\beta$ with the weight proportional to $Z(\beta_p)\langle\xi\rangle_{\beta_p}$, which is essentially equivalent to the configurational average of the correlation function with a similar form of the weight that appears frequently in the gauge theory of spin glasses, for example (4.43).

The overlap is closely related with the *Hamming distance* of the two messages (the number of different bits at the corresponding positions). For closer messages, the overlap is larger and the Hamming distance is smaller. When the two messages coincide, the overlap is one and the Hamming distance is zero, while, for two messages completely inverted from each other, the overlap is -1 and the Hamming distance is N.

5.3.2 Upper bound on the overlap

An interesting feature of the overlap $M(\beta)$ is that it is a non-monotonic function of β with its maximum at $\beta = \beta_p$:

$$M(\beta) \leq M(\beta_p). \qquad (5.21)$$

In other words the MPM at the correct parameter value $\beta = \beta_p$ gives the optimal result in the sense of maximization of the overlap (Ruján 1993; Nishimori 1993; Sourlas 1994; Iba 1999). The MPM is sometimes called the *Bayes-optimal* strategy.

To prove (5.21), we first take the absolute value of both sides of (5.20) and exchange the absolute value operation and the sum over $\boldsymbol{J}$ to obtain

$$M(\beta) \leq \frac{1}{2^N(2\cosh\beta_p)^{N_B}} \sum_{\boldsymbol{J}} \left| \text{Tr}_{\boldsymbol{\xi}} \, \xi_i \exp(\beta_p \sum J_{i_1\dots i_r}\xi_{i_1}\dots\xi_{i_r}) \right|, \qquad (5.22)$$

where we have used $|\text{sgn}\langle\sigma_i\rangle_\beta| = 1$. By rewriting the right hand side as follows, we can derive (5.21):

$$M(\beta) \leq \frac{1}{2^N(2\cosh\beta_p)^{N_B}} \sum_{\boldsymbol{J}} \frac{(\text{Tr}_{\boldsymbol{\xi}} \, \xi_i \exp(\beta_p \sum J_{i_1\dots i_r}\xi_{i_1}\dots\xi_{i_r}))^2}{|\text{Tr}_{\boldsymbol{\xi}} \, \xi_i \exp(\beta_p \sum J_{i_1\dots i_r}\xi_{i_1}\dots\xi_{i_r})|}$$

$$= \frac{1}{2^N (2 \cosh \beta_p)^{N_B}} \sum_{\boldsymbol{J}} \text{Tr}_{\boldsymbol{\xi}} \, \xi_i \exp(\beta_p \sum J_{i_1 \ldots i_r} \xi_{i_1} \cdots \xi_{i_r})$$

$$\cdot \frac{\text{Tr}_{\boldsymbol{\xi}} \, \xi_i \exp(\beta_p \sum J_{i_1 \ldots i_r} \xi_{i_1} \cdots \xi_{i_r})}{|\text{Tr}_{\boldsymbol{\xi}} \, \xi_i \exp(\beta_p \sum J_{i_1 \ldots i_r} \xi_{i_1} \cdots \xi_{i_r})|}$$

$$= \frac{1}{2^N (2 \cosh \beta_p)^{N_B}} \sum_{\boldsymbol{J}} \text{Tr}_{\boldsymbol{\xi}} \, \xi_i \exp(\beta_p \sum J_{i_1 \ldots i_r} \xi_{i_1} \cdots \xi_{i_r}) \text{sgn} \langle \sigma_i \rangle_{\beta_p}$$

$$= M(\beta_p). \tag{5.23}$$

Almost the same manipulations lead to the following inequality for the Gaussian channel:

$$M(\beta) = \frac{1}{2^N} \text{Tr}_{\boldsymbol{\xi}} \int \prod dJ_{i_1 \ldots i_r} \, P(\boldsymbol{J}|\boldsymbol{\xi}) \, \xi_i \, \text{sgn} \langle \sigma_i \rangle_{\beta} \leq M \left(\frac{J_0}{J^2} \right). \tag{5.24}$$

We have shown that the bit-wise overlap defined in (5.19) is maximized by the MPM with the correct parameter ($\beta = \beta_p$ for the BSC). This is natural in that the MPM at $\beta = \beta_p$ was introduced to maximize the bit-wise (marginalized) posterior. The MAP maximizes the posterior of the whole bit sequence $\boldsymbol{\sigma}$, but its probability of error for a given single bit is larger than the MPM with the correct parameter value. This observation is also confirmed from the viewpoint of Bayesian statistics (Sourlas 1994; Iba 1999).

The inequalities (5.21) and (5.24) are essentially identical to (4.63) derived by the gauge theory in §4.6.4. To understand this, we note that generality is not lost by the assumption $\xi_i = 1$ ($\forall i$) in the calculation of $M(\beta)$ for a uniform information source. This may be called the *ferromagnetic gauge*. Indeed, the gauge transformation $J_{i_1 \ldots i_r} \rightarrow J_{i_1 \ldots i_r} \xi_{i_1} \cdots \xi_{i_r}$ and $\sigma_i \rightarrow \sigma_i \xi_i$ in (5.19) removes ξ_i from the equation. Then M defined in (5.19) is seen to be identical to (4.63) with the two-point correlation replaced with a single-point spin expectation value. The argument in §4.6.4 applies not only to two-point correlations but to any correlations, and thus the result of §4.6.4 agrees with that of the present section. Therefore the overlap M of the decoded bit and the original bit becomes a maximum on the Nishimori line as a function of the decoding temperature with fixed error rate p. For the Gaussian channel, (5.24) corresponds to the fact that the Nishimori line is represented as $J_0/J^2 = \beta$ as observed in §4.3.3.

5.4 Infinite-range model

Inequality (5.21) shows that $M(\beta)$ is a non-monotonic function, but the inequality does not give the explicit β-dependence. It may also happen that it is not easy to adjust β exactly to β_p in practical situations where one may not know the noise rate p very precisely. One is then forced to estimate β_p but should always be careful of errors in the parameter estimation. It is therefore useful if one can estimate the effects of errors in the parameter estimation on the overlap $M(\beta)$.

A solvable model, for which we can calculate the explicit form of $M(\beta)$, would thus be helpful as a guide for non-solvable cases. The infinite-range model serves as an important prototype for this and other purposes.

5.4.1 Infinite-range model

The Sourlas code explained in §5.1.3 is represented as the infinite-range model. In the Sourlas code, the sum in the Hamiltonian

$$H = - \sum_{i_1 < \cdots < i_r} J_{i_1 \ldots i_r} \sigma_{i_1} \ldots \sigma_{i_r} \tag{5.25}$$

runs over all possible combinations of r spins out of N spins. Then the number of terms is $N_B = \binom{N}{r}$. This infinite-range model with r-body interactions can be solved explicitly by the replica method (Derrida 1981; Gross and Mézard 1984; Gardner 1985; Nishimori and Wong 1999). We show the solution for the Gaussian channel. The BSC is expected to give the same result in the thermodynamic limit according to the central limit theorem.

The parameters J_0 and J in the Gaussian distribution (5.17) must be scaled appropriately with N so that the expectation value of the infinite-range Hamiltonian (5.25) is extensive (proportional to N) in the limit $N \to \infty$. If we also demand that physical quantities remain finite in the limit $r \to \infty$ after $N \to \infty$, then r should also be appropriately scaled. The Gaussian distribution satisfying these requirements is

$$P(J_{i_1 \ldots i_r} | \xi_{i_1} \ldots \xi_{i_r}) = \left(\frac{N^{r-1}}{J^2 \pi r!} \right)^{1/2} \exp \left\{ - \frac{N^{r-1}}{J^2 r!} \left(J_{i_1 \ldots i_r} - \frac{j_0 r!}{N^{r-1}} \xi_{i_1} \ldots \xi_{i_r} \right)^2 \right\}, \tag{5.26}$$

where J and j_0 are independent of N and r. The appropriateness of (5.26) is justified by the expressions of various quantities to be derived below that have non-trivial limits as $N \to \infty$ and $r \to \infty$.

5.4.2 Replica calculations

Following the general prescription of the replica method, we first calculate the configurational average of the nth power of the partition function and take the limit $n \to 0$. Order parameters naturally emerge in due course. The overlap M is expressed as a function of these order parameters.

The configurational average of the nth power of the partition function of the infinite-range model is written for the uniform prior $(P = 2^{-N})$ as

$$[Z^n] = \text{Tr}_{\xi} \int \prod_{i_1 < \cdots < i_r} dJ_{i_i \ldots i_r} \, P(\xi) P(\boldsymbol{J} | \xi) \, Z^n$$

$$= \frac{1}{2^N} \text{Tr}_{\xi} \int \prod_{i_1 < \cdots < i_r} dJ_{i_i \ldots i_r} \left(\frac{N^{r-1}}{J^2 \pi r!} \right)^{1/2}$$

$$\cdot \exp \left\{ -\frac{N^{r-1}}{J^2 r!} \sum_{i_1 < \cdots < i_r} \left(J_{i_1 \ldots i_r} - \frac{j_0 r!}{N^{r-1}} \xi_{i_1} \cdots \xi_{i_r} \right)^2 \right\}$$

$$\cdot \mathrm{Tr}_{\sigma} \exp \left(\beta \sum_{i_1 < \cdots < i_r} J_{i_1 \ldots i_r} \sum_{\alpha} \sigma_{i_1}^{\alpha} \cdots \sigma_{i_r}^{\alpha} \right), \tag{5.27}$$

where $\alpha \, (= 1, \ldots, n)$ is the replica index. A gauge transformation

$$J_{i_1 \ldots i_r} \to J_{i_1 \ldots i_r} \xi_{i_1} \cdots \xi_{i_r}, \quad \sigma_i \to \sigma_i \xi_i \tag{5.28}$$

in (5.27) removes $\boldsymbol{\xi}$ from the integrand. The problem is thus equivalent to the case of $\xi_i = 1 \; (\forall i)$, the ferromagnetic gauge. We mainly use this gauge in the present chapter. The sum over $\boldsymbol{\xi}$ in (5.27) then simply gives 2^N, and the factor 2^{-N} in front of the whole expression disappears.

It is straightforward to carry out the Gaussian integral in (5.27). If we ignore the trivial overall constant and terms of lower order in N, the result is

$$[Z^n]$$

$$= \mathrm{Tr}_{\sigma} \exp \left\{ \frac{\beta^2 J^2 r!}{4 N^{r-1}} \sum_{i_1 < \cdots < i_r} \left(\sum_{\alpha} \sigma_{i_1}^{\alpha} \cdots \sigma_{i_r}^{\alpha} \right)^2 + \frac{\beta r! j_0}{N^{r-1}} \sum_{i_1 < \cdots < i_r} \sum_{\alpha} \sigma_{i_1}^{\alpha} \cdots \sigma_{i_r}^{\alpha} \right\}$$

$$= \mathrm{Tr}_{\sigma} \exp \left\{ \frac{\beta^2 J^2 r!}{4 N^{r-1}} \sum_{i_1 < \cdots < i_r} \left(\sum_{\alpha \neq \beta} \sigma_{i_1}^{\alpha} \cdots \sigma_{i_r}^{\alpha} \sigma_{i_1}^{\beta} \cdots \sigma_{i_r}^{\beta} + n \right) \right.$$

$$\left. + \frac{\beta r! j_0}{N^{r-1}} \sum_{i_1 < \cdots < i_r} \sum_{\alpha} \sigma_{i_1}^{\alpha} \cdots \sigma_{i_r}^{\alpha} \right\}$$

$$= \mathrm{Tr}_{\sigma} \exp \left\{ \frac{\beta^2 J^2 N}{2} \sum_{\alpha < \beta} \left(\frac{1}{N} \sum_i \sigma_i^{\alpha} \sigma_i^{\beta} \right)^r + \frac{\beta^2 J^2}{4} N n \right.$$

$$\left. + j_0 \beta N \sum_{\alpha} \left(\frac{1}{N} \sum_i \sigma_i^{\alpha} \right)^r \right\}. \tag{5.29}$$

Here Tr_{σ} is the sum over $\boldsymbol{\sigma}$. In deriving the final expression, we have used a relation that generalizes the following relation to the r-body case:

$$\frac{1}{N} \sum_{i_1 < i_2} \sigma_{i_1} \sigma_{i_2} = \frac{1}{2} \left(\frac{1}{N} \sum_i \sigma_i \right)^2 + \mathcal{O}(N^0)$$

$$\frac{1}{N^2} \sum_{i_1 < i_2 < i_3} \sigma_{i_1} \sigma_{i_2} \sigma_{i_3} = \frac{1}{3!} \left(\frac{1}{N} \sum_i \sigma_i \right)^3 + \mathcal{O}(N^0). \tag{5.30}$$

It is convenient to introduce the variables

$$q_{\alpha\beta} = \frac{1}{N} \sum_i \sigma_i^{\alpha} \sigma_i^{\beta}, \quad m_{\alpha} = \frac{1}{N} \sum_i \sigma_i^{\alpha} \tag{5.31}$$

to replace the expressions inside the parentheses in the final expression of (5.29) by $q_{\alpha\beta}$ and m_α so that we can carry out the Tr_σ operation. The condition to satisfy (5.31) is imposed by the Fourier-transformed expressions of delta functions with integration variables $\hat{q}_{\alpha\beta}$ and $\hat{m}_\alpha$:

$$
[Z^n] = \mathrm{Tr}_\sigma \int \prod_{\alpha<\beta} \mathrm{d}q_{\alpha\beta}\mathrm{d}\hat{q}_{\alpha\beta} \int \prod_\alpha \mathrm{d}m_\alpha \mathrm{d}\hat{m}_\alpha \exp\Bigg\{ \frac{\beta^2 J^2 N}{2} \sum_{\alpha<\beta}(q_{\alpha\beta})^r
$$
$$
- N\sum_{\alpha<\beta} q_{\alpha\beta}\hat{q}_{\alpha\beta} + N\sum_{\alpha<\beta}\hat{q}_{\alpha\beta}\left(\frac{1}{N}\sum_i \sigma_i^\alpha \sigma_i^\beta\right) + j_0\beta N\sum_\alpha (m_\alpha)^r
$$
$$
- N\sum_\alpha m_\alpha \hat{m}_\alpha + N\sum_\alpha \hat{m}_\alpha\left(\frac{1}{N}\sum_i \sigma_i^\alpha\right) + \frac{1}{4}\beta^2 J^2 N n \Bigg\}. \tag{5.32}
$$

We can now operate Tr_σ independently at each i to find

$$
[Z^n] = \int \prod_{\alpha<\beta} \mathrm{d}q_{\alpha\beta}\mathrm{d}\hat{q}_{\alpha\beta} \int \prod_\alpha \mathrm{d}m_\alpha \mathrm{d}\hat{m}_\alpha \exp\Bigg\{ \frac{\beta^2 J^2 N}{2} \sum_{\alpha<\beta}(q_{\alpha\beta})^r
$$
$$
- N\sum_{\alpha<\beta} q_{\alpha\beta}\hat{q}_{\alpha\beta} + \frac{1}{4}\beta^2 J^2 N n + j_0\beta N\sum_\alpha (m_\alpha)^r
$$
$$
- N\sum_\alpha m_\alpha \hat{m}_\alpha + N\log \mathrm{Tr}\exp\left(\sum_{\alpha<\beta}\hat{q}_{\alpha\beta}\sigma^\alpha\sigma^\beta + \sum_\alpha \hat{m}_\alpha \sigma^\alpha\right)\Bigg\}. \tag{5.33}
$$

Here Tr denotes sums over single-site replica spins $\{\sigma^1,\ldots,\sigma^n\}$.

5.4.3   Replica-symmetric solution

Further calculations are possible under the assumption of replica symmetry:

$$
q = q_{\alpha\beta},\quad \hat{q} = \hat{q}_{\alpha\beta},\quad m = m_\alpha,\quad \hat{m} = \hat{m}_\alpha. \tag{5.34}
$$

We fix n and take the thermodynamic limit $N \to \infty$ to evaluate the integral by steepest descent. The result is

$$
[Z^n] \approx \exp\Bigg[N\Big\{ \beta^2 J^2 \frac{n(n-1)}{4} q^r - \frac{n(n-1)}{2}q\hat{q} + j_0\beta n m^r - nm\hat{m} + \frac{1}{4}n\beta^2 J^2
$$
$$
+ \log\mathrm{Tr}\int \mathrm{D}u \exp\left(\sqrt{\hat{q}}\,u\sum_\alpha \sigma^\alpha + \hat{m}\sum_\alpha \sigma^\alpha - \frac{n}{2}\hat{q}\right)\Big\}\Bigg], \tag{5.35}
$$

where $\mathrm{D}u = \mathrm{e}^{-u^2/2}\mathrm{d}u/\sqrt{2\pi}$. This u has been introduced to reduce the double sum over α and β $(\sum_{\alpha<\beta})$ to a single sum. The trace operation can be performed for each replica independently so that the free energy βf defined by $[Z^n] = e^{-Nn\beta f}$ becomes in the limit $n \to 0$

$$-\beta f = -\frac{1}{4}\beta^2 J^2 q^r + \frac{1}{2}q\hat{q} + \beta j_0 m^r - m\hat{m}$$

$$+ \frac{1}{4}\beta^2 J^2 - \frac{1}{2}\hat{q} + \int Du \log 2\cosh(\sqrt{\hat{q}}\,u + \hat{m}). \qquad (5.36)$$

The equations of state for the order parameters are determined by the saddle-point condition. By variation of (5.36), we obtain

$$\hat{q} = \frac{1}{2}r\beta^2 J^2 q^{r-1}, \quad \hat{m} = \beta j_0 rm^{r-1} \qquad (5.37)$$

$$q = \int Du \tanh^2(\sqrt{\hat{q}}\,u + \hat{m}), \quad m = \int Du \tanh(\sqrt{\hat{q}}\,u + \hat{m}). \qquad (5.38)$$

Eliminating $\hat{q}$ and $\hat{m}$ from (5.38) using (5.37), we can write the equations for q and m in closed form:

$$q = \int Du \tanh^2 \beta G, \quad m = \int Du \tanh \beta G, \qquad (5.39)$$

where

$$G = J\sqrt{\frac{rq^{r-1}}{2}}\,u + j_0 rm^{r-1}. \qquad (5.40)$$

These reduce to the equations of state for the conventional SK model, (2.28) and (2.30), when $r = 2$. One should remember that $2j_0$ here corresponds to J_0 in the conventional notation of the SK model as is verified from (5.27) with $r = 2$.

5.4.4 Overlap

The next task is to derive the expression of the overlap M. An argument similar to §2.2.5 leads to formulae expressing the physical meaning of q and m:

$$q = \left[\langle\sigma_i^\alpha \sigma_i^\beta\rangle\right] = \left[\langle\sigma_i\rangle^2\right], \quad m = [\langle\sigma_i^\alpha\rangle] = [\langle\sigma_i\rangle]. \qquad (5.41)$$

Comparison of (5.41) and (5.39) suggests that $\tanh^2 \beta(\cdot)$ and $\tanh \beta(\cdot)$ in the integrands of the latter may be closely related with $\langle\sigma_i\rangle^2$ and $\langle\sigma_i\rangle$ in the former. To confirm this, we add $h\sum_i \sigma_i^\alpha \sigma_i^\beta$ to the final exponent of (5.27) and follow the calculations of the previous section to find a term $h\sigma^\alpha \sigma^\beta$ in the exponent of the integrand of (5.35). We then differentiate $-\beta nf$ with respect to h and let $n \to 0, h \to 0$ to find that σ^α and σ^β are singled out in the trace operation of replica spins, leading to the factor $\tanh^2 \beta(\cdot)$. A similar argument using an external field term $h\sum_i \sigma_i^\alpha$ and differentiation by h leads to $\tanh \beta(\cdot)$.

It should now be clear that the additional external field with the product of k spins, $h \sum_i \sigma_i^\alpha \sigma_i^\beta \ldots$, yields

$$[\langle \sigma_i \rangle_\beta^k] = \int Du \, \tanh^k \beta G. \tag{5.42}$$

Thus, for an arbitrary function $F(x)$ that can be expanded around $x = 0$, the following identity holds:

$$[F(\langle \sigma_i \rangle_\beta)] = \int Du \, F(\tanh \beta G). \tag{5.43}$$

The overlap is, in the ferromagnetic gauge, $M(\beta) = [\text{sgn}\langle \sigma_i \rangle_\beta]$. If we therefore take as $F(x)$ a function that approaches $\text{sgn}(x)$ (e.g. $\tanh(ax)$ with $a \to \infty$), we obtain the desired relation for the overlap:

$$M(\beta) = [\text{sgn}\langle \sigma_i \rangle_\beta] = \int Du \, \text{sgn} \, G. \tag{5.44}$$

It has thus been established that $M(\beta)$ is determined as a function of q and m through G.

5.5　Replica symmetry breaking

The system (5.25) and (5.26) in the ferromagnetic gauge is a spin glass model with r-body interactions. It is natural to go further to investigate the properties of the RSB solution (Derrida 1981; Gross and Mézard 1984; Gardner 1985; Nishimori and Wong 1999; Gillin *et al.* 2001). We shall show that, for small values of the centre of distribution j_0, a 1RSB phase appears after the RS paramagnetic phase as the temperature is lowered. A full RSB phase follows at still lower temperature. For larger j_0, paramagnetic phase, ferromagnetic phase without RSB, and then ferromagnetic phase with RSB phases appear sequentially as the temperature is decreased.

5.5.1　*First-step RSB*

The free energy with 1RSB can be derived following the method described in §3.2:

$$-\beta f = -\hat{m}m + \frac{1}{2}x\hat{q}_0 q_0 + \frac{1}{2}(1-x)\hat{q}_1 q_1 + \beta j_0 m^r$$
$$-\frac{1}{4}x\beta^2 J^2 q_0^r - \frac{1}{4}(1-x)\beta^2 J^2 q_1^r + \frac{1}{4}\beta^2 J^2 - \frac{1}{2}\hat{q}_1$$
$$+\frac{1}{x}\int Du \log \int Dv \cosh^x(\hat{m} + \sqrt{\hat{q}_0}\,u + \sqrt{\hat{q}_1 - \hat{q}_0}\,v) + \log 2, \tag{5.45}$$

where x ($0 \leq x \leq 1$) is the boundary of q_0 and q_1 in the matrix block, denoted by m_1 in §3.2. Extremization of (5.45) with respect to $q_0, q_1, \hat{q}_0, \hat{q}_1, m, \hat{m}, x$ leads to the equations of state. For m, q_0, q_1,

$$\hat{m} = \beta j_0 r m^{r-1}, \quad \hat{q}_0 = \frac{1}{2}\beta^2 J^2 r q_0^{r-1}, \quad \hat{q}_1 = \frac{1}{2}\beta^2 J^2 r q_1^{r-1}. \tag{5.46}$$

Elimination of $\hat{m}, \hat{q}_0, \hat{q}_1$ using (5.46) from the equations of extremization with respect to $\hat{q}_0, \hat{q}_1, \hat{m}$ leads to

$$m = \int Du \frac{\int Dv \cosh^x \beta G_1 \tanh \beta G_1}{\int Dv \cosh^x \beta G_1} \tag{5.47}$$

$$q_0 = \int Du \left(\frac{\int Dv \cosh^x \beta G_1 \tanh \beta G_1}{\int Dv \cosh^x \beta G_1} \right)^2 \tag{5.48}$$

$$q_1 = \int Du \frac{\int Dv \cosh^x \beta G_1 \tanh^2 \beta G_1}{\int Dv \cosh^x \beta G_1} \tag{5.49}$$

$$G_1 = J\sqrt{\frac{r q_0^{r-1}}{2}}\, u + J\sqrt{\frac{r}{2}(q_1^{r-1} - q_0^{r-1})}\, v + j_0 r m^{r-1}. \tag{5.50}$$

These equations coincide with (3.32)–(3.34) for $r = 2$ if we set $h = 0$ and $2j_0 = J_0$. The equation of extremization by x does not have an intuitively appealing compact form so that we omit it here.

The AT stability condition of the RS solution is

$$\frac{2T^2 q^{2-r}}{r(r-1)} > J^2 \int Du \operatorname{sech}^4 \beta G. \tag{5.51}$$

The criterion of stability for 1RSB is expressed similarly. For the replica pair $(\alpha\beta)$ in the same diagonal block, the stability condition for small deviations of $q_{\alpha\beta}$ and $\hat{q}_{\alpha\beta}$ from 1RSB is

$$\frac{2T^2 q_1^{2-r}}{r(r-1)} > J^2 \int Du \frac{\int Dv \cosh^{x-4} \beta G_1}{\int Dv \cosh^x \beta G_1}. \tag{5.52}$$

Further steps of RSB take place with the diagonal blocks breaking up into smaller diagonal and off-diagonal blocks. Thus it is sufficient to check the intra-block stability condition (5.52) only.

5.5.2 Random energy model

The model in the limit $r \to \infty$ is known as the *random energy model* (REM) (Derrida 1981). The problem can be solved completely in this limit and the spin glass phase is characterized by 1RSB.

As its name suggests, the REM has an independent distribution of energy. Let us demonstrate this fact for the case of $j_0 = 0$. The probability that the system has energy E will be denoted by $P(E)$,

$$P(E) = [\delta(E - H(\boldsymbol{\sigma}))]. \tag{5.53}$$

The average $[\cdots]$ over the distribution of $\boldsymbol{J}$, (5.26), can be carried out if we express the delta function by Fourier transformation. The result is

$$P(E) = \frac{1}{\sqrt{N\pi J^2}} \exp\left(-\frac{E^2}{J^2 N}\right).$$ (5.54)

The simultaneous distribution function of the energy values E_1 and E_2 of two independent spin configurations $\boldsymbol{\sigma}^{(1)}$ and $\boldsymbol{\sigma}^{(2)}$ with the same set of interactions can be derived similarly:

$$P(E_1, E_2) = \left[\delta(E_1 - H(\boldsymbol{\sigma}^{(1)}))\delta(E_2 - H(\boldsymbol{\sigma}^{(2)}))\right]$$

$$= \frac{1}{N\pi J^2 \sqrt{(1+q^r)(1-q^r)}} \exp\left(-\frac{(E_1+E_2)^2}{2N(1+q^r)J^2} - \frac{(E_1-E_2)^2}{2N(1-q^r)J^2}\right)$$

$$q = \frac{1}{N}\sum_i \sigma_i^{(1)}\sigma_i^{(2)}.$$ (5.55)

It is easy to see in the limit $r \to \infty$ that

$$P(E_1, E_2) \to P(E_1)P(E_2),$$ (5.56)

which implies independence of the energy distributions of two spin configurations. Similar arguments hold for three (and more) energy values.

The number of states with energy E is, according to the independence of energy levels,

$$n(E) = 2^N P(E) = \frac{1}{\sqrt{\pi N J^2}} \exp N \left\{\log 2 - \left(\frac{E}{NJ}\right)^2\right\}.$$ (5.57)

This expression shows that there are very many energy levels for $|E| < NJ\sqrt{\log 2} \equiv E_0$ but none in the other range $|E| > E_0$ in the limit $N \to \infty$. The entropy for $|E| < E_0$ is

$$S(E) = N\left[\log 2 - \left(\frac{E}{NJ}\right)^2\right].$$ (5.58)

We then have, from $\mathrm{d}S/\mathrm{d}E = 1/T$,

$$E = -\frac{NJ^2}{2T}.$$ (5.59)

The free energy is therefore

$$f = \begin{cases} -T\log 2 - \dfrac{J^2}{4T} & (T > T_c) \\ -J\sqrt{\log 2} & (T < T_c), \end{cases}$$ (5.60)

where $T_c/J = (2\sqrt{\log 2})^{-1}$. Equation (5.60) indicates that there is a phase transition at $T = T_c$ and the system freezes out completely ($S = 0$) below the transition temperature.

5.5.3 *Replica solution in the limit $r \to \infty$*

It is instructive to rederive the results of the previous subsection by the replica method. We first discuss the case $j_0 = 0$. It is quite reasonable to expect no RSB in the paramagnetic (P) phase at high temperature. We thus set $q = \hat{q} = m = \hat{m} = 0$ in (5.36) to obtain

$$f_{\mathrm{P}} = -T \log 2 - \frac{J^2}{4T}. \tag{5.61}$$

This agrees with (5.60) for $T > T_{\mathrm{c}}$.

It is necessary to introduce RSB in the spin glass (SG) phase. We try 1RSB and confirm that the result agrees with that of the previous subsection. For the 1RSB to be non-trivial (i.e. different from the RS), it is required that $q_0 < q_1 \leq 1$ and $\hat{q}_0 < \hat{q}_1$. Then, if $q_1 < 1$, we find $\hat{q}_0 = \hat{q}_1 = 0$ in the limit $r \to \infty$ from (5.46). We therefore have $q_1 = 1$, and $\hat{q}_1 = \beta^2 J^2 r / 2$ from (5.46). Then, in (5.48), we find $G_1 = J\sqrt{r/2v}$ for $r \gg 1$ and the v-integral in the numerator vanishes, leading to $q_0 = 0$. Hence $\hat{q}_0 = 0$ from (5.46). From these results, the free energy (5.45) in the limit $r \to \infty$ is

$$-\beta f = \frac{\beta^2 J^2}{4} x + \frac{1}{x} \log 2. \tag{5.62}$$

Variation with respect to x gives

$$(x\beta J)^2 = 4 \log 2. \tag{5.63}$$

The highest temperature satisfying this equation is the following one for $x = 1$:

$$\frac{T_{\mathrm{c}}}{J} = \frac{1}{2\sqrt{\log 2}}, \tag{5.64}$$

and therefore, for $T < T_{\mathrm{c}}$,

$$f_{\mathrm{SG}} = -J\sqrt{\log 2}. \tag{5.65}$$

This agrees with (5.60), which confirms that 1RSB is exact in the temperature range $T < T_{\mathrm{c}}$. It is also possible to show by explicit k-step RSB calculations that the solution reduces to the 1RSB for any $k \, (\geq 1)$ (Gross and Mézard 1984).

It is easy to confirm that $x < 1$ for $T < T_{\mathrm{c}}$ from (5.63); generally, $x = T/T_{\mathrm{c}}$. The order parameter function $q(x)$ is equal to $1 \, (= q_1)$ above $x \, (= T/T_{\mathrm{c}})$ and $0 \, (= q_0)$ below x (see Fig. 5.3).

For j_0 exceeding some critical value, a ferromagnetic (F) phase exists. It is easy to confirm that the solution of (5.48), (5.49), and (5.47) in the limit $r \to \infty$ is $q_0 = q_1 = m = 1$ when $j_0 > 0, m > 0$. The ferromagnetic phase is therefore replica symmetric since $q_0 = q_1$. The free energy (5.36) is then

$$f_{\mathrm{F}} = -j_0. \tag{5.66}$$

The phase boundaries between these three phases are obtained by comparison of the free energies:

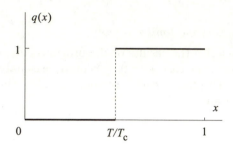

FIG. 5.3. Spin glass order parameter of the REM below the transition temperature

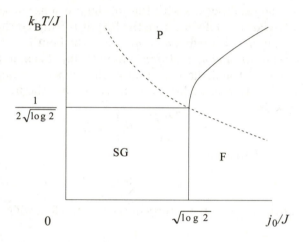

FIG. 5.4. Phase diagram of the REM. The Nishimori line is shown dashed.

1. P–SG transition: $T_c/J = (2\sqrt{\log 2})^{-1}$.
2. SG–F transition: $(j_0)_c/J = \sqrt{\log 2}$.
3. P–F transition: $j_0 = J^2/(4T) + T \log 2$.

The final phase diagram is depicted in Fig. 5.4.

Let us now turn to the interpretation of the above results in terms of error-correcting codes. The overlap M is one in the ferromagnetic phase ($j_0/J > \sqrt{\log 2}$) because the spin alignment is perfect ($m = 1$), implying error-free decoding.[6] To see the relation of this result to the Shannon bound (5.3), we first note that the transmission rate of information by the Sourlas code is

$$R = \frac{N}{\binom{N}{r}}. \tag{5.67}$$

As shown in Appendix C, the capacity of the Gaussian channel is

---

[6]Remember that we are using the ferromagnetic gauge.

$$C = \frac{1}{2} \log_2 \left(1 + \frac{J_0^2}{J^2} \right). \tag{5.68}$$

Here we substitute $J_0 = j_0 r!/N^{r-1}$ and $J^2 \to J^2 r!/2N^{r-1}$ according to (5.26) and take the limit $N \gg 1$ with r fixed to find

$$C \approx \frac{j_0^2 r!}{J^2 N^{r-1} \log 2}. \tag{5.69}$$

The transmission rate (5.67), on the other hand, reduces in the same limit to

$$R \approx \frac{r!}{N^{r-1}}. \tag{5.70}$$

It has thus been established that the transmission rate R coincides with the channel capacity C at the lower limit of the ferromagnetic phase $j_0/J = \sqrt{\log 2}$. In the context of error-correcting codes, j_0 represents the signal amplitude and J is the amplitude of the noise, and hence j_0/J corresponds to the S/N ratio. The conclusion is that the Sourlas code in the limit $r \to \infty$, equivalent to the REM, is capable of error-free decoding ($m = 1, M = 1$) for the S/N ratio exceeding some critical value and the Shannon bound is achieved at this critical value.

The general inequality (5.21) is of course satisfied. Both sides vanish if $j_0 < (j_0)_c$. For $j_0 > (j_0)_c$, the right hand side is one while the left hand side is zero in the paramagnetic phase and one in the ferromagnetic phase. In other words, the Sourlas code in the limit $r \to \infty$ makes it possible to transmit information without errors under the MAP as well as under the MPM. An important point is that the information transmission rate R is vanishingly small, impeding practical usefulness of this code.

The Nishimori line $\beta J^2 = J_0$ is in the present case $T/J = J/(2j_0)$ and passes through the point at $j_0/J = \sqrt{\log 2}$ and $T/J = 1/2\sqrt{\log 2}$ where three phases (P, SG, F) coexist. The exact energy on it, $E = -j_0$, derived from the gauge theory agrees with the above answer (5.66). One should remember here that the free energy coincides with the energy as the entropy vanishes.

5.5.4 Solution for finite r

It is necessary to solve the equations of state numerically for general finite r.[7] The result for the case of $r = 3$ is shown here as an example (Nishimori and Wong 1999; Gillin et al. 2001). If j_0 is close to zero, one finds a 1RSB SG solution with $q_1 > 0$ and $q_0 = m = 0$ below $T = 0.651J$. As the temperature is lowered, the stability condition of the 1RSB (5.52) breaks down at $T = 0.240J$, and the full RSB takes over.

The ferromagnetic phase is RS (5.39) in the high-temperature range but the RSB should be taken into account below the AT line (3.22); a mixed (M) phase with both ferromagnetic order and RSB exists at low temperatures. The

---

[7]Expansions from the large-r limit and from $r = 2$ are also possible (Gardner 1985).

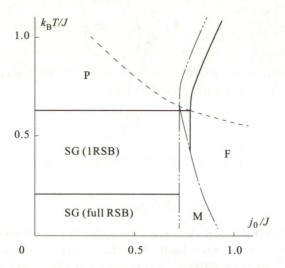

FIG. 5.5. Phase diagram of the model with $r = 3$. The double dotted line indicates the limit of metastability (spinodal) of the ferromagnetic phase. Error correction is possible to the right of this boundary. Thermodynamic phase boundaries are drawn in full lines. The Nishimori line is drawn dashed.

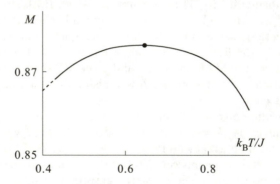

FIG. 5.6. Overlap for $r = 3, j_0 = 0.77$

ferromagnetic phase, with RS and/or RSB, continues to exist beyond the limit of thermodynamic stability as a *metastable state* (i.e. as a local minimum of the free energy). Figure 5.5 summarizes the result.

Dependence of the overlap $M(\beta)$ on $T (= 1/\beta)$ is depicted in Fig. 5.6 where j_0/J is fixed to 0.77 close to the boundary of the ferromagnetic phase. The overlap $M(\beta)$ is a maximum at the optimal temperature $T/J = J/2j_0 = 0.649$ appearing on the right hand side of (5.24) corresponding to the Nishimori line (the dot in Fig. 5.6). The ferromagnetic phase disappears above $T/J = 0.95$ even as a metastable state and one has $M = 0$: the paramagnetic phase has $\langle \sigma_i \rangle_\beta = 0$ at each site i, and $\text{sgn}\langle \sigma_i \rangle_\beta$ cannot be defined. It is impossible to decode the message there. For temperatures below $T/J = 0.43$, the RSB is observed (the dashed part). We have thus clarified in this example how much the decoded message agrees with the original one as the temperature is changed around the optimal value.

5.6 Codes with finite connectivity

The Sourlas code saturates the Shannon bound asymptotically with vanishing transmission rate. A mean-field model with finite connectivity has a more desirable property that the rate is finite yet the Shannon bound is achieved. We state some of the important results about this model in the present section. We refer the reader to the original papers cited in the text for details of the calculations.

5.6.1 *Sourlas-type code with finite connectivity*

The starting point is analogous to the Sourlas code described by the Hamiltonian (5.25) but with diluted binary interactions for the BSC,

$$H = - \sum_{i_1 < \cdots < i_r} A_{i_1 \ldots i_r} J_{i_1 \ldots i_r} \sigma_{i_1} \cdots \sigma_{i_r} - F \sum_i \sigma_i, \qquad (5.71)$$

where the element of the symmetric tensor $A_{i_1 \ldots i_r}$ (representing dilution) is either zero or one depending on the set of indices $(i_1, i_2, \ldots, i_r)$. The final term has been added to be prepared for biased messages in which 1 may appear more frequently than -1 (or vice versa). The connectivity is c; there are c non-zero elements randomly chosen for any given site index i:

$$\sum_{i_2, \ldots, i_r} A_{i\, i_2 \ldots i_r} = c. \qquad (5.72)$$

The code rate is $R = r/c$ because an encoded message has c bits per index i and carries r bits of the original information.

Using the methods developed for diluted spin glasses (Wong and Sherrington 1988), one can calculate the free energy under the RS ansatz as (Kabashima and Saad 1998, 1999; Vicente *et al.* 1999)

$$-\beta f^{(\mathrm{RS})} = \frac{c}{r} \log \cosh \beta + \frac{c}{r} \int \prod_{l=1}^{r} \mathrm{d}x_l\, \pi(x_l) \left[\log \left(1 + \tanh \beta J \prod_{j=1}^{r} \tanh \beta x_j \right) \right]_J$$

$$- c \int \mathrm{d}x \mathrm{d}y\, \pi(x)\hat{\pi}(y) \log(1 + \tanh \beta x \tanh \beta y) - c \int \mathrm{d}y\, \hat{\pi}(y) \log \cosh \beta y$$

$$+ \int \prod_{l=1}^{c} \mathrm{d}y_l\, \hat{\pi}(y_l) \left[\log \left(2 \cosh(\beta \sum_j y_j + \beta F \xi) \right) \right]_\xi. \qquad (5.73)$$

Here $[\cdots]_J$ and $[\cdots]_\xi$ denote the configurational averages over the distributions of J and ξ, respectively. The order functions $\pi(x)$ and $\hat{\pi}(y)$ represent distributions of the multi-replica spin overlap and its conjugate:

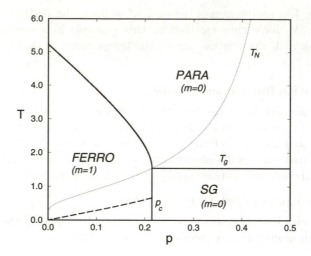

FIG. 5.7. Finite-temperature phase diagram of the unbiased diluted Sourlas code with $R = 1/4$ in the limit $r, c \to \infty$ (Vicente *et al.* 1999). The Shannon bound is achieved at p_c. The ferromagnetic phase is stable at least above the dashed line. (Copyright 1999 by the American Physical Society)

$$q_{\alpha\beta\ldots\gamma} = a \int dx\, \pi(x) \tanh^l \beta x, \quad \hat{q}_{\alpha\beta\ldots\gamma} = \hat{a} \int dy\, \hat{\pi}(y) \tanh^l \beta y, \qquad (5.74)$$

where a and $\hat{a}$ are normalization constants, and l is the number of replica indices on the left hand side. Extremization of the free energy gives paramagnetic and ferromagnetic solutions for the order functions. The spin glass solution should be treated with more care under the 1RSB scheme. The result becomes relatively simple in the limit where r and c tend to infinity with the ratio $R = r/c \, (\equiv 1/\alpha)$ kept finite and $F = 0$:

$$f_P = -T(\alpha \log \cosh \beta + \log 2)$$
$$f_F = -\alpha(1 - 2p) \qquad (5.75)$$
$$f_{1\text{RSB}-\text{SG}} = -T_g(\alpha \log \cosh \beta_g + \log 2),$$

where p is the noise probability of the BSC, and T_g is determined by the condition of vanishing paramagnetic entropy, $\alpha(\log \cosh \beta_g - \beta_g \tanh \beta_g) + \log 2 = 0$. The ferromagnetic and 1RSB–SG phases are completely frozen (vanishing entropy). The finite-temperature phase diagram for a given R is depicted in Fig. 5.7. Perfect decoding ($m = M = 1$) is possible in the ferromagnetic phase that extends to the limit p_c. It can be verified by equating f_F and $f_{1\text{RSB}-\text{SG}}$ that the Shannon bound is achieved at p_c,

$$R = 1 + p \log_2 p + (1 - p) \log_2(1 - p) \quad (p = p_c). \qquad (5.76)$$

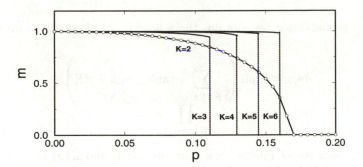

FIG. 5.8. Ground-state magnetization as a function of the noise probability for various r (written as K in the figure) (Vicente et $al.$ 1999). The rate R is $1/2$ and $F = 0$. Also shown by open circles are the numerical results from the TAP-like decoding algorithm. (Copyright 1999 by the American Physical Society)

The ferromagnetic solution appears as a metastable state at a very high temperature of $\mathcal{O}(r/\log r)$, but the thermodynamic transition takes place at T of $\mathcal{O}(1)$. This suggests that there exists a high energy barrier between the ferromagnetic and paramagnetic solutions. Consequently, it might be difficult to reach the correctly decoded (i.e. ferromagnetic) state starting from an arbitrary initial condition (which is almost surely a paramagnetic state) by some decoding algorithm. We are therefore lead to consider moderate r, in which case the ferromagnetic phase would have a larger basin of attraction although we have to sacrifice the final quality of the decoded result (magnetization smaller than unity). In Fig. 5.8 the ground-state magnetization (overlap) is shown as a function of the noise probability for various finite values of r (written as K in the figure) in the case of $R = 1/2$. The transition is of first order except for $r = 2$. It can be seen that the decoded result is very good (m close to one) for moderate values of r and p.

It is useful to devise a practical algorithm of decoding, given the channel output $\{J_\mu\}$, where μ denotes an appropriate combination of site indices. The following method based on an iterative solution of TAP-like equations is a powerful tool for this purpose (Kabashima and Saad 2001; Saad et $al.$ 2001) since its computational requirement is only of $\mathcal{O}(N)$. For a given site i and an interaction μ that includes i, one considers a set of conditional probabilities

$$P(\sigma_i|\{J_{\nu\neq\mu}\}) \equiv \frac{1 + m_{\mu i}\sigma_i}{2}, \quad P(J_\mu|\sigma_i, \{J_{\nu\neq\mu}\}) \equiv \text{const} \cdot (1 + \hat{m}_{\mu i}\sigma_i), \quad (5.77)$$

where ν also includes i.[8] Under an approximate mean-field-like decoupling of the

---

[8]We did not write out the normalization constant in the second expression of (5.77) because the left hand side is to be normalized with respect to J_μ in contrast to the first expression to be normalized for σ_i.

conditional probabilities, one obtains the following set of equations for $m_{\mu i}$ and $\hat{m}_{\mu i}$:

$$
\begin{aligned}
m_{\mu i} &= \tanh\left(\sum_{\nu \in \mathcal{M}(i)\backslash\mu} \tanh^{-1}\hat{m}_{\nu i} + \beta F\right) \\
\hat{m}_{\mu i} &= \tanh\beta J_\mu \cdot \prod_{l \in \mathcal{L}(\mu)\backslash i} m_{\mu l},
\end{aligned}
\tag{5.78}
$$

where $\mathcal{M}(i)$ is a set of interactions that include i, and $\mathcal{L}(\mu)$ is a set of sites connected by J_μ. After iteratively solving these equations for $m_{\mu i}$ and $\hat{m}_{\mu i}$, one determines the final decoded result of the ith bit as $\mathrm{sgn}(m_i)$, where

$$
m_i = \tanh\left(\sum_{\nu \in \mathcal{M}(i)} \tanh^{-1}\hat{m}_{\nu i} + \beta F\right).
\tag{5.79}
$$

This method is equivalent to the technique of *belief propagation* used in information theory. It is also called a TAP approach in the statistical mechanics literature owing to its similarity to the TAP equations in the sense that $m_{\mu i}$ and $\hat{m}_{\mu i}$ reflect the effects of removal of a bond μ from the system.

The resulting numerical data are shown in Fig. 5.8. One can see satisfactory agreement with the replica solution. It also turns out that the basin of attraction of the ferromagnetic solution is very large for $r = 2$ but not for $r \geq 3$.

5.6.2 *Low-density parity-check code*

Statistical-mechanical analysis is applicable also to other codes that are actively investigated from the viewpoint of information theory. We explain the *low-density parity-check code* (LDPC) here because of its formal similarity to the diluted Sourlas code treated in the previous subsection (Kabashima *et al.* 2000*a*; Murayama *et al.* 2000). In statistical mechanics terms, the LDPC is a diluted many-body Mattis model in an external field.[9]

Let us start the argument with the definition of the code in terms of a Boolean representation (0 and 1, instead of ± 1). The original message of length N is denoted by an N-dimensional Boolean vector $\boldsymbol{\xi}$ and the encoded message of length M by $\boldsymbol{z}_0$. The latter is generated from the former using two sparse matrices C_s and C_n according to the following modulo-2 operation of Boolean numbers:

$$
\boldsymbol{z}_0 = C_n^{-1}C_s\boldsymbol{\xi}.
\tag{5.80}
$$

The matrix C_s has the size $M \times N$ and the number of ones per row is r and that per column is c, located at random. Similarly, C_n is $M \times M$ and has l ones per

---

[9]There are several variations of the LDPC. We treat in this section the one discussed by MacKay and Neal (1997) and MacKay (1999). See also Vicente *et al.* (2000) for a slightly different code.

row and column randomly. The channel noise ζ is added to z_0, and the output is

$$z = z_0 + \zeta. \tag{5.81}$$

Decoding is carried out by multiplying z by C_n:

$$C_n z = C_n z_0 + C_n \zeta = C_s \xi + C_n \zeta. \tag{5.82}$$

One finds the most probable solution of this equation for the decoded message σ and the inferred noise τ

$$C_s \sigma + C_n \tau = C_s \xi + C_n \zeta. \tag{5.83}$$

The Ising spin representation corresponding to the above prescription of the LDPC, in particular (5.83), is

$$\prod_{i \in \mathcal{L}_s(\mu)} \sigma_i \prod_{j \in \mathcal{L}_n(\mu)} \tau_j = \prod_{i \in \mathcal{L}_s(\mu)} \xi_i \prod_{j \in \mathcal{L}_n(\mu)} \zeta_j \ (\equiv J_\mu), \tag{5.84}$$

where $\mathcal{L}_s(\mu)$ is a set of indices of non-zero elements in the μth row of C_s and similarly for $\mathcal{L}_n(\mu)$. Note that σ, τ, ξ, and ζ are all Ising variables (± 1) from now on. The Hamiltonian reflects the constraint (5.84) as well as the bias in the original message and the channel noise:

$$H = \sum A_{i_1 \ldots i_r; j_1 \ldots j_l} \delta[-1, J_{i_1 \ldots i_r; j_1 \ldots j_l} \sigma_{i_1} \ldots \sigma_{i_r} \tau_{j_1} \ldots \tau_{j_l}]$$
$$- TF_s \sum_i \sigma_i - TF_n \sum_j \tau_j. \tag{5.85}$$

Here A is a sparse tensor for choosing the appropriate combination of indices corresponding to C_s and C_n (or $\mathcal{L}_s(\mu)$ and $\mathcal{L}_s(\mu)$), F_s is the bias of the original message, and $F_n = \frac{1}{2} \log(1 - p)/p$ comes from the channel noise of rate p. The interaction $J_{i_1 \ldots i_r; j_1 \ldots j_l}$ is specified by the expressions involving ξ and ζ in (5.84). The problem is to find the ground state of this Hamiltonian to satisfy (5.84), given the output of the channel $\{J_\mu\}$ defined in (5.84).

The replica analysis of the present system works similarly to the diluted Sourlas code. The resulting RS free energy at $T = 0$ is

$$f = \frac{c}{r} \log 2 + c \int \mathrm{d}x \mathrm{d}\hat{x}\, \pi(x)\hat{\pi}(\hat{x}) \log(1 + x\hat{x}) + \frac{cl}{r} \int \mathrm{d}y \mathrm{d}\hat{y}\, \rho(y)\hat{\rho}(\hat{y}) \log(1 + y\hat{y})$$

$$- \frac{c}{r} \int \prod_{k=1}^{r} \mathrm{d}x_k \pi(x_k) \prod_{m=1}^{l} \mathrm{d}y_m \rho(y_m) \log \left(1 + \prod_k x_k \prod_m y_m \right)$$

$$- \int \prod_{k=1}^{r} \mathrm{d}\hat{x}_k \hat{\pi}(\hat{x}_k) \left[\log \left(e^{F_s \xi} \prod_k (1 + \hat{x}_k) + e^{-F_s \xi} \prod_k (1 - \hat{x}_k) \right) \right]_\xi$$

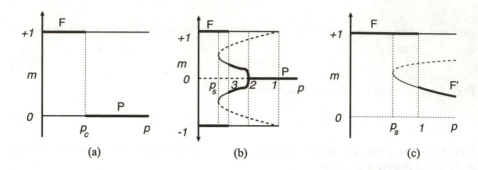

FIG. 5.9. Magnetization as a function of the channel-error probability in the LDPC (Murayama *et al.* 2000). Bold lines represent stable states. (a) $r \geq 3$ or $l \geq 3, r > 1$. (b) $r = l = 2$. (c) $r = 1$. (Copyright 2000 by the American Physical Society)

$$-\frac{c}{r} \int \prod_{m=1}^{l} \mathrm{d}\hat{y}_m \hat{\rho}(\hat{y}_m) \left[\log \left(\mathrm{e}^{F_n \zeta} \prod_m (1 + \hat{y}_m) + \mathrm{e}^{-F_n \zeta} \prod_m (1 - \hat{y}_m) \right) \right]_\zeta . \quad (5.86)$$

The order functions $\pi(x)$ and $\hat{\pi}(\hat{x})$ denote the distributions of the multi-replica overlaps and their conjugates for the σ-spins, and $\rho(y)$ and $\hat{\rho}(\hat{y})$ are for the τ-spins:

$$q_{\alpha\beta...\gamma} = a_q \int \mathrm{d}x \, \pi(x) x^l, \quad \hat{q}_{\alpha\beta...\gamma} = a_{\hat{q}} \int \mathrm{d}\hat{x} \, \hat{\pi}(\hat{x}) \hat{x}^l,$$

$$r_{\alpha\beta...\gamma} = a_r \int \mathrm{d}y \, \rho(y) y^l, \quad \hat{r}_{\alpha\beta...\gamma} = a_{\hat{r}} \int \mathrm{d}\hat{y} \, \hat{\rho}(\hat{y}) \hat{y}^l. \quad (5.87)$$

Extremization of the free energy (5.86) with respect to these order functions yields ferromagnetic and paramagnetic solutions. Since the interactions in the Hamiltonian (5.85) are of Mattis type without frustration, there is no spin glass phase. When $r \geq 3$ or $l \geq 3, r > 1$, the free energy for an unbiased message ($F_s = 0$) is

$$f_{\mathrm{F}} = -\frac{1}{R} F_n \tanh F_n, \quad f_{\mathrm{P}} = \frac{1}{R} \log 2 - \log 2 - \frac{1}{R} \log 2 \cosh F_n. \quad (5.88)$$

The spin alignment is perfect ($m = 1$) in the ferromagnetic phase. The magnetization as a function of the channel-error probability p is shown in Fig. 5.9(a). The ferromagnetic state has a lower free energy below p_c that coincides with the Shannon bound as can be verified by equating f_{F} and f_{P}. The paramagnetic solution loses its significance below p_c because its entropy is negative in this region. A serious drawback is that the basin of attraction of the ferromagnetic state is quite small in the present case.

If $r = l = 2$, the magnetization behaves as in Fig. 5.9(b). The perfect ferromagnetic state and its reversal are the only solutions below a threshold p_s. Any initial state converges to this perfect state under an appropriate decoding algorithm. Thus the code with $r = l = 2$ is quite useful practically although the threshold p_s lies below the Shannon bound.

The system with single-body interactions $r = 1$ has the magnetization as shown in Fig. 5.9(c). Again, the Shannon bound is not saturated, but the perfect ferromagnetic state is the only solution below p_s. An advantage of the present case is that there is no mirror image ($m = -1$).

Iterative solutions using TAP-like equations work also in the LDPC as a rapidly converging tool for decoding (Kabashima and Saad 2001; Saad *et al.* 2001). These equations have similar forms to (5.78) but with two types of parameters, one for the σ-spins and the other for τ. Iterative numerical solutions of these equations for given dilute matrices C_s and C_n show excellent agreement with the replica predictions.

5.6.3 *Cryptography*

The LDPC is also useful in *public-key cryptography* (Kabashima *et al.* 2000b). The N-dimensional Boolean *plaintext* $\boldsymbol{\xi}$ is *encrypted* to an M-dimensional *ciphertext* $\boldsymbol{z}$ by the public key $G \equiv C_n^{-1} C_s D$ (where D is an arbitrary invertible dense matrix of size $N \times N$) and the noise $\boldsymbol{\zeta}$ with probability p according to (5.81)

$$z = G\boldsymbol{\xi} + \boldsymbol{\zeta}. \tag{5.89}$$

Only the authorized user has the knowledge of C_n, C_s, and D separately, not just the product G. The authorized user then carries out the process of *decryption* equivalent to the decoding of the LDPC to infer $D\boldsymbol{\xi}$ and consequently the original plaintext $\boldsymbol{\xi}$. This user succeeds if $r = l = 2$ and $p < p_s$ as was discussed in the previous subsection.

The task of decomposing G into C_n, C_s, and D is NP complete[10] and is very difficult for an unauthorized user, who is therefore forced to find the ground state of the Hamiltonian, which is the Ising spin representation of (5.89):

$$H = -\sum \mathcal{G}_{i_1 \dots i_{r'}} \mathcal{J}_{i_1 \dots i_{r'}} \sigma_{i_1} \dots \sigma_{i_{r'}} - TF_s \sum_i \sigma_i, \tag{5.90}$$

where $\mathcal{G}$ is a dense tensor with elements 1 or 0 corresponding to G, and $\mathcal{J}$ is either 1 or -1 according to the noise added as $\boldsymbol{\zeta}$ in the Boolean representation (5.89). Thus the system is frustrated. For large N, the number r' in the above Hamiltonian and c' (the connectivity of the system described by (5.90)) tend to infinity (but are smaller than N itself) with the ratio c'/r' kept finite. The problem is thus equivalent to the Sourlas-type code in the same limit. We know, as mentioned in §5.6.1, that the basin of attraction of the correctly decrypted state in such a system is very narrow. Therefore the unauthorized user almost surely fails to decrypt.

This system of cryptography has the advantage that it allows for relatively high values of p, and thus an increased tolerance against noise in comparison

---

[10]See Chapter 9 for elucidation of the term 'NP completeness'.

with existing systems. The computational requirement for decryption is of $\mathcal{O}(N)$, which is much better than some of the commonly used methods.

5.7 Convolutional code

The *convolutional code* corresponds to a one-dimensional spin glass and plays important roles in practical applications. It also has direct relevance to the turbo code, to be elucidated in the next section, which is rapidly becoming the standard in practical scenes owing to its high capability of error correction. We explain the convolutional code and its decoding from a statistical-mechanical point of view following Montanari and Sourlas (2000).

5.7.1 *Definition and examples*

In a convolutional code, one first transforms the original message sequence $\boldsymbol{\xi} = \{\xi_1, \ldots, \xi_N\}$ ($\xi_i = \pm 1, \ \forall i$) into a *register sequence* $\boldsymbol{\tau} = \{\tau_1(\boldsymbol{\xi}), \ldots, \tau_N(\boldsymbol{\xi})\}$ ($\tau_i = \pm 1, \ \forall i$). In the *non-recursive convolutional code*, the register sequence coincides with the message sequence ($\tau_i = \xi_i, \ \forall i$), but this is not the case in the recursive convolutional code to be explained later in §5.7.3. To encode the message, one prepares r registers, the state of which at time t is described by $\Sigma_1(t), \Sigma_2(t), \ldots, \Sigma_r(t)$.[11] The number r is called the *memory order* of the code. The register sequence $\boldsymbol{\tau}$ is fed into the register sequentially (*shift register*):

$$\begin{aligned}
\Sigma_1(t+1) &= \Sigma_0(t) \equiv \tau_t \\
\Sigma_2(t+1) &= \Sigma_1(t) = \tau_{t-1} \\
&\vdots \\
\Sigma_r(t+1) &= \Sigma_{r-1}(t) = \tau_{t-r}.
\end{aligned} \tag{5.91}$$

The encoder thus carries the information of $(r+1)$ bits $\tau_t, \tau_{t-1}, \ldots, \tau_{t-r}$ at any moment t.

We restrict ourselves to the convolutional code with rate $R = 1/2$ for simplicity. Code words $\boldsymbol{J} = \{J_1^{(1)}, \ldots, J_N^{(1)}; J_1^{(2)}, \ldots, J_N^{(2)}\}$ are generated from the register bits by the rule

$$J_i^{(\alpha)} = \prod_{j=0}^{r} (\tau_{i-j})^{\kappa(j;\alpha)}. \tag{5.92}$$

Here, $\alpha = 1$ or 2, and we define $\tau_j = 1$ for $j \leq 0$. The superscript $\kappa(j;\alpha)$ is either 0 or 1 and characterizes a specific code. We define $\kappa(0;1) = \kappa(0;2) = 1$ to remove ambiguities in code construction. Two simple examples will be used frequently to illustrate the idea:

---

[11] The original source message is assumed to be generated sequentially from $i = 1$ to $i = N$. Consequently the time step denoted by $t (= 1, 2, \ldots, N)$ is identified with the bit number $i (= 1, 2, \ldots, N)$.

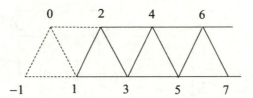

FIG. 5.10. A convolutional code with code rate 1/2 (example 2 in the text) expressed as a spin system. Interactions exist among three spins around each triangle and between two horizontally neighbouring spins. Two up spins are located at $i = -1$ and $i = 0$ to fix the initial condition.

1. $\kappa(0; 1) = \kappa(1; 1) = 1$, and the other $\kappa(j; 1) = 0$; $\kappa(0; 2) = 1$, and the other $\kappa(j; 2) = 0$. The memory order is $r = 1$. The code words are $J_i^{(1)} = \tau_i \tau_{i-1}$ and $J_i^{(2)} = \tau_i$. The corresponding spin Hamiltonian is

$$H = -\sum_{i=1}^{N} \tilde{J}_i^{(1)} \sigma_i \sigma_{i-1} - \sum_{i=1}^{N} \tilde{J}_i^{(2)} \sigma_i, \tag{5.93}$$

where $\tilde{J}_i^{(\alpha)}$ is the noisy version of $J_i^{(\alpha)}$ and σ_i is the dynamical variable used for decoding. This is a one-dimensional spin system with random interactions and random fields.

2. $\kappa(0; 1) = \kappa(1; 1) = \kappa(2; 1) = 1$, and the other $\kappa(j; 1) = 0$; $\kappa(0; 2) = \kappa(2; 2) = 1$, and the other $\kappa(j; 2) = 0$. The memory order is $r = 2$ and the code words are $J_i^{(1)} = \tau_i \tau_{i-1} \tau_{i-2}$ and $J_i^{(2)} = \tau_i \tau_{i-2}$. There are three-body and two-body interactions in the corresponding spin system

$$H = -\sum_{i=1}^{N} \tilde{J}_i^{(1)} \sigma_i \sigma_{i-1} \sigma_{i-2} - \sum_{i=1}^{N} \tilde{J}_i^{(2)} \sigma_i \sigma_{i-2}, \tag{5.94}$$

which can be regarded as a system of ladder-like structures shown in Fig. 5.10. A diagrammatic representation of the encoder is depicted in Fig. 5.11.

5.7.2 *Generating polynomials*

Exposition of the encoding procedure in terms of the Boolean (0 or 1) representation instead of the binary (± 1) representation is useful to introduce the recursive convolutional code in the next subsection. For this purpose, we express the original message sequence $\boldsymbol{\xi} = \{\xi_1, \ldots, \xi_N\}$ by its *generating polynomial* defined as

$$H(x) = \sum_{j=1}^{N} H_j x^j, \tag{5.95}$$

where $H_j (= 0$ or $1)$ is the Boolean form of ξ_j: $\xi_j = (-1)^{H_j}$. Similarly the generating polynomial for the register sequence τ is

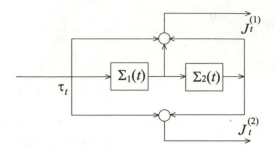

FIG. 5.11. Encoder corresponding to the code of Fig. 5.10. $J_t^{(1)}$ is formed from the three consecutive register bits and $J_t^{(2)}$ from two bits.

$$G(x) = \sum_{j=1}^{N} G_j x^j, \quad \tau_j = (-1)^{G_j}. \tag{5.96}$$

The non-recursive convolutional code has $G(x) = H(x)$, but this is not the case for the recursive code to be explained in the next subsection. The code word $\boldsymbol{J}^{(\alpha)}$ ($\alpha = 1, 2$) is written as

$$L^{(\alpha)}(x) = \sum_{j=1}^{N} L_j^{(\alpha)} x^j \tag{5.97}$$

with $J_j^{(\alpha)} = (-1)^{L_j^{(\alpha)}}$. The relation between $L^{(\alpha)}(x)$ and $G(x)$ is determined by (5.92) and is described by another polynomial

$$g_\alpha(x) = \sum_{j=0}^{r} \kappa(j; \alpha) x^j \tag{5.98}$$

as

$$L^{(\alpha)}(x) = g_\alpha(x) G(x) \tag{5.99}$$

or equivalently

$$L_i^{(\alpha)} = \sum_{j=0}^{r} \kappa(j; \alpha) G_{i-j} \quad (\text{mod } 2). \tag{5.100}$$

The right hand side is the convolution of κ and G, from which the name of convolutional code comes.

The examples 1 and 2 of §5.7.1 have the generating polynomials as (1) $g_1(x) = 1 + x$ and $g_2(x) = 1$, and (2) $g_1(x) = 1 + x + x^2$ and $g_2(x) = 1 + x^2$.

5.7.3 Recursive convolutional code

The relation between the source and register sequences $\boldsymbol{\xi}$ and $\boldsymbol{\tau}$ (or $H(x)$ and $G(x)$) is not simple in the *recursive convolutional code*. The register sequence of recursive convolutional code is defined by the generating polynomial as

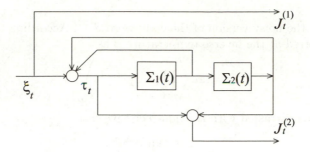

FIG. 5.12. Encoder of the recursive convolutional code to be compared with the non-recursive case of Fig. 5.11.

$$G(x) = \frac{1}{g_1(x)} H(x). \tag{5.101}$$

Code words satisfy $L^{(\alpha)}(x) = g_\alpha(x)G(x)$ and therefore we have

$$L^{(1)}(x) = H(x), \quad L^{(2)}(x) = \frac{g_2(x)}{g_1(x)} H(x). \tag{5.102}$$

The first relation means $\boldsymbol{J}^{(1)} = \boldsymbol{\xi}$ in the binary representation.

The relation between the source and register sequences (5.101) can be written in terms of the binary representation as follows. Equation (5.101) is seen to be equivalent to $G(x) = H(x) + (g_1(x) - 1)G(x)$ because $G(x) = -G(x)$ and $H(x) = -H(x)$ (mod 2). The coefficient of x^i in this relation is, if we recall $\kappa(0;1) = 1$, $G_i = H_i + \sum_{j=1}^r \kappa(j;1)G_{i-j}$, which has the binary representation

$$\tau_i = \xi_i \prod_{j=1}^r (\tau_{i-j})^{\kappa(j;1)} \quad \left(\Leftrightarrow \xi_i = \prod_{j=0}^r (\tau_{i-j})^{\kappa(j;1)} \right). \tag{5.103}$$

This equation allows us to determine τ_i recursively; that is, τ_i is determined if we know $\tau_1, \ldots, \tau_{i-1}$. From the definition $L^{(\alpha)}(x) = g_\alpha(x)G(x)$, code words are expressed in terms of the register sequence in the same form as in the case of the non-recursive convolutional code:

$$J_i^{(\alpha)} = \prod_{j=0}^r (\tau_{i-j})^{\kappa(j;\alpha)}. \tag{5.104}$$

The encoder for the code of example 2 of §5.7.1 is shown in Fig. 5.12. Decoding is carried out under the Hamiltonian

$$H = -\sum_{i=1}^N \left\{ \tilde{J}_i^{(1)} \prod_{j=0}^r (\sigma_{i-j})^{\kappa(j;1)} + \tilde{J}_i^{(2)} \prod_{j=0}^r (\sigma_{i-j})^{\kappa(j;2)} \right\}, \tag{5.105}$$

where $\tilde{J}_i^{(\alpha)}$ is the noisy version of the code word $J_i^{(\alpha)}$. According to (5.103), the ith bit is inferred at the inverse temperature β as

$$\hat{\xi}_i = \text{sgn}\left\langle \prod_{j=0}^{r}(\sigma_{i-j})^{\kappa(j;1)} \right\rangle_{\beta},\tag{5.106}$$

which is to be contrasted with the non-recursive case

$$\hat{\xi}_i = \text{sgn}\langle\sigma_i\rangle_{\beta}.\tag{5.107}$$

5.8 Turbo code

The *turbo code* is a powerful coding/decoding technique frequently used recently. In has near-optimal performance (i.e. the transmission rate can be made close to the Shannon bound under the error-free condition), which is exceptional in a practicable code. We explain its statistical-mechanical formulation and some of the results (Montanari and Sourlas 2000; Montanari 2000).

The turbo code is a variant of the recursive convolutional code with the source message sequence $\boldsymbol{\xi} = \{\xi_1, \ldots, \xi_N\}$ and the permuted sequence $\boldsymbol{\xi}^P = \{\xi_{P(1)}, \ldots, \xi_{P(N)}\}$ as the input to the encoder. The permutation P operates on the set $\{1, 2, \ldots, N\}$ and is fixed arbitrarily for the moment. Correspondingly, two register sequences are generated according to the prescription of the recursive convolutional code (5.103):

$$\tau_i^{(1)} = \tau_i(\boldsymbol{\xi}) = \xi_i \prod_{j=1}^{r}(\tau_{i-j}(\boldsymbol{\xi}))^{\kappa(j;1)}\tag{5.108}$$

$$\tau_i^{(2)} = \tau_i(\boldsymbol{\xi}^P) = \xi_{P(i)} \prod_{j=1}^{r}(\tau_{i-j}(\boldsymbol{\xi}^P))^{\kappa(j;1)}\tag{5.109}$$

or, equivalently,

$$\xi_i = \prod_{j=0}^{r}(\tau_{i-j}^{(1)})^{\kappa(j;1)} \equiv \epsilon_i(\boldsymbol{\tau}^{(1)})\tag{5.110}$$

$$\xi_i^P = \prod_{j=0}^{r}(\tau_{i-j}^{(2)})^{\kappa(j;1)} \equiv \epsilon_i(\boldsymbol{\tau}^{(2)}).\tag{5.111}$$

The code words are comprised of three sequences and the rate is $R = 1/3$:

$$J_i^{(0)} = \prod_{j=0}^{r}(\tau_{i-j}^{(1)})^{\kappa(j;1)}, \quad J_i^{(1)} = \prod_{j=0}^{r}(\tau_{i-j}^{(1)})^{\kappa(j;2)}, \quad J_i^{(2)} = \prod_{j=0}^{r}(\tau_{i-j}^{(2)})^{\kappa(j;2)}.\tag{5.112}$$

The posterior to be used at the receiving end of the channel has the following expression:

$$P(\boldsymbol{\sigma}^{(1)}, \boldsymbol{\sigma}^{(2)} | \tilde{\boldsymbol{J}}^{(0)}, \tilde{\boldsymbol{J}}^{(1)}, \tilde{\boldsymbol{J}}^{(2)}) = \frac{1}{Z} \prod_{i=1}^{N} \delta(\epsilon_{P(i)}(\boldsymbol{\sigma}^{(1)}), \epsilon_i(\boldsymbol{\sigma}^{(2)}))$$

$$\cdot \exp\{-\beta H(\boldsymbol{\sigma}^{(1)}, \boldsymbol{\sigma}^{(2)})\}, \qquad (5.113)$$

where the Hamiltonian is, corresponding to (5.112),

$$H(\boldsymbol{\sigma}^{(1)}, \boldsymbol{\sigma}^{(2)}) = -\sum_{i=1}^{N} \left\{ \tilde{J}_i^{(0)} \prod_{j=0}^{r} (\sigma_{i-j}^{(1)})^{\kappa(j;1)} \right.$$

$$\left. + \tilde{J}_i^{(1)} \prod_{j=0}^{r} (\sigma_{i-j}^{(1)})^{\kappa(j;2)} + \tilde{J}_i^{(2)} \prod_{j=0}^{r} (\sigma_{i-j}^{(2)})^{\kappa(j;2)} \right\}. \qquad (5.114)$$

The interactions $\tilde{J}_i^{(0)}$, $\tilde{J}_i^{(1)}$, and $\tilde{J}_i^{(2)}$ are the noisy versions of the code words $J_i^{(0)}$, $J_i^{(1)}$, and $J_i^{(2)}$, respectively. The system (5.114) is a one-dimensional spin glass composed of two chains ($\boldsymbol{\sigma}^{(1)}$ and $\boldsymbol{\sigma}^{(2)}$) interacting via the constraint $\epsilon_{P(i)}(\boldsymbol{\sigma}^{(1)}) = \epsilon_i(\boldsymbol{\sigma}^{(2)})$ ($\forall i$). In decoding, one calculates the thermal expectation value of the variable representing the original bit, (5.110), using the posterior (5.113):

$$\hat{\xi}_i = \text{sgn}\langle \epsilon_i(\boldsymbol{\sigma}^{(1)}) \rangle_\beta. \qquad (5.115)$$

The finite-temperature (MPM) decoding with the appropriate β is used in practice because an efficient TAP-like finite-temperature iterative algorithm exists as explained later briefly.

To understand the effectiveness of turbo code intuitively, it is instructive to express the spin variable $\sigma_i^{(1)}$ in terms of the other set $\boldsymbol{\sigma}^{(2)}$. In example 1 of §5.7.1, we have $\kappa(0;1) = \kappa(1;1) = 1$ and therefore, from the constraint $\epsilon_{P(i)}(\boldsymbol{\sigma}^{(1)}) = \epsilon_i(\boldsymbol{\sigma}^{(2)})$, $\sigma_i^{(2)}\sigma_{i-1}^{(2)} = \sigma_{P(i)}^{(1)}\sigma_{P(i)-1}^{(1)}$, see (5.110) and (5.111). We thus have $\sigma_i^{(1)} = \prod_{j=1}^{i} \sigma_j^{(1)}\sigma_{j-1}^{(1)} = \prod_{j=1}^{i} \sigma_{P^{-1}(j)}^{(2)}\sigma_{P^{-1}(j)-1}^{(2)}$ with $\sigma_j^{(\alpha)} = 1$ for $j \leq 0$. If i is of $\mathcal{O}(N)$ and the permutation P is random, it is very plausible that this final product of the $\boldsymbol{\sigma}^{(2)}$ is composed of $\mathcal{O}(N)$ different $\sigma^{(2)}$. This means that the Hamiltonian (5.114) has long-range interactions if expressed only in terms of $\boldsymbol{\sigma}^{(2)}$, and the ferromagnetic phase (in the ferromagnetic gauge) is likely to have an enhanced stability compared to simple one-dimensional systems. We may therefore expect that good performance is achieved in a turbo code with random permutation P, which is indeed confirmed to be the case in numerical experiments.

The decoding algorithm of the turbo code is described as follows. One prepares two chains labelled by $\alpha = 1, 2$ with the Hamiltonian

$$H^{(\alpha)}(\boldsymbol{\sigma}^{(\alpha)}) = -\sum_i (\tilde{J}_i^{(0)} + \Gamma_i^{(\alpha)}) \prod_{j=0}^{r} (\sigma_{i-j}^{(\alpha)})^{\kappa(j;1)} - \sum_i \tilde{J}_i^{(\alpha)} \prod_{j=0}^{r} (\sigma_{i-j}^{(\alpha)})^{\kappa(j;2)}.$$

$$(5.116)$$

Then one iteratively solves a set of TAP-like equations for the effective fields $\Gamma_i^{(\alpha)}$ that represent the effects of the other chain:

$$\Gamma_i^{(1)}(k+1) = \beta^{-1} \tanh^{-1} \langle \epsilon_{P^{-1}(i)}(\boldsymbol{\sigma}^{(2)}) \rangle^{(2)} - \Gamma_{P^{-1}(i)}^{(2)}(k) \qquad (5.117)$$

$$\Gamma_i^{(2)}(k+1) = \beta^{-1}\tanh^{-1}\langle\epsilon_{P(i)}(\boldsymbol{\sigma}^{(1)})\rangle^{(1)} - \Gamma_{P(i)}^{(1)}(k), \tag{5.118}$$

where $\langle\cdots\rangle^{(\alpha)}$ is the thermal average with the Hamiltonian $H^{(\alpha)}(\boldsymbol{\sigma}^{(\alpha)})$ and k denotes the iteration step. The process (5.116)–(5.118) is an approximation to the full system (5.114) and yet yields excellent performance numerically.

Detailed statistical-mechanical analysis of the system $H^{(1)}(\boldsymbol{\sigma}^{(1)})+H^{(2)}(\boldsymbol{\sigma}^{(2)})$ with $\epsilon_{P(i)}(\boldsymbol{\sigma}^{(1)}) = \epsilon_i(\boldsymbol{\sigma}^{(2)})$ has been carried out (Montanari 2000). We describe some of its important results. Let us suppose that the channel is Gaussian and β is adjusted to the optimal value (MPM). The S/N ratio is denoted as $1/w^2$. There exists a phase of error-free decoding (overlap $M = 1$) that is locally unstable in the high-noise region $w^2 > w_c^2$. The numerical values w_c^2 are $1/\log 4 = 0.721$ for the code 1 of §5.7.1 and 1.675 for the code 2. The latter is very close to the Shannon limit $w_S^2 = 1/(2^{2/3} - 1) = 1.702$ derived by equating the capacity of the Gaussian channel with the rate $R = 1/3$:

$$\frac{1}{2}\log_2\left(1 + \frac{1}{w_S^2}\right) = \frac{1}{3}. \tag{5.119}$$

The limit of the first example ($w_c = 0.721$) is found to be close to numerical results whereas the second ($w_c = 1.675$) shows some deviation from numerical results. The stability analysis leading to these values may not give the correct answer if a first-order phase transition takes place in the second example.

5.9   CDMA multiuser demodulator

In this section we present a statistical-mechanical analysis of signal transmission by modulation (T. Tanaka 2001). This topic deviates somewhat from the other parts of this chapter. The signal is *not* encoded and decoded but is modulated and demodulated as described below. Nevertheless, the goal is very similar to error-correcting codes: to extract the best possible information from a noisy output using the idea of Bayesian inference.

5.9.1   *Basic idea of CDMA*

Code-division multiple access (*CDMA*) is an important standard of modern mobile communications (Simon *et al.* 1994; Viterbi 1995). The digital signal of a user is modulated and transmitted to a base station through a channel that is shared by multiple users. At the base station, the original digital signal is retrieved by demodulation of the received signal composed of the superposition of multiple original signals and noise. An important problem is therefore to design an efficient method to modulate and demodulate signals.

In CDMA, one modulates a signal in the following way. Let us focus our attention to a *signal interval*, which is the time interval carrying a single digital signal, with a signal $\xi_i \,(= \pm 1)$ for the ith user. The signal interval is divided into p *chip intervals* ($p = 4$ in Fig. 5.13). User i is assigned a *spreading code sequence* $\eta_i^t \,(= \pm 1) \,(t = 1, \ldots, p)$. The signal ξ_i is modulated in each chip interval t by the

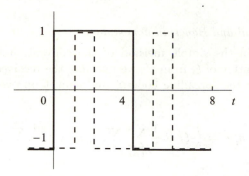

FIG. 5.13. Modulation of the signal of a single user in CDMA. A signal interval is composed of four chip intervals in this example. The full line represents the original signal and the dashed line denotes the spreading code sequence.

spreading code sequence according to the multiplication $\eta_i^t \xi_i$. Modulated signals of N users are superimposed in a channel and are further disturbed by noise. At the base station, one receives the signal

$$y^t = \sum_{i=1}^{N} \eta_i^t \xi_i + \nu^t \tag{5.120}$$

at the chip interval t and is asked to retrieve the original signals $\xi_i \,(i = 1, \dots, N)$ from $y_t \,(t = 1, \dots, p)$ with the knowledge of the spreading code sequence η_i^t $(t = 1, \dots, p; i = 1, \dots, N)$.

Before proceeding to the problem of demodulation, we list a few points of idealization that lead to the simple formula (5.120): modulated signals of N users are assumed to be transmitted under perfect synchronization at each chip interval t throughout an information signal interval. This allows us simply to sum up $\eta_i^t \xi_i$ over all $i\,(= 1, \dots, N)$ at any given chip interval t. Furthermore, all signals are supposed to have the same amplitude (normalized to unity in (5.120)), a perfect power control. Other complications (such as the effects of reflections) are ignored in the present formulation. These aspects would have to be taken into account when one applies the theory to realistic situations.

The measure of performance is the overlap of the original (ξ_i) and demodulated ($\hat{\xi}_i$) signals

$$M = \frac{1}{N} \sum_{i=1}^{N} \xi_i \hat{\xi}_i, \tag{5.121}$$

averaged over the distributions of ξ_i, η_i^t, and ν^t. Equivalently, one may try to minimize the bit-error rate (the average probability of error per bit) $(1 - M)/2$. We show in the following that the CDMA multiuser demodulator, which uses Bayesian inference, gives a larger overlap than the conventional demodulator.

5.9.2   *Conventional and Bayesian demodulators*

Let us first explain the simple method of the conventional demodulator. To extract the information of ξ_i from y^t, we multiply the received signal at the t th chip interval y^t by the spreading code η_i^t and sum it up over the whole signal interval:

$$h_i \equiv \frac{1}{N} \sum_{t=1}^{p} \eta_i^t y^t = \frac{p}{N} \xi_i + \frac{1}{N} \sum_{t=1}^{p} \sum_{k(\neq i)} \eta_i^t \eta_k^t \xi_k + \frac{1}{N} \sum_{t=1}^{p} \eta_i^t \nu^t. \tag{5.122}$$

The first term on the right hand side is the original signal, the second represents multiuser interference, and the third is the channel noise (which is assumed to be Gaussian). We then demodulate the signal by taking the sign of this quantity

$$\hat{\xi}_i = \text{sgn}(h_i). \tag{5.123}$$

It is easy to analyse the performance of this conventional demodulator in the limit of large N and p with $\alpha = p/N$ fixed. We also assume that the noise power σ_s^2, the variance of ν^t, scales with N such that $\beta_s \equiv N/\sigma_s^2$ is of $\mathcal{O}(1)$, and that η_i^t and ξ_k are all independent. Then the second and third terms on the right hand side of (5.122) are Gaussian variables, resulting in the overlap

$$M = \frac{2}{\sqrt{\pi}} \text{Erf} \left(\sqrt{\frac{\alpha}{2(1 + \beta_s^{-1})}} \right), \tag{5.124}$$

where $\text{Erf}(x)$ is the error function $\int_0^x \mathrm{e}^{-t^2} \, \mathrm{d}t$. This represents the performance of the conventional demodulator as a function of the number of chip intervals per signal interval α and the noise power β_s.

To improve the performance, it is useful to construct the posterior of the original signal, given the noisy signal, following the method of Bayesian inference. Let us denote the set of original signals by $\boldsymbol{\xi} = {}^t(\xi_1, \ldots, \xi_N)$ and the corresponding dynamical variables for demodulation by $\boldsymbol{S} = {}^t(S_1, \ldots, S_N)$. The sequence of received signals within p chip intervals is also written as a vector in a p-dimensional space $\boldsymbol{y} = {}^t(y^1, \ldots, y^p)$. Once the posterior $P(\boldsymbol{S}|\boldsymbol{y})$ is given, one demodulates the signal by the MAP or MPM:

$$\text{MAP}: \quad \hat{\boldsymbol{\xi}} = \arg \max_{\boldsymbol{S}} P(\boldsymbol{S}|\boldsymbol{y}), \tag{5.125}$$

$$\text{MPM}: \quad \hat{\xi}_i = \arg \max_{S_i} \text{Tr}_{\boldsymbol{S} \backslash S_i} P(\boldsymbol{S}|\boldsymbol{y}). \tag{5.126}$$

To construct the posterior, we first write the distribution of Gaussian noise $\nu^t = y^t - \sum_i \eta_i^t \xi_i$, (5.120), as

$$\prod_{t=1}^{p} P(y^t|\boldsymbol{\xi}) \propto \exp \left\{ -\frac{\beta_s}{2N} \sum_{t=1}^{p} \left(y^t - \sum_{i=1}^{N} \eta_i^t \xi_i \right)^2 \right\} \propto \exp\{-\beta_s H(\boldsymbol{\xi})\}, \tag{5.127}$$

where the effective Hamiltonian has been defined as

$$H(\xi) = \frac{1}{2} \sum_{i,j=1}^{N} J_{ij}\xi_i\xi_j - \sum_{i=1}^{N} h_i\xi_i, \quad J_{ij} = \frac{1}{N}\sum_{t=1}^{p}\eta_i^t\eta_j^t. \tag{5.128}$$

The field h_i has already been defined in (5.122). If we assume that the prior is uniform, $P(\xi) = \text{const}$, the posterior is seen to be directly proportional to the prior according to the Bayes formula:

$$P(S|y) \propto \exp\{-\beta_s H(S)\}. \tag{5.129}$$

The Hamiltonian (5.128) looks very similar to the Hopfield model to be discussed in Chapter 7, (7.7) with (7.4), the only difference being that the sign of the interaction is the opposite (Miyajima *et al.* 1993).

5.9.3 *Replica analysis of the Bayesian demodulator*

The replica method is useful to analyse the performance of the Bayesian demodulator represented by the posterior (5.129).

Since we usually do not know the noise power of the channel β_s, it is appropriate to write the normalized posterior with an arbitrary noise parameter β in place of β_s, the latter being the true value. From (5.127)–(5.129), we then find

$$P(S|r) = \frac{2^{-N}}{Z(r)} \exp\left\{-\frac{\beta}{2}\sum_{t=1}^{p}\left(r^t - N^{-1/2}\sum_{i=1}^{N}\eta_i^t S_i\right)^2\right\}, \tag{5.130}$$

where the vector r denotes $^t(r^1,\ldots,r^p)$ with $r^t = y^t/\sqrt{N}$, and the normalization factor (or the partition function) is given as

$$Z(r) = 2^{-N}\text{Tr}_S \exp\left\{-\frac{\beta}{2}\sum_{t=1}^{p}\left(r^t - N^{-1/2}\sum_{i=1}^{N}\eta_i^t S_i\right)^2\right\}. \tag{5.131}$$

The factor 2^{-N} is the uniform prior for ξ. The macroscopic behaviour of the system is determined by the free energy averaged over the distributions of the spreading code sequence, which is assumed to be completely random, and the channel noise. The latter distribution of noise is nothing more than the partition function (5.131) with the true hyperparameter $\beta = \beta_s$, which we denote by $Z_0(r)$. The replica average is therefore expressed as

$$[Z^n] = \int \prod_{t=1}^{p} dr^t [Z_0(r)Z^n(r)]_\eta, \tag{5.132}$$

where the configurational average on the right hand side is over the spreading code sequence. It is convenient to separate the above quantity into the spin-dependent part g_1 and the rest g_2 for further calculations:

$$[Z^n] = \int \prod_{0 \le \alpha < \beta \le n} dQ_{\alpha\beta}\, e^{N(g_1 + \alpha g_2)}, \tag{5.133}$$

where $\alpha = p/N$ in the exponent should not be confused with the replica index. The zeroth replica ($\alpha = 0$) corresponds to the probability weight Z_0. The two functions g_1 and g_2 are defined by

$$e^{Ng_1} = \mathrm{Tr}_{\boldsymbol{S}} \prod_{0 \le \alpha < \beta \le n} \delta(\boldsymbol{S}_\alpha \cdot \boldsymbol{S}_\beta - N Q_{\alpha\beta}) \tag{5.134}$$

$$e^{g_2} = \int dr \left[\exp\left\{ -\frac{\beta_s}{2}(r - v_0)^2 - \frac{\beta}{2} \sum_{\alpha=1}^{n}(r - v_\alpha)^2 \right\} \right]_\eta, \tag{5.135}$$

where the following notations have been used:

$$v_0 = \frac{1}{\sqrt{N}} \sum_{i=1}^{N} \eta_i S_{i0}, \quad v_\alpha = \frac{1}{\sqrt{N}} \sum_{i=1}^{N} \eta_i S_{i\alpha} \quad (\alpha = 1, \ldots, n). \tag{5.136}$$

In the thermodynamic limit $p, N \to \infty$ with their ratio α fixed, these v_0 and v_α become Gaussian variables with vanishing mean and covariance given by the overlap of spin variables, under the assumption of a random distribution of the spreading code sequence:

$$Q_{\alpha\beta} = [v_\alpha v_\beta]_\eta = \frac{\boldsymbol{S}_\alpha \cdot \boldsymbol{S}_\beta}{N} \quad (\alpha, \beta = 0, \ldots, n). \tag{5.137}$$

To proceed further, we assume symmetry between replicas ($\alpha = 1, \ldots, n$): $Q_{0\alpha} = m, Q_{\alpha\beta} = q\, (\alpha, \beta \ge 1)$. Then v_0 and v_α are more conveniently written in terms of independent Gaussian variables u, t, and z_α with vanishing mean and unit variance,

$$v_0 = u\sqrt{1 - \frac{m^2}{q} - \frac{tm}{\sqrt{q}}}, \quad v_\alpha = z_\alpha\sqrt{1 - q} - t\sqrt{q} \quad (\alpha \ge 1). \tag{5.138}$$

We are now ready to evaluate the factor e^{g_2} explicitly as

$$e^{g_2} = \int dr \int Dt \int Du \exp\left\{ -\frac{\beta_s}{2}\left(u\sqrt{1 - \frac{m^2}{q} - \frac{tm}{\sqrt{q}}} - r \right)^2 \right\}$$

$$\cdot \left\{ \int Dz \exp\left(-\frac{\beta}{2}(z\sqrt{1 - q} - t\sqrt{q} - r)^2 \right) \right\}^n$$

$$= \{1 + \beta_s(1 - m^2/q)\}^{-1/2}\{1 + \beta(1 - q)\}^{-n/2}$$

$$\cdot \int dr \int Dt \exp\left\{ -\frac{\beta_s(tm/\sqrt{q} + r)^2}{2\{1 + \beta_s(1 - m^2/q)\}} - \frac{n\beta(t\sqrt{q} + r)^2}{2\{1 + \beta(1 - q)\}} \right\}$$

$$= \sqrt{2\pi}\{1 + \beta(1 - q)\}^{-(n-1)/2}$$

$$\cdot \left[\beta_s \{ 1 + \beta(1 - q) \} + n\beta \{ 1 + \beta_s(1 - 2m + q) \} \right]^{-1/2}. \qquad (5.139)$$

The other factor e^{Ng_1} (5.134) can be evaluated using the Fourier representation of the delta function

$$e^{Ng_1} = \int \prod_{0 \le \alpha < \beta \le n} \frac{dM_{\alpha\beta}}{2\pi i}$$

$$\cdot \exp N \left\{ \log G(M) - \sum_{0 \le \alpha < \beta \le n} M_{\alpha\beta} Q_{\alpha\beta} \right\}, \qquad (5.140)$$

$$G(M) = \text{Tr}_S \exp \left(\sum_{0 \le \alpha < \beta \le n} M_{\alpha\beta} S_\alpha S_\beta \right)$$

$$= 2 \int Dz \, (2 \cosh(\sqrt{F} z + E))^n e^{-nF/2}, \qquad (5.141)$$

where we have used the RS form of the matrix $M_{0\alpha} = E$ and $M_{\alpha\beta} = F$ ($\alpha \ne \beta \ge 1$). In the thermodynamic limit, the leading contribution is

$$g_1 = \log \int Dz \, (2 \cosh(\sqrt{F} z + E))^n - \frac{n}{2} F - nEm - \frac{1}{2} n(n-1) Fq. \qquad (5.142)$$

From (5.139) and (5.142), the total free energy $g_1 + \alpha g_2$ is given in the limit $n \to 0$ as

$$-\beta f = \int Dz \, \log 2 \cosh(\sqrt{F} z + E) - Em - \frac{1}{2} F(1 - q)$$

$$- \frac{\alpha}{2} \left\{ \log\{1 + \beta(1 - q)\} + \frac{\beta \{ 1 + \beta_s(1 - 2m + q) \}}{\beta_s \{ 1 + \beta(1 - q) \}} \right\}. \qquad (5.143)$$

Extremization of the free energy yields the equations of state for the order parameters as

$$m = \int Dz \, \tanh(\sqrt{F} z + E), \quad q = \int Dz \, \tanh^2(\sqrt{F} z + E) \qquad (5.144)$$

$$E = \frac{\alpha\beta}{1 + \beta(1 - q)}, \quad F = \frac{\alpha\beta^2(\beta_s^{-1} + 1 - 2m + q)}{\{1 + \beta(1 - q)\}^2}. \qquad (5.145)$$

The overlap is determined from these quantities by

$$M = \int Dz \, \text{sgn}(\sqrt{F} z + E). \qquad (5.146)$$

The stability limit of the RS solution, the AT line, is expressed as

$$\alpha = E^2 \int Dz \, \text{sech}^4(\sqrt{F} z + E). \qquad (5.147)$$

The optimum demodulation by MPM is achieved at the parameter $\beta = \beta_s$ whereas the MAP corresponds to $\beta \to \infty$.

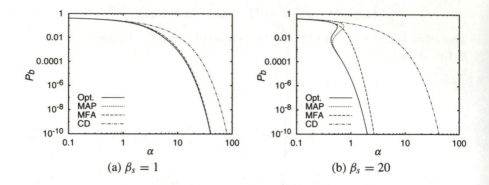

FIG. 5.14. Bit-error rate of the CDMA demodulators. The left one (a) is for
the noise power $\beta_s = 1$ and the right (b) is for $\beta_s = 20$. The symbols are:
Opt. for the MPM, MFA for the mean-field demodulator with $\beta = \beta_s$, and
CD for the conventional demodulator (T. Tanaka 2001; Copyright 2001 by
the Massachusetts Institute of Technology).

5.9.4   *Performance comparison*

The results of the previous analysis in terms of the bit-error rate $(1 - M)/2$
are plotted in Fig. 5.14 for (a) $\beta_s = 1$ and (b) $\beta_s = 20$ for the conventional
demodulator (CD), MPM ('Opt.'), and MAP demodulators. Also shown is the
mean-field demodulator in which one uses the mean-field equation of state for
local magnetization

$$m_i = \tanh\{\beta(-\sum_j J_{ij}m_j + h_i)\} \tag{5.148}$$

in combination with $\hat{\xi}_i = \mathrm{sgn}(m_i)$. This method has the advantage that it serves
as a demodulating algorithm of direct practical usefulness.

It is observed that the MAP and MPM show much better performance than
the conventional demodulator. The curve for the MAP almost overlaps with the
MPM curve when the noise power is high, $\beta_s = 1$, but a clear deviation is found
in the low-noise case $\beta_s = 20$. The MPM result has been confirmed to be stable
for RSB. By contrast, one should take RSB into account for the MAP except in
a region with small α and large β_s.

Bibliographical note

General expositions of information theory and error-correcting codes are found
in textbooks on these subjects (McEliece 1977; Clark and Cain 1981; Lin and
Costello 1983; Arazi 1988; Rhee 1989; Ash 1990; Wicker 1995). The present
form of statistical-mechanical analysis of error-correcting codes was proposed by
Sourlas (1989) and has been expanding rapidly as described in the text. Some

of the recent papers along this line of development (but not cited in the text) include Kanter and Saad (2000), Nakamura *et al.* (2000), and Kabashima *et al.* (2000*c*). See also Heegard and Wicker (1999) for the turbo code.

6

IMAGE RESTORATION

The problem of statistical inference of the original image given a noisy image can be formulated in a similar way to error-correcting codes. By the Bayes formula the problem reduces to a form of random spin systems, and methods of statistical mechanics apply. It will be shown that image restoration using statistical fluctuations (finite-temperature restoration or MPM) gives better performance than the MAP if we are to maximize the pixel-wise similarity of the restored image to the original image. This is the same situation as in error-correcting codes. Mean-field treatments and the problem of parameter estimation will also be discussed.

6.1 Stochastic approach to image restoration

Let us consider the problem of inference of the original image from a given digital image corrupted by noise. This problem would seem to be very difficult without any hints about which part has been corrupted by the noise. In the stochastic approach to image restoration, therefore, one usually makes use of empirical knowledge on images in general (a priori knowledge) to facilitate reasonable restoration. The Bayes formula plays an important role in the argument.

6.1.1 Binary image and Bayesian inference

We formulate the stochastic method of *image restoration* for the simple case of a binary ('black and white') image represented by a set of Ising spins $\boldsymbol{\xi} = \{\xi_i\}$. The index i denotes a lattice site in the spin system and corresponds to the pixel index of an image. The set of pixel states $\boldsymbol{\xi}$ is called the *Markov random field* in the literature of image restoration.

Suppose that the image is corrupted by noise, and one receives a *degraded* (corrupted) *image* with the state of the pixel τ_i inverted from the original value ξ_i with probability p. This conditional probability is written as

$$P(\tau_i|\xi_i) = \frac{\exp(\beta_p \tau_i \xi_i)}{2\cosh\beta_p}, \tag{6.1}$$

where β_p is the same function of p as in (5.7). Under the assumption of independent noise at each pixel, the conditional probability for the whole image is the product of (6.1):

$$P(\boldsymbol{\tau}|\boldsymbol{\xi}) = \frac{1}{(2\cosh\beta_p)^N} \exp(\beta_p \sum_i \tau_i \xi_i), \tag{6.2}$$

where N is the total number of pixels.

116

The problem is to infer the original image $\boldsymbol{\xi}$, given a degraded image $\boldsymbol{\tau}$. For this purpose, it is useful to use the Bayes formula (5.10) to exchange the entries $\boldsymbol{\tau}$ and $\boldsymbol{\xi}$ in the conditional probability (6.2). We use the notation $\boldsymbol{\sigma} = \{\sigma_i\}$ for dynamical variables to restore the image which are to be distinguished from the true original image $\boldsymbol{\xi}$. Then the desired conditional probability (posterior) is

$$P(\boldsymbol{\sigma}|\boldsymbol{\tau}) = \frac{\exp(\beta_p \sum_i \tau_i \sigma_i) P(\boldsymbol{\sigma})}{\mathrm{Tr}_{\boldsymbol{\sigma}} \exp(\beta_p \sum_i \tau_i \sigma_i) P(\boldsymbol{\sigma})}. \tag{6.3}$$

Here the original image is assumed to have been generated with the probability (prior) $P(\boldsymbol{\sigma})$.

One usually does not know the correct prior $P(\boldsymbol{\sigma})$. Nevertheless (6.3) shows that it is necessary to use $P(\boldsymbol{\sigma})$ in addition to the given degraded image $\boldsymbol{\tau}$ to restore the original image. In error-correcting codes, it was reasonable to assume a uniform prior. This is not the case in image restoration where non-trivial structures (such as local smoothness) are essential. We therefore rely on our knowledge on images in general to construct a model prior to be used in place of the true prior.

Let us consider a degraded image in which a black pixel is surrounded by white pixels. It then seems natural to infer that the black pixel is likely to have been caused by noise than to have existed in the original image because real images often have extended areas of smooth parts. This leads us to the following model prior that gives a larger probability to neighbouring pixels in the same state than in different states:

$$P(\boldsymbol{\sigma}) = \frac{\exp(\beta_m \sum_{\langle ij \rangle} \sigma_i \sigma_j)}{Z(\beta_m)}, \tag{6.4}$$

where the sum $\langle ij \rangle$ runs over neighbouring pixels. The normalization factor $Z(\beta_m)$ is the partition function of the ferromagnetic Ising model at temperature $T_m = 1/\beta_m$. Equation (6.4) represents our general knowledge that meaningful images usually tend to have large areas of smooth parts rather than rapidly changing parts. The β_m is the parameter to control smoothness. Larger β_m means a larger probability of the same state for neighbouring pixels.

6.1.2   MAP and MPM

With the model prior (6.4) inserted in the Bayes formula (6.3), we have the explicit form of the posterior,

$$P(\boldsymbol{\sigma}|\boldsymbol{\tau}) = \frac{\exp(\beta_p \sum_i \tau_i \sigma_i + \beta_m \sum_{\langle ij \rangle} \sigma_i \sigma_j)}{\mathrm{Tr}_{\boldsymbol{\sigma}} \exp(\beta_p \sum_i \tau_i \sigma_i + \beta_m \sum_{\langle ij \rangle} \sigma_i \sigma_j)}. \tag{6.5}$$

The numerator is the Boltzmann factor of an Ising ferromagnet in random fields represented by $\boldsymbol{\tau}$. We have thus reduced the problem of image restoration to the statistical mechanics of a random-field Ising model.

If one follows the idea of MAP, one should look for the ground state of the random-field Ising model because the ground state maximizes the Boltzmann factor (6.5). Note that the set τ, the degraded image, is given and fixed, which in other words represents quenched randomness. Another strategy (MPM) is to minimize the pixel-wise error probability as described in §5.2.3 and accept $\mathrm{sgn}\langle\sigma_i\rangle$ as the restored value of the ith pixel calculated through the finite-temperature expectation value. It should also be noted here that, in practical situations of restoration of grey-scale natural images, one often uses multivalued spin systems, which will be discussed in §§6.4 and 6.5.

6.1.3   Overlap

The parameter β_p in (6.5) represents the noise rate in the degraded image. One does not know this noise rate beforehand, so that it makes sense to replace it with a general variable h to be estimated by some method. We therefore use the posterior (6.5) with β_p replaced by h. Our theoretical analysis will be developed for a while for the case where the original image has been generated according to the Boltzmann factor of the ferromagnetic Ising model:

$$P(\boldsymbol{\xi}) = \frac{\exp(\beta_s \sum_{\langle ij \rangle} \xi_i \xi_j)}{Z(\beta_s)}, \qquad (6.6)$$

where β_s is the inverse of the temperature T_s of the prior.

We next define the average overlap of the original and restored images as in (5.19)

$$\begin{aligned}
M(\beta_m, h) &= \mathrm{Tr}_{\boldsymbol{\xi}} \mathrm{Tr}_{\boldsymbol{\tau}} P(\boldsymbol{\xi}) P(\boldsymbol{\tau}|\boldsymbol{\xi}) \left\{ \xi_i \, \mathrm{sgn}\langle\sigma_i\rangle \right\} \\
&= \frac{1}{(2 \cosh \beta_p)^N Z(\beta_s)} \\
&\quad \cdot \mathrm{Tr}_{\boldsymbol{\xi}} \mathrm{Tr}_{\boldsymbol{\tau}} \exp\left(\beta_s \sum_{\langle ij \rangle} \xi_i \xi_j + \beta_p \sum_i \tau_i \xi_i \right) \left\{ \xi_i \mathrm{sgn}\langle\sigma_i\rangle \right\}. \quad (6.7)
\end{aligned}$$

Here $\langle\sigma_i\rangle$ is the average by the Boltzmann factor with β_p replaced by h in (6.5). The dependence of M on β_m and h is in the quantity $\mathrm{sgn}\langle\sigma_i\rangle$. The overlap $M(\beta_m, h)$ assumes the largest value when β_m and h are equal to the true values, β_s and β_p, respectively:

$$M(\beta_m, h) \leq M(\beta_s, \beta_p). \qquad (6.8)$$

This inequality can be proved in the same way as in §5.3.2.

The inequality (6.8) has been derived for the artificial image generated by the Ising model prior (6.6). It is not possible to prove comparable results for general natural images because the prior is different from one image to another. Nevertheless it may well happen in many images that the maximization of pixel-wise overlap is achieved at finite values of the parameters β_m and h.

We have been discussing noise of type (6.2), a simple reversal of the binary value. Similar arguments can be developed for the Gaussian noise

$$P(\boldsymbol{\tau}|\boldsymbol{\xi}) = \frac{1}{(\sqrt{2\pi}\tau)^N} \exp\left\{-\sum_i \frac{(\tau_i - \tau_0\xi_i)^2}{2\tau^2}\right\}. \tag{6.9}$$

The inequality for the maximum overlap between pixels, corresponding to (6.8), is then

$$M(\beta_m, h) \le M\left(\beta_s, \frac{\tau_0}{\tau^2}\right). \tag{6.10}$$

6.2 Infinite-range model

The true values of the parameters β_m and h (β_s and β_p, respectively) are not known beforehand. One should estimate them to bring the overlap M close to the largest possible value. It is therefore useful to have information on how the overlap $M(\beta_m, h)$ depends upon the parameters near the best values $\beta_m = \beta_s$ and $h = \beta_p$. The infinite-range model serves as a prototype to clarify this point. In the present section we calculate the overlap for the infinite-range model (Nishimori and Wong 1999).

6.2.1 *Replica calculations*

Let us consider the infinite-range model with the following priors (the real and model priors):

$$P(\boldsymbol{\xi}) = \frac{\exp\left((\beta_s/2N)\sum_{i\neq j}\xi_i\xi_j\right)}{Z(\beta_s)}, \quad P(\boldsymbol{\sigma}) = \frac{\exp\left((\beta_m/2N)\sum_{i\neq j}\sigma_i\sigma_j\right)}{Z(\beta_m)}. \tag{6.11}$$

This model is very artificial in the sense that all pixels are neighbours to each other, and it cannot be used to restore the original image of a realistic two-dimensional degraded image. However, our aim here is not to establish a model of practical usefulness but to understand the generic features of macroscopic variables such as the overlap M. It is well established in statistical mechanics that the infinite-range model is suited for such a purpose.

For the Gaussian noise (6.9), we can calculate the overlap $M(\beta_m, h)$ by the replica method. The first step is the evaluation of the configurational average of the nth power of the partition function:

$$[Z^n] = \int \prod_i d\tau_i \frac{1}{(\sqrt{2\pi}\tau)^N} \exp\left(-\frac{1}{2\tau^2}\sum_i(\tau_i^2 + \tau_0^2)\right)$$

$$\cdot \mathrm{Tr}_{\boldsymbol{\xi}} \frac{\exp\left((\beta_s/2N)\sum_{i\neq j}\xi_i\xi_j + (\tau_0/\tau^2)\sum_i \tau_i\xi_i\right)}{Z(\beta_s)}$$

$$\cdot \mathrm{Tr}_{\boldsymbol{\sigma}} \exp\left(\frac{\beta_m}{2N}\sum_{i\neq j}\sum_\alpha \sigma_i^\alpha\sigma_j^\alpha + h\sum_i\sum_\alpha \tau_i\sigma_i^\alpha\right)$$

$$= \frac{1}{Z(\beta_s)(\sqrt{2\pi}\tau)^N} \int \prod_i d\tau_i \, \exp\left(-\frac{1}{2\tau^2}\sum_i (\tau_i^2 + \tau_0^2)\right)$$

$$\cdot \left(\frac{N\beta_s}{2\pi}\right)^{1/2} \left(\frac{N\beta_m}{2\pi}\right)^{n/2} \int dm_0 \int \prod_\alpha dm_\alpha \mathrm{Tr}_\xi \mathrm{Tr}_\sigma$$

$$\cdot \exp\left\{-\frac{N\beta_s m_0^2}{2} - \frac{\beta_s}{2} - \frac{n\beta_m}{2} + \beta_s m_0 \sum_i \xi_i\right.$$

$$\left. - \frac{N\beta_m}{2}\sum_\alpha m_\alpha^2 + \beta_m \sum_\alpha m_\alpha \sum_i \sigma_i^\alpha + \sum_i \left(\frac{\tau_0}{\tau^2}\xi_i + h\sum_\alpha \sigma_i^\alpha\right)\tau_i\right\}$$

$$\propto \frac{1}{Z(\beta_s)} \int dm_0 \int \prod_\alpha dm_\alpha \exp N\left\{-\frac{\beta_s m_0^2}{2} - \frac{\beta_m}{2}\sum_\alpha m_\alpha^2\right.$$

$$+ \log \mathrm{Tr} \int Du \exp\left(\beta_s m_0 \xi + \beta_m \sum_\alpha m_\alpha \sigma^\alpha\right.$$

$$\left.\left. + \tau_0 h\xi \sum_\alpha \sigma^\alpha + h\tau u \sum_\alpha \sigma^\alpha\right)\right\}, \tag{6.12}$$

where Tr denotes the sums over σ^α and ξ. We write $[Z^n] = \exp(-\beta_m n N f)$, and evaluate $-\beta_m n f$ to first order in n by steepest descent assuming replica symmetry,

$$-\beta_m n f = -\frac{1}{2}\beta_s m_0^2 + \log 2\cosh \beta_s m_0 - \frac{1}{2}n\beta_m m^2$$

$$+ n\frac{\mathrm{Tr}_\xi \int Du \, e^{\beta_s m_0 \xi} \log 2\cosh(\beta_m m + \tau_0 h\xi + \tau h u)}{2\cosh \beta_s m_0}, \tag{6.13}$$

where Tr_ξ is the sum over $\xi = \pm 1$.

By extremization of the free energy at each order of n, we obtain the equations of state for order parameters. From the n-independent terms, we find

$$m_0 = \tanh \beta_s m_0. \tag{6.14}$$

This represents the ferromagnetic order parameter $m_0 = [\xi_i]$ in the original image. It is natural to have a closed equation for m_0 because the original image should not be affected by degraded or restored images.

The terms of $\mathcal{O}(n)$ give

$$m = \frac{\mathrm{Tr}_\xi \int Du \, e^{\beta_s m_0 \xi} \tanh(\beta_m m + \tau_0 h\xi + \tau h u)}{2\cosh \beta_s m_0}. \tag{6.15}$$

This is the equation for the ferromagnetic order parameter $m = [\langle\sigma_i\rangle]$ of the restored image. The overlap M can be calculated by replacing $\tanh(\cdot)$ in (6.15)

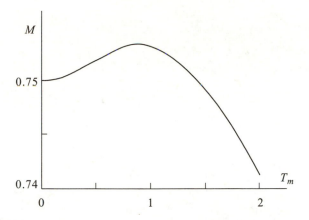

FIG. 6.1. The overlap as a function of the restoration temperature

by $\xi\mathrm{sgn}(\cdot)$ as in §5.4.4. Here, we should remember that we cannot use the ferro-magnetic gauge in image restoration (because the prior is not a constant) and the value ξ remains explicitly in the formulae.

$$M = \frac{\mathrm{Tr}_\xi \int Du \, e^{\beta_s m_0 \xi} \xi \mathrm{sgn}(\beta_m m + \tau_0 h\xi + \tau h u)}{2\cosh\beta_s m_0}. \tag{6.16}$$

The information on the original image (6.14) determines the order parameter of the restored image (6.15) and then we have the overlap (6.16).

6.2.2 *Temperature dependence of the overlap*

It is straightforward to investigate the temperature dependence of M by numerically solving the equations for m_0, m, and M in (6.14), (6.15), and (6.16). In Fig. 6.1 we have drawn M as a function of $T_m = 1/\beta_m$ by fixing the ratio of β_m and h to the optimum value $\beta_s/(\tau_0/\tau^2)$ determined in (6.10). We have set $T_s = 0.9, \tau_0 = \tau = 1$. The overlap is seen to be a maximum at the optimal parameter $T_m = 0.9 \, (= T_s)$. The MAP corresponds to $T_m \to 0$ and the overlap there is smaller than the maximum value. It is clear that the annealing process (in which one tries to reach equilibrium by decreasing the temperature from a high value) gives a smaller overlap if one lowers the temperature beyond the optimal value.

6.3 Simulation

It is in general difficult to discuss quantitatively the behaviour of the overlap M for two-dimensional images by the infinite-range model. We instead use Monte Carlo simulations and compare the results with those for the infinite-range model (Nishimori and Wong 1999).

In Fig. 6.2 is shown the overlap M of the original and restored images by finite-temperature restoration. The original image has 400×400 pixels and was

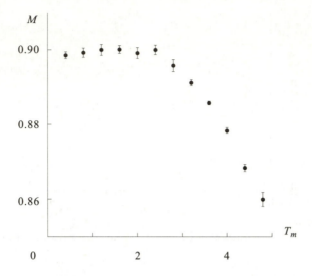

FIG. 6.2. The overlap as a function of the restoration temperature for a
two-dimensional image.

generated by the prior (6.11) ($T_s = 2.15$). Degradation was caused by the binary
noise with $p = 0.1$. The overlap M assumes a maximum when the restoration
temperature T_m is equal to the original $T_s = 2.15$ according to the inequality
(6.8), which is seen to be true within statistical uncertainties. In this example,
the parameter h has been changed with β_m so that the ratio of β_m and h is kept
to the optimum value β_s/β_p.

Comparison with the case of the infinite-range model in Fig. 6.1 indicates
that M depends relatively mildly on the temperature in the two-dimensional
case below the optimum value. One should, however, be aware that this result
is for the present specific values of T_s and p, and it still has to be clarified how
general this conclusion is.

An explicit illustration is given in Fig. 6.3 that corresponds to the situation
of Fig. 6.2. Figure 6.3(a) is the original image ($T_s = 2.15$), (b) is the degraded
image ($p = 0.1$), (c) is the result of restoration at a low temperature ($T_m = 0.5$),
and (d) has been obtained at the optimum temperature ($T_m = 2.15$). It is clear
that (d) is closer to the original image than (c). The MAP has $T_m = 0$ and is
expected to give an even less faithful restored image than (c), in particular in
fine structures. It has thus become clear that the MPM, the finite-temperature
restoration with correct parameter values, gives better results than the MAP for
two-dimensional images generated by the ferromagnetic Ising prior (6.11).

6.4   Mean-field annealing

In practical implementations of image restoration by the MAP as well as by
the MPM, the required amount of computation is usually very large because

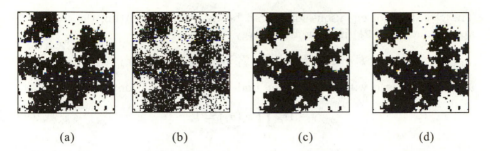

(a)   (b)   (c)   (d)

FIG. 6.3. Restoration of an image generated by the two-dimensional Ising model: (a) original image, (b) degraded image, (c) restored image at a very low temperature (close to MAP), and (d) restored image at the optimum temperature (MPM).

there are 2^N degrees of freedom for a binary image. Therefore one often makes use of approximations, and a typical example is mean-field annealing in which one looks for the optimum solution numerically using the idea of the mean-field approximation (Geiger and Girosi 1991; Zhang 1992; Bilbro *et al.* 1992).

6.4.1 *Mean-field approximation*

We now generalize the argument from binary to grey-scale images to be represented by the *Potts model*. Generalization of (6.5) to the Potts model is

$$P(\boldsymbol{\sigma}|\boldsymbol{\tau}) = \frac{\exp(-\beta_p H(\boldsymbol{\sigma}|\boldsymbol{\tau}))}{Z} \tag{6.17}$$

$$H(\boldsymbol{\sigma}|\boldsymbol{\tau}) = -\sum_i \delta(\sigma_i, \tau_i) - J\sum_{\langle ij\rangle} \delta(\sigma_i, \sigma_j) \tag{6.18}$$

$$Z = \mathrm{Tr}_{\boldsymbol{\sigma}} \exp(-\beta_p H(\boldsymbol{\sigma}|\boldsymbol{\tau})), \tag{6.19}$$

where $\boldsymbol{\tau}$ and $\boldsymbol{\sigma}$ are Q-state Potts spins ($\tau_i, \sigma_i = 0, 1, \ldots, Q-1$) to denote grey scales of degraded and restored images, respectively. In the ferromagnetic Potts model, (6.18) with $J > 0$, the interaction energy is $-J$ if the neighbouring spins (pixels) are in the same state $\sigma_i = \sigma_j$ and zero otherwise. Thus the neighbouring spins tend to be in the same state. The Ising model corresponds to $Q = 2$. The MAP evaluates the ground state of (6.18), and the MPM calculates the thermal average of each spin σ_i at an appropriate temperature.

Since it is difficult to evaluate (6.17) explicitly, we approximate it by the product of marginal distributions

$$\rho_i(n) = \mathrm{Tr}_{\boldsymbol{\sigma}} P(\boldsymbol{\sigma}|\boldsymbol{\tau})\delta(n, \sigma_i) \tag{6.20}$$

as

$$P(\boldsymbol{\sigma}|\boldsymbol{\tau}) \approx \prod_i \rho_i(\sigma_i). \tag{6.21}$$

The closed set of equations for ρ_i can be derived by inserting the mean-field approximation (6.21) into the free energy

$$F = \mathrm{Tr}_{\sigma}\{H(\boldsymbol{\sigma}|\boldsymbol{\tau}) + T_p \log P(\boldsymbol{\sigma}|\boldsymbol{\tau})\}P(\boldsymbol{\sigma}|\boldsymbol{\tau}) \tag{6.22}$$

and minimizing it with respect to ρ_i under the normalization condition

$$\mathrm{Tr}_{\sigma} \prod_i \rho_i(\sigma_i) = 1. \tag{6.23}$$

Simple manipulations then show that ρ_i satisfies the following equation:

$$\rho_i(\sigma) = \frac{\exp\big(-\beta_p H_i^{\mathrm{MF}}(\sigma)\big)}{\sum_{n=0}^{Q-1} \exp\big(-\beta_p H_i^{\mathrm{MF}}(n)\big)} \tag{6.24}$$

$$H_i^{\mathrm{MF}}(n) = -\delta(n, \tau_i) - J \sum_{\mathrm{n.n.}\in i} \rho_j(n), \tag{6.25}$$

where the sum in the second term on the right hand side of (6.25) runs over nearest neighbours of i.

6.4.2 Annealing

A numerical solution of (6.24) can be obtained relatively straightforwardly by iteration if the parameters β_p and J are given. In practice, one iterates not for the function ρ_i itself but for the coefficients $\{m_i^{(l)}\}$

$$\rho_i(\sigma) = \sum_{l=0}^{Q-1} m_i^{(l)} \Phi_l(\sigma) \tag{6.26}$$

of the expansion of the function in terms of the complete orthonormal system of polynomials

$$\sum_{\sigma=0}^{Q-1} \Phi_l(\sigma)\Phi_{l'}(\sigma) = \delta(l, l'). \tag{6.27}$$

The following discrete Tchebycheff polynomials are useful for this purpose (Tanaka and Morita 1996):

$$\Psi_0(\sigma) = 1, \ \Psi_1(\sigma) = 1 - \frac{2}{Q-1}\sigma,$$

$$(l+1)(Q-1-l)\Psi_{l+1}(\sigma)$$
$$= -(2\sigma - Q + 1)(2l+1)\Psi_l(\sigma) - l(Q+l)\Psi_{l-1}(\sigma) \tag{6.28}$$

$$\Phi_l(\sigma) = \frac{\Psi_l(\sigma)}{\sqrt{\sum_{\sigma=0}^{Q-1} \Psi_l(\sigma)^2}}.$$

Multiplying both sides of (6.24) by $\Phi_l(\sigma)$ and summing the result over σ, we find from (6.27) and (6.26)

$$m_i^{(l)} = \frac{\text{Tr}_\sigma \Phi_l(\sigma) \exp \left\{ \beta_p \delta(\sigma, \tau_i) + \beta_p J \sum_{\text{n.n.} \in i} \sum_{l'} m_j^{(l')} \Phi_{l'}(\sigma) \right\}}{Z_{\text{MF}}}, \qquad (6.29)$$

where Z_{MF} is the denominator of (6.24). The set of coefficients $\{m_i^{(l)}\}$ can thus be calculated by iteration. In practice, one usually does not know the correct values of β_p and J, and therefore it is necessary to estimate them by the methods explained below.

Equation (6.29) is a generalization of the usual mean-field approximation to the Potts model. To confirm this explicitly, we apply (6.24) and (6.25) to the Ising model ($Q = 2$):

$$\rho_i(\sigma) = m_i^{(0)} + m_i^{(1)}(1 - 2\sigma) \qquad (6.30)$$

$$H_i^{\text{MF}}(\sigma) = -\delta(\sigma, \tau_i) - J \sum_{\text{n.n.} \in i} \{ m_j^{(0)} + m_j^{(1)}(1 - 2\sigma) \}, \qquad (6.31)$$

where $\sigma = 0$ or 1. Using the first two Tchebycheff polynomials Ψ_0 and Ψ_1, we find from (6.29) that $m_i^{(0)} = 1$ and the mean-field equation in a familiar form

$$m_i^{(1)} = \tanh \left(\beta_p J \sum_{\text{n.n.} \in i} m_j^{(1)} + \frac{\beta_p}{2} \tau_i \right), \qquad (6.32)$$

where we have used the conventional Ising variables (± 1 instead of 0 and 1).

In the MAP ($\beta_p \to \infty$) as well as in the MPM, one has to lower the temperature gradually starting from a sufficiently high temperature ($\beta_p \approx 0$) to obtain a reliable solution of (6.29). This is the process of *mean-field annealing*.

6.5 Edges

For non-binary images, it is useful to introduce variables representing discontinuous changes of pixel values between neighbouring positions to restore the edges of surfaces as faithfully as possible (Geman and Geman 1984; Marroquin *et al.* 1987). Such an edge variable u_{ij} takes the value 0 (no edge between pixels i and j) or 1 (existence of an edge). In the present section, we solve a Gaussian model of image restoration with edges using the mean-field approximation (Geiger and Girosi 1991; Zerubia and Chellappa 1993; Zhang 1996; K. Tanaka 2001b).

Let us consider a Q-state grey-scale image. The model prior is assumed to have a Gaussian form

$$P(\sigma, u) = \frac{1}{Z} \exp \left\{ -\beta_m \sum_{\langle ij \rangle} (1 - u_{ij}) \{ (\sigma_i - \sigma_j)^2 - \gamma^2 \} \right\}, \qquad (6.33)$$

where $\sigma_i = 0, 1, \ldots, Q - 1$ and $u_{ij} = 0, 1$. Note that we are considering a Q-state Ising model here, which is different from the Potts model. The difference between neighbouring pixel values $|\sigma_i - \sigma_j|$ is constrained to be small (less than γ) if $u_{ij} = 0$ (no edge) to reflect the smoothness of the grey scale, whereas the same difference $|\sigma_i - \sigma_j|$ can take arbitrary values if $u_{ij} = 1$ (edge). Thus this prior favours the existence of an edge if the neighbouring pixel values differ by a large amount.[12] Noise is also supposed to be Gaussian

$$P(\boldsymbol{\tau}|\boldsymbol{\xi}) = \prod_i \frac{1}{\sqrt{2\pi}w} \exp\left\{-\frac{(\tau_i - \xi_i)^2}{2w^2}\right\}, \tag{6.34}$$

where the true original image $\boldsymbol{\xi}$ and degraded image $\boldsymbol{\tau}$ both have Q values at each pixel. The posterior is therefore of the form

$$P(\boldsymbol{\sigma}, \boldsymbol{u}|\boldsymbol{\tau}) = \frac{\exp(-H(\boldsymbol{\sigma}, \boldsymbol{u}|\boldsymbol{\tau}))}{\sum_{\boldsymbol{u}} \text{Tr}_{\boldsymbol{\sigma}} \exp(-H(\boldsymbol{\sigma}, \boldsymbol{u}|\boldsymbol{\tau}))}, \tag{6.35}$$

where

$$H(\boldsymbol{\sigma}, \boldsymbol{u}|\boldsymbol{\tau}) = -\beta_m \sum_{\langle ij \rangle} (1 - u_{ij})\{(\sigma_i - \sigma_j)^2 - \gamma^2\} - (2w^2)^{-1} \sum_i (\tau_i - \sigma_i)^2. \tag{6.36}$$

In the finite-temperature (MPM) estimation, we accept the value n_i that maximizes the marginalized posterior

$$P(n_i|\boldsymbol{\tau}) = \sum_{\boldsymbol{u}} \text{Tr}_{\boldsymbol{\sigma}} P(\boldsymbol{\sigma}, \boldsymbol{u}|\boldsymbol{\tau}) \delta(\sigma_i, n_i) \tag{6.37}$$

as the restored pixel state at i.

It is usually quite difficult to carry out the above procedure explicitly. A convenient yet powerful approximation is the mean-field method discussed in the previous section. The central quantities in the mean-field approximation are the marginal probabilities

$$\begin{aligned} \rho_i(n) &= \sum_{\boldsymbol{u}} \text{Tr}_{\boldsymbol{\sigma}} P(\boldsymbol{\sigma}, \boldsymbol{u}|\boldsymbol{\tau}) \delta(\sigma_i, n) \\ \rho_{ij}(u) &= \sum_{\boldsymbol{u}} \text{Tr}_{\boldsymbol{\sigma}} P(\boldsymbol{\sigma}, \boldsymbol{u}|\boldsymbol{\tau}) \delta(u_{ij}, u). \end{aligned} \tag{6.38}$$

The full probability distribution is approximated as

$$P(\boldsymbol{\sigma}, \boldsymbol{u}|\boldsymbol{\tau}) \approx \prod_i \rho_i(\sigma_i) \prod_{\langle ij \rangle} \rho_{ij}(u_{ij}). \tag{6.39}$$

The marginal probabilities are determined by minimization of the free energy

$$F = \sum_{\boldsymbol{u}} \text{Tr}_{\boldsymbol{\sigma}} \{H(\boldsymbol{\sigma}, \boldsymbol{u}|\boldsymbol{\tau}) + \log P(\boldsymbol{\sigma}, \boldsymbol{u}|\boldsymbol{\tau})\} P(\boldsymbol{\sigma}, \boldsymbol{u}|\boldsymbol{\tau}) \tag{6.40}$$

---

[12]Interactions between edges represented by the products of the u_{ij} are often included to take into account various types of straight and crossing edges of extended lengths in real images. The set $\{u_{ij}\}$ is called the *line field* in such cases.

with respect to $\rho_i(\sigma_i)$ and $\rho_{ij}(u_{ij})$ under the normalization condition

$$\sum_{n=0}^{Q-1} \rho_i(n) = 1, \qquad \sum_{u=0,1} \rho_{ij}(u) = 1. \tag{6.41}$$

The result is

$$\rho_i(n) = \frac{e^{-E_i(n)}}{\sum_{m=0}^{Q-1} e^{-E_i(m)}}$$

$$\rho_{ij}(l) = \frac{e^{-E_{ij}(l)}}{\sum_{k=0,1} e^{-E_{ij}(k)}}$$

$$E_i(n) = \frac{(n-\tau_i)^2}{2w^2} \tag{6.42}$$

$$+ \sum_{j \in G_i} \sum_{k=0,1} \sum_{m=0}^{Q-1} \beta_m (1-k)\{(n-m)^2 - \gamma^2\}\rho_j(m)\rho_{ij}(k)$$

$$E_{ij}(l) = \beta_m \sum_{m,m'=0}^{Q-1} (1-l)\{(m-m')^2 - \gamma^2\}\rho_i(m)\rho_j(m'),$$

where G_i is the set of neighbours of i. By solving these equations iteratively, we obtain $\rho_i(n)$, from which it is possible to determine the restored value of the ith pixel as the n that gives the largest value of $\rho_i(n)$.

For practical implementation of the iterative solution of the set of equations (6.42) for large Q, it is convenient to approximate the sum over Q pixel values $\sum_{m=0}^{Q-1}$ by the integral $\int_{-\infty}^{\infty} dm$ because the integrals are analytically calculated to give

$$\rho_i(n) = \frac{1}{\sqrt{2\pi}w_i} \exp\left(-\frac{(n-\mu_i)^2}{2w_i^2}\right) \tag{6.43}$$

$$\mu_i = \frac{(2\sigma^2)^{-1}\tau_i + \beta_m \sum_{j \in G_i}(1-\lambda_{ij})\mu_j}{(2\sigma^2)^{-1} + \beta_m \sum_{j \in G_i}(1-\lambda_{ij})} \tag{6.44}$$

$$\frac{1}{2w_i^2} = \frac{1}{2w^2} + \beta_m \sum_{j \in G_i}(1-\lambda_{ij}) \tag{6.45}$$

$$\lambda_{ij} = \frac{1}{1 + \exp[-\beta_m\{w_i^2 + w_j^2 + (\mu_i - \mu_j)^2 - \gamma^2\}]}. \tag{6.46}$$

It is straightforward to solve (6.44), (6.45), and (6.46) by iteration. Using the result in (6.43), the final estimation of the restored pixel value is obtained.

An example is shown in Fig. 6.5 for $Q = 256$. The original image (a) has been degraded by Gaussian noise of vanishing mean and variance 900 into (b). The image restored by the set of equations (6.43) to (6.46) is shown in (c) together with a restored image (d) obtained by a more sophisticated cluster variation method in which the correlation effects of neighbouring sites are taken

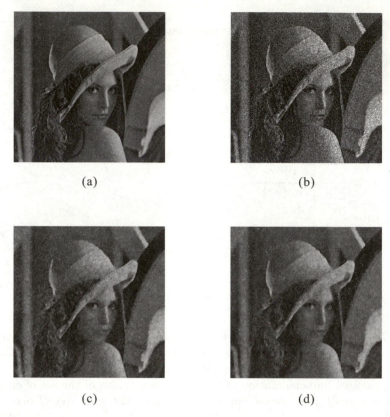

(a)   (b)

(c)   (d)

FIG. 6.4. Restoration of 256 grey-scale image by the Gaussian prior and edges
(line process). Degradation was by Gaussian noise of vanishing mean and
variance 900: (a) original image, (b) degraded images, (c) restored image by
mean-field annealing, and (d) restored image by cluster variation method.
Courtesy of Kazuyuki Tanaka (copyright 2001).

into account in the Bethe-like approximation (K. Tanaka 2001b). Even in the
mean-field level (c), discontinuous changes of pixel values (edges) around the
eyes are well reproduced. The edges would have been blurred without the u_{ij}-
term in (6.33).

6.6   Parameter estimation

It is necessary to use appropriate values of the parameters β_p and J to restore
the image using the posterior (6.17). However, one usually has only the degraded
image and no explicit knowledge of the degradation process characterized by
β_p or the parameter J of the original image. We therefore have to estimate
these parameters (*hyperparameters*) from the degraded image only (Besag 1986;

Lakshmanan and Derin 1989; Pryce and Bruce 1995; Zhou *et al.* 1997; Molina *et al.* 1999).

The following procedure is often used for this purpose. We first marginalize the probability of the given degraded image τ, erasing the original image

$$P(\tau|\beta_p, J) = \mathrm{Tr}_\xi P(\tau|\xi, \beta_p)P(\xi, J). \tag{6.47}$$

The above notation denotes the probability of degraded image τ given the parameters β_p and J. Since we know τ, it is possible to estimate the parameters β_p and J as the ones that maximize the marginalized likelihood function (6.47) or the *evidence*. However, the computational requirement for the sum in (6.47) is exponentially large, and one should resort to simulations or the mean-field approximation to implement this idea.

A different strategy is to estimate σ that maximizes $P(\tau|\sigma, \beta_p)P(\sigma, J)$ as a function of β_p and J without marginalization of ξ in (6.47). One denotes the result as $\{\hat{\sigma}(\beta_p, J)\}$ and estimates β_p and J that maximize the product

$$P(\tau|\{\hat{\sigma}(\beta_p, J)\}, \beta_p)P(\{\hat{\sigma}(\beta_p, J)\}, J).$$

This method is called the *maximum likelihood estimation*.

Another idea is useful when one knows the number of neighbouring pixel pairs L having different grey scales (Tanaka and Morita 1995; Morita and Tanaka 1996, 1997; Tanaka and Morita 1997; K. Tanaka 2001*a*)

$$L = \sum_{\langle ij \rangle} \{1 - \delta(\xi_i, \xi_j)\}. \tag{6.48}$$

One then accepts the image nearest to the degraded image under this constraint (6.48). By taking account of the constraint using the Lagrange multiplier, we see that the problem is to find the ground state of

$$H = -\sum_i \delta(\sigma_i, \tau_i) - J\left(L - \sum_{\langle ij \rangle}\{1 - \delta(\sigma_i, \sigma_j)\}\right).$$

The Potts model in random fields (6.18) has thus been derived naturally. The parameter J is chosen such that the solution satisfies the constraint (6.48). Figure 6.5 is an example of restoration of a 256-level grey scale image by this method of *constrained optimization*.[13]

---

[13]Precisely speaking, the restored image (c) has been obtained by reducing the 256-level degraded image to an eight-level image and then applying the constrained optimization method and mean-field annealing. The result has further been refined by a method called conditional maximization with respect to the grey scale of 256 levels.

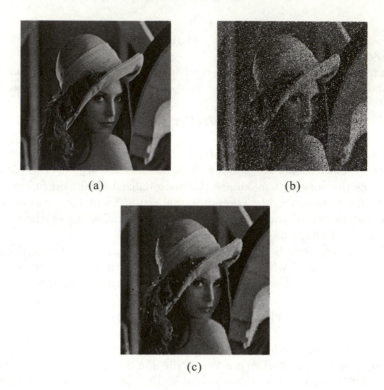

FIG. 6.5. Restoration of 256-level image by the Potts model: (a) original image,
(b) degraded image, and (c) restored image. Courtesy of Kazuyuki Tanaka
(1999). (Copyright 1999 by the Physical Society of Japan)

Bibliographical note

The papers by Geman and Geman (1984), Derin *et al.* (1984), Marroquin *et
al.* (1987), and Pryce and Bruce (1995) are important original contributions on
stochastic approaches to image restoration and, at the same time, are useful to
obtain an overview of the field. For reviews mainly from an engineering point
of view, see Chellappa and Jain (1993). Some recent topics using statistical-
mechanical ideas include dynamics of restoration (Inoue and Carlucci 2000), state
search by quantum fluctuations (Tanaka and Horiguchi 2000; Inoue 2001), hyperparameter estimation in a solvable model (Tanaka and Inoue 2000), segmentation
by the XY model (Okada *et al.* 1999), and the cluster variation method to improve the naïve mean-field approach (Tanaka and Morita 1995, 1996; K. Tanaka
2001*b*).

7

ASSOCIATIVE MEMORY

The scope of the theory of neural networks has been expanding rapidly, and statistical-mechanical techniques stemming from the theory of spin glasses have been playing important roles in the analysis of model systems. We summarize basic concepts in the present chapter and study the characteristics of networks with interneuron connections given by a specific prescription. The next chapter deals with the problem of learning where the connections gradually change according to some rules to achieve specified goals.

7.1 Associative memory

The states of processing units (*neurons*) in an associative memory change with time autonomously and, under certain circumstances, reach an equilibrium state that reflects the initial condition. We start our argument by elucidating the basic concepts of an associative memory, a typical neural network. Note that the emphasis is, in the present book, on mathematical analyses of information processing systems with engineering applications in mind (however remote they might be), rather than on understanding the functioning of the real brain. We nevertheless use words borrowed from neurobiology (neuron, synapse, etc.) because of their convenience to express various basic building blocks of the theory.

7.1.1 *Model neuron*

The structure of a neuron in the real brain is schematically drawn in Fig. 7.1. A neuron receives inputs from other neurons through *synapses*, and if the weighted sum of the input signals exceeds a threshold, the neuron starts to emit its own signal. This signal is transmitted through an *axon* to many other neurons.

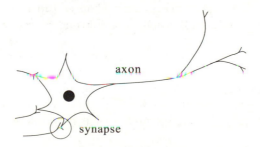

FIG. 7.1. Schematic structure of a neuron

To construct a system of information processing, it is convenient to model the functioning of a neuron in a very simple way. We label the state of a neuron by the variable $S_i = 1$ if the neuron is *excited* (transmitting a signal) and by $S_i = -1$ when it is *at rest*. The *synaptic efficacy* from neuron j to neuron i will be denoted by $2J_{ij}$. Then the sum of signals to the ith neuron, h_i, is written as

$$h_i = \sum_j J_{ij}(S_j + 1). \tag{7.1}$$

Equation (7.1) means that the input signal from j to i is $2J_{ij}$ when $S_j = 1$ and zero if $S_j = -1$. The synaptic efficacy J_{ij} (which will often be called the connection or interaction) can be both positive and negative. In the former case, the signal from S_j increases the value of the right hand side of (7.1) and tends to excite neuron i; a positive connection is thus called an *excitatory* synapse. The negative case is the *inhibitory* synapse.

Let us assume that the neuron i becomes excited if the input signal (7.1) exceeds a threshold θ_i at time t and is not excited otherwise:

$$S_i(t + \Delta t) = \operatorname{sgn}\left(\sum_j J_{ij}(S_j(t) + 1) - \theta_i\right). \tag{7.2}$$

We focus our argument on the simple case where the threshold θ_i is equal to $\sum_j J_{ij}$ so that there is no constant term in the argument on the right hand side of the above equation:

$$S_i(t + \Delta t) = \operatorname{sgn}\left(\sum_j J_{ij}S_j(t)\right). \tag{7.3}$$

7.1.2   *Memory and stable fixed point*

The capability of highly non-trivial information processing emerges in a neural network when very many neurons are connected with each other by synapses, and consequently the properties of connections determine the characteristics of the network. The first half of the present chapter (up to §7.5) discusses how memory and its *retrieval* (recall) become possible under a certain rule for synaptic connections.

A pattern of excitation of a neural network will be denoted by $\{\xi_i^\mu\}$. Here $i\,(= 1, \ldots, N)$ is the neuron index, $\mu\,(= 1, \ldots, p)$ denotes the excitation pattern index, and ξ_i^μ is an Ising variable (± 1). For example, if the μth pattern has the ith neuron in the excited state, we write $\xi_i^\mu = 1$. The μth excitation pattern can be written as $\{\xi_1^\mu, \xi_2^\mu, \ldots, \xi_N^\mu\}$, and p such patterns are assumed to exist ($\mu = 1, 2, \ldots, p$).

Let us suppose that a specific excitation pattern of a neural network corresponds to a memory and investigate the problem of memorization of p patterns in a network of N neurons. We identify memorization of a pattern $\{\xi_i^\mu\}_{i=1,...,N}$ with the fact that the pattern is a stable fixed point of the time evolution rule (7.3) that $S_i(t) = \xi_i^\mu \rightarrow S_i(t + \Delta t) = \xi_i^\mu$ holds at all i. We investigate the condition for this stability.

To facilitate our theoretical analysis as well as to develop arguments independent of a specific pattern, we restrict ourselves to random patterns in which each ξ_i^μ takes ± 1 at random. For random patterns, each pattern is a stable fixed point as long as p is not too large if we choose J_{ij} as follows:

$$J_{ij} = \frac{1}{N} \sum_{\mu=1}^{p} \xi_i^\mu \xi_j^\mu. \tag{7.4}$$

The diagonal term J_{ii} is assumed to be vanishing ($J_{ii} = 0$). This is called the *Hebb rule*. In fact, if the state of the system is in perfect coincidence with the pattern μ at time t (i.e. $S_i(t) = \xi_i^\mu, \forall i$), the time evolution (7.3) gives the state of the ith neuron at the next time step as

$$\mathrm{sgn}\left(\sum_j J_{ij}\xi_j^\mu\right) = \mathrm{sgn}\left(\frac{1}{N}\sum_j\sum_\nu \xi_i^\nu \xi_j^\nu \xi_j^\mu\right) = \mathrm{sgn}\left(\sum_\nu \xi_i^\nu \delta_{\nu\mu}\right) = \mathrm{sgn}\left(\xi_i^\mu\right),$$
$$\tag{7.5}$$

where we have used the approximate orthogonality between random patterns

$$\frac{1}{N}\sum_j \xi_j^\mu \xi_j^\nu = \delta_{\nu\mu} + \mathcal{O}\left(\frac{1}{\sqrt{N}}\right). \tag{7.6}$$

Consequently we have $S_i(t+\Delta t) = \xi_i^\mu$ at all i for sufficiently large N. Drawbacks of this argument are, first, that we have not checked the contribution of the $\mathcal{O}(1/\sqrt{N})$ term in the orthogonality relation (7.6), and, second, that the stability of the pattern is not clear when one starts from a state slightly different from the *embedded* (memorized) pattern (i.e. $S_i(t) = \xi_i^\mu$ at most, but not all, i). These points will be investigated in the following sections.

7.1.3 *Statistical mechanics of the random Ising model*

The time evolution described by (7.3) is equivalent to the zero-temperature dynamics of the Ising model with the Hamiltonian

$$H = -\frac{1}{2}\sum_{i,j} J_{ij}S_iS_j = -\frac{1}{2}\sum_i S_i \sum_j J_{ij}S_j. \tag{7.7}$$

The reason is that $\sum_j J_{ij}S_j$ is the local field h_i to the spin S_i,[14] and (7.3) aligns the spin (neuron state) S_i to the direction of the local field at the next time step

[14]Note that the present local field is different from (7.1) by a constant $\sum_j J_{ij}$.

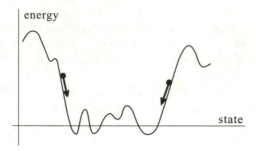

FIG. 7.2. Energy landscape and time evolution

$t + \Delta t$, leading to a monotonic decrease of the energy (7.7). Figure 7.2 depicts this situation intuitively where the network reaches a minimum of the energy closest to the initial condition and stops its time evolution there. Consequently, if the system has the property that there is a one-to-one correspondence between memorized patterns and energy minima, the system starting from the initial condition with a small amount of noise (i.e. an initial pattern slightly different from an embedded pattern) will evolve towards the closest memorized pattern, and the noise in the initial state will thereby be erased. The system therefore works as an information processor to remove noise by autonomous and distributed dynamics. Thus the problem is to find the conditions for this behaviour to be realized under the Hebb rule (7.4).

We note here that the dynamics (7.3) definitely determines the state of the neuron, given the input $h_i(t) = \sum_j J_{ij} S_j(t)$. However, the functioning of real neurons may not be so deterministic, which suggests the introduction of a stochastic process in the time evolution. For this purpose, it is convenient to assume that $S_i(t + \Delta t)$ becomes 1 with probability $1/(1 + e^{-2\beta h_i(t)})$ and is -1 with probability $e^{-2\beta h_i(t)}/(1 + e^{-2\beta h_i(t)})$. Here β is a parameter introduced to control uncertainty in the functioning of the neuron. This stochastic dynamics reduces to (7.3) in the limit $\beta \to \infty$ because $S_i(t + \Delta t) = 1$ if $h_i(t) > 0$ and is $S_i(t + \Delta t) = -1$ otherwise. The network becomes perfectly random if $\beta = 0$ (see Fig. 7.3).

This stochastic dynamics is equivalent to the kinetic Ising model (4.93) and (4.94). To confirm this, we consider, for example, the process of the second term on the right hand side of (4.94). Since $\boldsymbol{S}''$ is a spin configuration with S_i inverted to $-S_i$ in $\boldsymbol{S} = \{S_i\}$, we find $\Delta(\boldsymbol{S}'', \boldsymbol{S}) = H(\boldsymbol{S}'') - H(\boldsymbol{S}) = 2S_i h_i$. The transition probability of this spin inversion process is, according to (4.94),

$$w = \frac{1}{1 + \exp(\beta\Delta(\boldsymbol{S}'', \boldsymbol{S}))} = \frac{1}{1 + e^{2\beta S_i h_i}}. \tag{7.8}$$

If $S_i = 1$ currently, the possible new state is $S_i = -1$ and the transition probability for such a process is $w = (1 + e^{2\beta h_i})^{-1} = e^{-2\beta h_i}/(1 + e^{-2\beta h_i})$, which

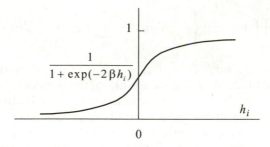

FIG. 7.3. The probability that the neuron i becomes excited.

coincides with the above-mentioned transition probability of neuron update to $S_i = -1$.

If we now insert the equilibrium Gibbs–Boltzmann distribution into $P_t(\boldsymbol{S})$ of (4.93), the right hand side vanishes, and the equilibrium distribution does not change with time as expected. It is also known that, under the kinetic Ising model, the state of a system approaches the equilibrium Gibbs–Boltzmann distribution at an appropriate temperature $\beta^{-1} = T$ even when the initial condition is away from equilibrium. Thus the problem of memory retrieval after a sufficiently long time starting from an initial condition in a neural network can be analysed by equilibrium statistical mechanics of the Ising model (7.7) with random interactions specified by the Hebb rule (7.4). The model Hamiltonian (7.7) with random patterns embedded by the Hebb rule (7.4) is usually called the *Hopfield model* (Hopfield 1982).

7.2 Embedding a finite number of patterns

It is relatively straightforward to analyse statistical-mechanical properties of the Hopfield model when the number of embedded patterns p is finite (Amit *et al.* 1985).

7.2.1 *Free energy and equations of state*

The partition function of the system described by the Hamiltonian (7.7) with the Hebb rule (7.4) is

$$Z = \mathrm{Tr} \exp\left(\frac{\beta}{2N} \sum_\mu (\sum_i S_i \xi_i^\mu)^2 \right),\tag{7.9}$$

where Tr is the sum over $\boldsymbol{S}$. The effect of the vanishing diagonal ($J_{ii} = 0$) has been ignored here since it is of lower order in N. By introducing a new integration variable m^μ to linearize the square in the exponent, we have

$$Z = \mathrm{Tr} \int \prod_{\mu=1}^p \mathrm{d}m^\mu \exp\left\{ -\frac{1}{2} N\beta \sum_\mu m_\mu^2 + \beta \sum_\mu m_\mu \sum_i S_i \xi_i^\mu \right\}$$

$$= \int \prod_{\mu} dm^{\mu} \exp \left\{ -\frac{1}{2} N \beta \boldsymbol{m}^2 + \sum_i \log(2 \cosh \beta \boldsymbol{m} \cdot \boldsymbol{\xi}_i) \right\}, \qquad (7.10)$$

where $\boldsymbol{m} = {}^t(m^1, \ldots, m^p)$, $\boldsymbol{\xi}_i = {}^t(\xi_i^1, \ldots, \xi_i^p)$, and the overall multiplicative constant has been omitted as it does not affect the physical properties.

In consideration of the large number of neurons in the brain (about 10^{11}) and also from the viewpoint of designing computational equipment with highly non-trivial information processing capabilities, it makes sense to consider the limit of large-size systems. It is also interesting from a statistical-mechanical point of view to discuss phase transitions that appear only in the thermodynamic limit $N \to \infty$. Hence we take the limit $N \to \infty$ in (7.10), so that the integral is evaluated by steepest descent. The free energy is

$$f = \frac{1}{2} \boldsymbol{m}^2 - \frac{T}{N} \sum_i \log(2 \cosh \beta \boldsymbol{m} \cdot \boldsymbol{\xi}_i). \qquad (7.11)$$

The extremization condition of the free energy (7.11) gives the equation of state as

$$\boldsymbol{m} = \frac{1}{N} \sum_i \boldsymbol{\xi}_i \tanh(\beta \boldsymbol{m} \cdot \boldsymbol{\xi}_i). \qquad (7.12)$$

In the limit of large N, the sum over i in (7.12) becomes equivalent to the average over the random components of the vector $\boldsymbol{\xi} = {}^t(\xi^1, \ldots, \xi^p)$ $(\xi^1 = \pm 1, \ldots, \xi^p = \pm 1)$ according to the self-averaging property. This averaging corresponds to the configurational average in the theory of spin glasses, which we denote by $[\cdots]$, and we write the free energy and the equation of state as

$$f = \frac{1}{2} \boldsymbol{m}^2 - T \left[\log(2 \cosh \beta \boldsymbol{m} \cdot \boldsymbol{\xi}) \right] \qquad (7.13)$$

$$\boldsymbol{m} = \left[\boldsymbol{\xi} \tanh \beta \boldsymbol{m} \cdot \boldsymbol{\xi} \right]. \qquad (7.14)$$

The physical significance of the order parameter $\boldsymbol{m}$ is revealed by the saddle-point condition of the first expression of (7.10) for large N:

$$m^{\mu} = \frac{1}{N} \sum_i S_i \xi_i^{\mu}. \qquad (7.15)$$

This equation shows that m^{μ} is the overlap between the μth embedded pattern and the state of the system. If the state of the system is in perfect coincidence with the μth pattern $(S_i = \xi_i^{\mu}, \forall i)$, we have $m^{\mu} = 1$. In the total absence of correlation, on the other hand, S_i assumes ± 1 independently of ξ_i^{μ}, and consequently $m^{\mu} = 0$. It should then be clear that the success of retrieval of the μth pattern is measured by the order parameter m^{μ}.

7.2.2   Solution of the equation of state

Let us proceed to the solution of the equation of state (7.14) and discuss the macroscopic properties of the system. We first restrict ourselves to the case of a

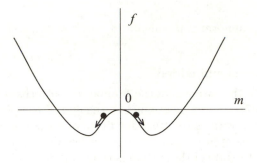

FIG. 7.4. Free energy of the Hopfield model when a single pattern is retrieved.

single-pattern retrieval. There is no loss of generality by assuming that the first pattern is to be retrieved: $m_1 = m, m_2 = \cdots = m_p = 0$. Then the equation of state (7.14) is

$$m = [\xi^1 \tanh(\beta m \xi^1)] = \tanh \beta m. \tag{7.16}$$

This is precisely the mean-field equation of the usual ferromagnetic Ising model, (1.19) with $h = 0$, which has a stable non-trivial solution $m \neq 0$ for $T = \beta^{-1} < 1$. The free energy, accordingly, has minima at $m \neq 0$ if $T < 1$ as shown in Fig. 7.4. It is seen in this figure that an initial condition away from a stable state will evolve towards the nearest stable state with gradually decreasing free energy under the dynamics of §7.1.3 if the uncertainty parameter of the neuron functioning T (temperature in the physics terms) is smaller than 1. In particular, if $T = 0$, then the stable state is $m = \pm 1$ from (7.16), and a perfect retrieval of the embedded pattern (or its complete reversal) is achieved. It has thus been shown that the Hopfield model, if the number of embedded patterns is finite and the temperature is not very large, works as an *associative memory* that retrieves the appropriate embedded pattern when a noisy version of the pattern is given as the initial condition.

The equation of state (7.14) has many other solutions. A simple example is the solution to retrieve l patterns simultaneously with the same amplitude

$$\boldsymbol{m} = (m_l, m_l, m_l, \ldots, m_l, 0, \ldots, 0). \tag{7.17}$$

It can be shown that the ground state ($T = 0$) energy E_l of this solution satisfies the following inequality:

$$E_1 < E_3 < E_5 < \ldots . \tag{7.18}$$

The single-retrieval solution is the most stable one, and other odd-pattern retrieval cases follow. All even-pattern retrievals are unstable. At finite temperatures, the retrieval solution with $l = 1$ exists stably below the critical temperature $T_c = 1$ as has been shown already. This is the unique solution in the range $0.461 < T < 1$. The $l = 3$ solution appears at $T = 0.461$. Other solutions with $l = 5, 7, \ldots$ show up one after another as the temperature is further decreased.

Solutions with non-uniform components in contrast to (7.17) also exist at low temperatures.

7.3 Many patterns embedded

We next investigate the case where the number of patterns p is proportional to N (Amit *et al.* 1987). As was mentioned at the end of the previous section, the equation of state has various solutions for finite p. For larger p, more and more complicated solutions appear, and when p reaches $\mathcal{O}(N)$, a spin glass solution emerges, in which the state of the system is randomly frozen in the sense that the state has no correlation with embedded patterns. If the ratio $\alpha \equiv p/N$ exceeds a threshold, the spin glass state becomes the only stable state at low temperatures. This section is devoted to a detailed account of this scenario.

7.3.1 *Replicated partition function*

It is necessary to calculate the configurational average of the nth power of the partition function to derive the free energy $F = -T[\log Z]$ averaged over the pattern randomness $\{\xi_i^\mu\}$. It will be assumed for simplicity that only the first pattern ($\mu = 1$) is to be retrieved. The configurational average of the replicated partition function is, by (7.4) and (7.7),

$$
[Z^n] = \left[\mathrm{Tr}\exp\left(\frac{\beta}{2N} \sum_{i,j} \sum_\mu \sum_{\rho=1}^n \xi_i^\mu \xi_j^\mu S_i^\rho S_j^\rho \right) \right]
$$

$$
= \mathrm{Tr} \int \prod_{\mu\rho} dm_\rho^\mu \left[\exp\left\{ \beta N \left(-\frac{1}{2}\sum_{\mu\geq 2}\sum_\rho (m_\rho^\mu)^2 + \frac{1}{N}\sum_{\mu\geq 2}\sum_\rho m_\rho^\mu \sum_i \xi_i^\mu S_i^\rho \right. \right. \right.
$$

$$
\left. \left. \left. -\frac{1}{2}\sum_\rho (m_\rho^1)^2 + \frac{1}{N}\sum_\rho m_\rho^1 \sum_i \xi_i^1 S_i^\rho \right) \right\} \right]. \tag{7.19}
$$

It is convenient to separate the contribution of the first pattern from the rest.

7.3.2 *Non-retrieved patterns*

The overlap between the state of the system and a pattern other than the first one ($\mu \geq 2$) is due only to coincidental contributions from the randomness of $\{\xi_i^\mu\}$, and is

$$
m_\rho^\mu = \frac{1}{N}\sum_i \xi_i^\mu S_i^\rho \approx \mathcal{O}\left(\frac{1}{\sqrt{N}}\right). \tag{7.20}
$$

The reason is that, if ξ_i^μ and S_i^ρ assume 1 or -1 randomly and independently, the stochastic variable $\sum_i \xi_i^\mu S_i^\rho$ has average 0 and variance N, since $\overline{(\sum_i \xi_i^\mu S_i^\rho)^2} = N + \overline{\sum_{i\neq j} \xi_i^\mu \xi_j^\mu S_i^\rho S_j^\rho} = N$. In the case of finite p treated in the previous section, the number of terms with $\mu \geq 2$ in (7.19) is finite, and we could ignore these contributions in the limit of large N. Since the number of such terms is proportional to N in the present section, a more careful treatment is needed.

The first step is the rescaling of the variable $m_\rho^\mu \to m_\rho^\mu/\sqrt{\beta N}$ to reduce m_ρ^μ to $\mathcal{O}(1)$. We then perform configurational averaging $[\cdots]$ for the terms with $\mu \geq 2$ in (7.19) to find

$$\exp\left\{-\frac{1}{2}\sum_{\mu\rho}(m_\rho^\mu)^2 + \sum_{i\mu}\log\cosh\left(\sqrt{\frac{\beta}{N}}\sum_\rho m_\rho^\mu S_i^\rho\right)\right\}. \qquad (7.21)$$

The sum over μ is for the range $\mu \geq 2$. In the limit of large N, we may expand $\log\cosh(\cdot)$ and keep only the leading term:

$$\exp\left(-\frac{1}{2}\sum_\mu\sum_{\rho\sigma}m_\rho^\mu K_{\rho\sigma}m_\sigma^\mu\right), \qquad (7.22)$$

where

$$K_{\rho\sigma} = \delta_{\rho\sigma} - \frac{\beta}{N}\sum_i S_i^\rho S_i^\sigma. \qquad (7.23)$$

The term (7.22) is quadratic in m_ρ^μ, and the integral over m_ρ^μ can be carried out by the multivariable Gaussian integral. The result is, up to a trivial overall constant,

$$(\det K)^{-(p-1)/2} = \exp\left(-\frac{p-1}{2}\operatorname{Tr}\log K\right) = \int\prod_{(\rho\sigma)}dq_{\rho\sigma}\delta\left(q_{\rho\sigma} - \frac{1}{N}\sum_i S_i^\rho S_i^\sigma\right)$$

$$\cdot\exp\left\{-\frac{p-1}{2}\operatorname{Tr}_n\log\{(1-\beta)I - \beta Q\}\right\}. \qquad (7.24)$$

Here $(\rho\sigma)$ is the set of $n(n-1)$ replica pairs. We have also used the fact that the diagonal element $K_{\rho\rho}$ of the matrix K is equal to $1 - \beta$ and the off-diagonal part is

$$K_{\rho\sigma} = -\frac{\beta}{N}\sum_i S_i^\rho S_i^\sigma = -\beta q_{\rho\sigma}. \qquad (7.25)$$

The expression Q is a matrix with the off-diagonal elements $q_{\rho\sigma}$ and 0 along the diagonal. The operation Tr_n in the exponent is the trace of the $n \times n$ matrix. Fourier representation of the delta function in (7.24) with integral variable $r_{\rho\sigma}$ and insertion of the result into (7.19) yield

$$[Z^n] = \operatorname{Tr}\int\prod_\rho dm_\rho^1\int\prod_{(\rho\sigma)}dq_{\rho\sigma}dr_{\rho\sigma}\exp\left(-\frac{N}{2}\alpha\beta^2\sum_{(\rho\sigma)}r_{\rho\sigma}q_{\rho\sigma}\right.$$

$$\left. + \frac{\alpha\beta^2}{2}\sum_{i,(\rho\sigma)}r_{\rho\sigma}S_i^\rho S_i^\sigma\right)\exp\left\{-\frac{p-1}{2}\operatorname{Tr}_n\log\{(1-\beta)I - \beta Q\}\right\}$$

$$\cdot\left[\exp\beta N\left(-\frac{1}{2}\sum_\rho(m_\rho^1)^2 + \frac{1}{N}\sum_\rho m_\rho^1\sum_i\xi_i^1 S_i^\rho\right)\right]. \qquad (7.26)$$

7.3.3 Free energy and order parameter

The S_i-dependent parts in (7.26) are written as

$$
\left[\mathrm{Tr}\exp\left(\beta\sum_{i,\rho} m_\rho^1 \xi_i^1 S_i^\rho + \frac{1}{2}\alpha\beta^2 \sum_{i,(\rho\sigma)} r_{\rho\sigma} S_i^\rho S_i^\sigma \right) \right]
$$

$$
= \left[\exp\left\{ \sum_i \log\mathrm{Tr}\exp\left(\beta\sum_\rho m_\rho^1 \xi_i^1 S^\rho + \frac{1}{2}\alpha\beta^2 \sum_{(\rho\sigma)} r_{\rho\sigma} S^\rho S^\sigma \right) \right\} \right]
$$

$$
= \exp N \left[\log\mathrm{Tr}\exp\left(\beta\sum_\rho m_\rho^1 \xi^1 S^\rho + \frac{1}{2}\alpha\beta^2 \sum_{(\rho\sigma)} r_{\rho\sigma} S^\rho S^\sigma \right) \right]. \qquad (7.27)
$$

In going from the second expression to the last one, we have used the fact that the sum over i is equivalent to the configurational average $[\cdots]$ by self-averaging in the limit of large N. Using this result in (7.26), we obtain

$$
[Z^n] = \int \prod \mathrm{d}m_\rho \int \prod \mathrm{d}r_{\rho\sigma} \prod \mathrm{d}q_{\rho\sigma}
$$

$$
\cdot \exp N \left\{ -\frac{\beta}{2}\sum_\rho (m_\rho^1)^2 - \frac{\alpha}{2}\mathrm{Tr}_n \log\{(1-\beta)I - \beta Q\} - \frac{1}{2}\alpha\beta^2 \sum_{(\rho\sigma)} r_{\rho\sigma} q_{\rho\sigma} \right.
$$

$$
\left. + \left[\log\mathrm{Tr}\exp\left(\frac{1}{2}\alpha\beta^2 \sum_{(\rho\sigma)} r_{\rho\sigma} S^\rho S^\sigma + \beta\sum_\rho m_\rho^1 \xi^1 S^\rho \right) \right] \right\}, \qquad (7.28)
$$

where we have set $(p-1)/N = \alpha$ assuming $N, p \gg 1$. In the thermodynamic limit $N \to \infty$, the free energy is derived from the term proportional to $n\beta$ in the exponent:

$$
f = \frac{1}{2n}\sum_\rho (m_\rho^1)^2 + \frac{\alpha}{2n\beta}\mathrm{Tr}_n \log\{(1-\beta)I - \beta Q\}
$$

$$
+ \frac{\alpha\beta}{2n}\sum_{(\rho\sigma)} r_{\rho\sigma} q_{\rho\sigma} - \frac{1}{\beta n}\left[\log\mathrm{Tr}\,e^{\beta H_\xi} \right], \qquad (7.29)
$$

where βH_ξ is the quantity in the exponent after the operator $\log\mathrm{Tr}$ in (7.28).

The order parameters have the following significance. The parameter m_ρ^μ is the overlap between the state of the system and the μth pattern. This can be confirmed by extremizing the exponent of (7.19) with respect to m_ρ^μ to find

$$
m_\rho^\mu = \frac{1}{N}\sum_i \xi_i^\mu S_i^\rho. \qquad (7.30)
$$

Next, $q_{\alpha\beta}$ is the spin glass order parameter from comparison of (7.24) and (2.20). As for $r_{\rho\sigma}$, extremization of the exponent of (7.26) with respect to $q_{\alpha\beta}$ gives

$$r_{\rho\sigma} = \frac{1}{\alpha} \sum_{\mu \geq 2} m_\rho^\mu m_\sigma^\mu. \tag{7.31}$$

To calculate the variation with respect to the components of Q in the $\mathrm{Tr}_n \log$ term, we have used the facts that the $\mathrm{Tr}_n \log$ term is written by the integral over m_ρ^μ of (7.22) with $K_{\rho\sigma}$ replaced by $(1 - \beta)I - \beta Q$ and that we have performed scaling of m_ρ^μ by $\sqrt{\beta N}$ just before (7.22). From (7.31), $r_{\rho\sigma}$ is understood as the sum of effects of non-retrieved patterns.

7.3.4   Replica-symmetric solution

The assumption of replica symmetry leads to an explicit form of the free energy (7.29). From $m_\rho^1 = m, q_{\rho\sigma} = q$, and $r_{\rho\sigma} = r$ for $\rho \neq \sigma$, the first term of (7.29) is $m^2/2$ and the third term in the limit $n \to 0$ is $-\alpha\beta rq/2$. To calculate the second term, it should be noted that the eigenvectors of the matrix $(1 - \beta)I - \beta Q$ are, first, the uniform one ${}^t(1, 1, \ldots, 1)$ and, second, the sequence of the nth roots of unity ${}^t(e^{2\pi i k/n}, e^{4\pi i k/n}, \ldots, e^{2(n-1)\pi i k/n})$ $(k = 1, 2, \ldots, n - 1)$. The eigenvalue of the first eigenvector is $1 - \beta + \beta q - n\beta q$ without degeneracy, and for the second eigenvector, it is $1 - \beta + \beta q$ (degeneracy $n - 1$). Thus in the limit $n \to 0$,

$$\frac{1}{n}\mathrm{Tr}_n \log\{(1 - \beta)I - \beta Q\} = \frac{1}{n} \log(1 - \beta + \beta q - n\beta q) + \frac{n-1}{n} \log(1 - \beta + \beta q)$$

$$\to \log(1 - \beta + \beta q) - \frac{\beta q}{1 - \beta + \beta q}. \tag{7.32}$$

The final term of (7.29) is, as $n \to 0$,

$$\frac{1}{n}\left[\log \mathrm{Tr} \exp\left(\frac{1}{2}\alpha\beta^2 r(\sum_\rho S^\rho)^2 - \frac{1}{2}\alpha\beta^2 rn - \beta\sum_\rho m\xi^1 S^\rho \right) \right]$$

$$= -\frac{1}{2}\alpha r \beta^2 + \frac{1}{n}\left[\log \mathrm{Tr} \int Dz \exp\left(-\beta\sqrt{\alpha r}z\sum_\rho S^\rho - \beta\sum_\rho m\xi S^\rho \right) \right]$$

$$\to -\frac{1}{2}\alpha r \beta^2 + \left[\int Dz \log 2\cosh\beta(\sqrt{\alpha r}z + m\xi) \right]. \tag{7.33}$$

The integral over z has been introduced to reduce the quadratic form $(\sum_\rho S^\rho)^2$ to a linear expression. Collecting everything together, we find the RS free energy as

$$f = \frac{1}{2}m^2 + \frac{\alpha}{2\beta}\left(\log(1 - \beta + \beta q) - \frac{\beta q}{1 - \beta + \beta q} \right) + \frac{\alpha\beta}{2}r(1 - q)$$

$$-T \int Dz \left[\log 2\cosh\beta(\sqrt{\alpha r}z + m\xi) \right]. \tag{7.34}$$

The equations of state are derived by extremization of (7.34). The equation for m is

$$m = \int Dz \left[\xi \tanh \beta(\sqrt{\alpha r} z + m\xi)\right] = \int Dz \tanh \beta(\sqrt{\alpha r} z + m). \tag{7.35}$$

Next, a slight manipulation of the extremization condition with respect to r gives

$$q = \int Dz \left[\tanh^2 \beta(\sqrt{\alpha r} z + m\xi)\right] = \int Dz \tanh^2 \beta(\sqrt{\alpha r} z + m), \tag{7.36}$$

and from extremization by q,

$$r = \frac{q}{(1 - \beta + \beta q)^2}. \tag{7.37}$$

It is necessary to solve these three equations simultaneously.

It is easy to check that there is a paramagnetic solution $m = q = r = 0$. At low temperatures, the spin glass solution $m = 0, q > 0, r > 0$ exists. The critical temperature for the spin glass solution to appear is $T = 1 + \sqrt{\alpha}$, as can be verified by combining (7.37) and the leading term in the expansion of the right hand side of (7.36) with $m = 0$.

The retrieval solution $m > 0, q > 0, r > 0$ appears discontinuously by a first-order transition. Numerical solution of the equations of state (7.35), (7.36), and (7.37) should be employed to draw phase boundaries.[15]

The final phase diagram is shown in Fig. 7.5. For α smaller than 0.138, three phases (paramagnetic, spin glass, and metastable retrieval phases) appear in this order as the temperature is decreased. If $\alpha < 0.05$, the retrieval phase is stable (i.e. it is the global, not local, minimum of the free energy) at low temperatures. The RS solution of the retrieval phase is unstable for RSB at very low temperatures. However, the RSB region is very small and the qualitative behaviour of the order parameter m and related quantities is expected to be relatively well described by the RS solution.

7.4 Self-consistent signal-to-noise analysis

The replica method is powerful but is not easily applicable to some problems. For instance, when the input–output relation of a neuron, which is often called the *activation function*, is not represented by a simple monotonic function like (7.3), it is difficult (and often impossible) to use the replica method. *Self-consistent signal-to-noise analysis* (SCSNA) is a convenient approximation applicable to many of such cases (Shiino and Fukai 1993).

7.4.1 *Stationary state of an analogue neuron*

The idea of SCSNA is most easily formulated for analogue neurons. The membrane potential of a real neuron is an analogue quantity and its change with time is modelled by the following equation:

$$\frac{dh_i}{dt} = -h_i + \sum_j J_{ij} F(h_j). \tag{7.38}$$

---

[15]Analytical analysis is possible in the limits $\alpha \to 0$ and $T \to 0$.

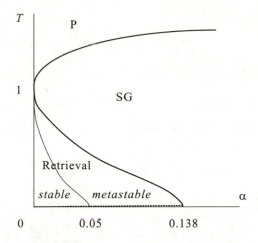

FIG. 7.5. Phase diagram of the Hopfield model. The dashed line near the α axis is the AT line and marks the instability of the RS solution.

The first term on the right hand side is for a natural decay of h_i, and the second term is the input signal from other neurons. $F(h)$ is the activation function of an analogue neuron, $S_j = F(h_j)$ in the notation of (7.3). The variable S_j has continuous values, and the connection J_{ij} is assumed to be given by the Hebb rule (7.4) with random patterns $\xi_i^\mu = \pm 1$. The self-interaction J_{ii} is vanishing.

In the stationary state, we have from (7.38)

$$h_i = \sum_j J_{ij} S_j. \tag{7.39}$$

Let us look for a solution which has the overlap

$$m^\mu = \frac{1}{N} \sum_j \xi_j^\mu S_j \tag{7.40}$$

of order $m^1 = m = \mathcal{O}(1)$ and $m^\mu = \mathcal{O}(1/\sqrt{N})$ $(\mu \geq 2)$ as in §§7.2 and 7.3.

7.4.2 Separation of signal and noise

To evaluate the overlap (7.40), we first insert the Hebb rule (7.4) into the definition of h_i, (7.39), to find the following expression in which the first and the μth patterns $(\mu \geq 2)$ are treated separately:

$$h_i = \xi_i^1 m + \xi_i^\mu m^\mu + \sum_{\nu \neq 1, \mu} \xi_i^\nu m^\nu - \alpha S_i. \tag{7.41}$$

The last term $-\alpha S_i$ is the correction due to $J_{ii} = 0$. The possibility that a term proportional to S_i may be included in the third term of (7.41) is taken into account by separating this into a term proportional to S_i and the rest,

$$\sum_{\nu \neq 1, \mu} \xi_i^\nu m^\nu = \gamma S_i + z_{i\mu}. \tag{7.42}$$

From (7.41) and (7.42),

$$S_i = F(h_i) = F(\xi_i^1 m + \xi_i^\mu m^\mu + z_{i\mu} + \Gamma S_i), \tag{7.43}$$

where we have written $\Gamma = \gamma - \alpha$. Now, suppose that the following solution has been obtained for S_i from (7.43):

$$S_i = \tilde{F}(\xi_i^1 m + \xi_i^\mu m^\mu + z_{i\mu}). \tag{7.44}$$

Then the overlap for $\mu \geq 2$ is

$$m^\mu = \frac{1}{N} \sum_j \xi_j^\mu \tilde{F}(\xi_j^1 m + z_{j\mu}) + \frac{m^\mu}{N} \sum_j \tilde{F}'(\xi_j^1 m + z_{j\mu}), \tag{7.45}$$

where we have expanded the expression to first order in $m^\mu = \mathcal{O}(1/\sqrt{N})$. The solution of (7.45) for m^μ is

$$m^\mu = \frac{1}{KN} \sum_j \xi_j^\mu \tilde{F}(\xi_j^1 m + z_{j\mu}), \quad K = 1 - \frac{1}{N} \sum_j \tilde{F}'(\xi_j^1 m + z_{j\mu}). \tag{7.46}$$

Here we have dropped the μ-dependence of K in the second relation defining K, since $z_{j\mu}$ will later be assumed to be a Gaussian random variable with vanishing mean and μ-independent variance. Replacement of this equation in the left hand side of (7.42) and separation of the result into the $(j = i)$-term and the rest give

$$\sum_{\nu \neq 1, \mu} \xi_i^\nu m^\nu = \frac{1}{KN} \sum_{\nu \neq 1, \mu} \tilde{F}(\xi_i^1 m + z_{i\nu}) + \frac{1}{KN} \sum_{j \neq i} \sum_{\nu \neq 1, \mu} \xi_i^\nu \xi_j^\nu \tilde{F}(\xi_j^1 m + z_{j\nu}). \tag{7.47}$$

Comparison of this equation with the right hand side of (7.42) reveals that the first term on the right hand side of (7.47) is γS_i and the second $z_{i\mu}$. Indeed, the first term is $(p/KN)S_i = \alpha S_i / K$ according to (7.44). Small corrections of $\mathcal{O}(1/\sqrt{N})$ can be ignored in this correspondence. From $\alpha S_i / K = \gamma S_i$, we find $\gamma = \alpha / K$.

The basic assumption of the SCSNA is that the second term on the right hand side of (7.47) is a Gaussian random variable $z_{i\mu}$ with vanishing mean. Under the assumption that various terms in the second term on the right hand side of (7.47) are independent of each other, the variance is written as

$$\sigma^2 = \langle z_{i\mu}^2 \rangle = \frac{1}{K^2 N^2} \sum_{j \neq i} \sum_{\nu \neq 1, \mu} \langle \tilde{F}(\xi_j^1 m + z_{j\nu})^2 \rangle = \frac{\alpha}{K^2} \langle \tilde{F}(\xi^1 m + z)^2 \rangle_{\xi, z}. \tag{7.48}$$

Here $\langle \cdots \rangle_{\xi, z}$ is the average by the Gaussian variable z and the random variable $\xi (= \pm 1)$. The above equation holds because in the large-N limit the sum over j (and division by N) is considered to be equivalent to the average over ξ and z.

A similar manipulation to rewrite (7.46) yields

$$K = 1 - \langle \tilde{F}'(\xi m + z) \rangle_{\xi,z}. \tag{7.49}$$

Finally, m is seen to satisfy

$$m = \langle \xi \tilde{F}(\xi m + z) \rangle_{\xi,z}. \tag{7.50}$$

Solution of (7.48), (7.49), and (7.50) determines the properties of the system.

7.4.3 Equation of state

Let us rewrite (7.48), (7.49), and (7.50) in a more compact form. Equation (7.41) is expressed as a relation between random variables according to (7.42):

$$h = \xi m + z + \Gamma Y(\xi, z) \quad \left(\Gamma = \frac{\alpha}{K} - \alpha \right), \tag{7.51}$$

where $Y(\xi, z) = F(\xi m + z + \Gamma Y(\xi, z))$. Introducing new symbols

$$q = \frac{K^2 \sigma^2}{\alpha}, \quad \sqrt{\alpha r} = \sigma, \quad U = 1 - K, \quad x\sigma = z, \tag{7.52}$$

we have the following equations in place of (7.50),(7.48), and (7.49)

$$m = \left\langle \int Dx\, \xi Y(\xi, x) \right\rangle_\xi \tag{7.53}$$

$$q = \left\langle \int Dx Y(\xi, x)^2 \right\rangle_\xi \tag{7.54}$$

$$U\sqrt{\alpha r} = \left\langle \int Dx\, x Y(\xi, x) \right\rangle_\xi \tag{7.55}$$

and auxiliary relations

$$Y(\xi, x) = F(\xi m + \sqrt{\alpha r}\, x + \Gamma Y(\xi, x)), \quad \Gamma = \frac{\alpha U}{1 - U}, \quad q = (1 - U)^2 r. \tag{7.56}$$

Equations (7.53)–(7.56) are the equations of state of the SCSNA.

7.4.4 Binary neuron

Let us exemplify the idea of the SCSNA using the case of a binary neuron $F(x) = \text{sgn}(x)$. For an odd function $F(x)$, it is seen from (7.56) that $Y(-1, x) = -Y(1, -x)$. Accordingly, we may drop the average over ξ in (7.53)–(7.55) for $m, q, U\sqrt{\alpha r}$ and replace the integrands by their values at $\xi = 1$.

The equation for Y, (7.56), reads

$$Y(x) = \text{sgn}(m + \sqrt{\alpha r}\, x + \Gamma Y(x)). \tag{7.57}$$

The stable solution of this equation is, according to Fig. 7.6, $Y(x) = 1$ when $\sqrt{\alpha r}\, x + m > 0$ and is $Y(x) = -1$ for $\sqrt{\alpha r}\, x + m < 0$. When two solutions exist

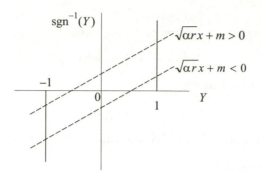

FIG. 7.6. Solutions of the equation for Y

simultaneously as in Fig. 7.6, one accepts the one with a larger area between $\sqrt{\alpha r}\, x + m$ and the Y axis, corresponding to the Maxwell rule of phase selection. From this and (7.54), we have $q = 1$, and from (7.53) and (7.55),

$$m = 2 \int_0^{m/\sqrt{\alpha r}} \mathrm{D}x, \quad \sqrt{\alpha r} = \sqrt{\alpha} + \sqrt{\frac{2}{\pi}} \mathrm{e}^{-m^2/2\alpha r}. \tag{7.58}$$

The solution obtained this way agrees with the limit $T \to 0$ of the RS solution (7.35)–(7.37). This is trivial for m and q. As for r, (7.56) coincides with (7.37) using the correspondence $U = \beta(1 - q)$.

The SCSNA generally gives the same answer as the RS solution when the latter is known. The SCSNA is applicable also to many problems for which the replica method is not easily implemented. A future problem is to clarify the limit of applicability of the SCSNA, corresponding to the AT line in the replica method.

7.5   Dynamics

It is an interesting problem how the overlap m changes with time in neural networks. An initial condition close to an embedded pattern μ would develop towards a large value of m_μ at $T = 0$. It is necessary to introduce a theoretical framework beyond equilibrium theory to clarify the details of these time-dependent phenomena.

It is not difficult to construct a dynamical theory when the number of embedded patterns p is finite (Coolen and Ruijgrok 1988; Riedel *et al.* 1988; Shiino *et al.* 1989). However, for p proportional to N, the problem is very complicated and there is no closed exact theory to describe the dynamics of the macroscopic behaviour of a network. This aspect is closely related to the existence of a complicated structure of the phase space of the spin glass state as mentioned in §7.3; it is highly non-trivial to describe the system properties rigorously when the state of the system evolves in a phase space with infinitely many free energy minima. The present section gives a flavour of approximation theory using the example of the dynamical theory due to Amari and Maginu (1988).

7.5.1 *Synchronous dynamics*

Let us denote the state of neuron i at time t by S_i^t $(= \pm 1)$. We assume in the present section that t is an integer and consider *synchronous dynamics* in which all neurons update their states according to

$$S_i^{t+1} = \text{sgn}\left(\sum_j J_{ij}S_j^t\right) \tag{7.59}$$

simultaneously at each discrete time step $t (= 0, 1, 2, \ldots)$. In other words, we apply (7.59) to all i simultaneously and the new states thus obtained $\{S_i^{t+1}\}$ are inserted in the right hand side in place of $\{S_i^t\}$ for the next update.[16] Synchronous dynamics is usually easier to treat than its asynchronous counterpart. Also it is often the case that systems with these two types of dynamics share very similar equilibrium properties (Amit *et al.* 1985; Fontanari and Köberle 1987).

7.5.2 *Time evolution of the overlap*

Let us investigate how the overlap m_μ changes with time under the synchronous dynamics (7.59) when the synaptic couplings J_{ij} are given by the Hebb rule (7.4) with $J_{ii} = 0$. The goal is to express the time evolution of the system in terms of a few macroscopic variables (order parameters). It is in general impossible to carry out this programme with only a finite number of order parameters including the overlap (Gardner *et al.* 1987). Some approximations should be introduced. In the *Amari–Maginu dynamics*, one derives dynamical equations in an approximate but closed form for the overlap and the variance of noise in the input signal.

Let us consider the case where the initial condition is close to the first pattern only, $m_1 = m > 0$ and $m_\mu = 0$ for $\mu \neq 1$. The input signal to S_i is separated into the true signal part (contributing to retrieval of the first pattern) and the rest (noise):

$$h_i^t = \sum_{j \neq i} J_{ij}S_j^t = \frac{1}{N} \sum_{\mu=1}^{p} \sum_{j \neq i} \xi_i^\mu \xi_j^\mu S_j^t$$

$$= \frac{1}{N} \sum_{j \neq i} \xi_i^1 \xi_j^1 S_j^t + \frac{1}{N} \sum_{\mu \neq 1} \sum_{j \neq i} \xi_i^\mu \xi_j^\mu S_j^t$$

$$= \xi_i^1 m_t + N_i^t. \tag{7.60}$$

The term with $\mu = 1$ gives the true signal $\xi_i^1 m_t$, and the rest is the noise N_i^t. Self-feedback $J_{ii}S_i^t$ vanishes and is omitted here as are also terms of $\mathcal{O}(N^{-1})$. The time evolution of the overlap is, from (7.59) and (7.60),

---

[16]We implicitly had *asynchronous dynamics* in mind so far in which (7.59) is applied to a single i and the new state of this ith neuron, S_i^{t+1}, is inserted in the right hand side with all other neuron states (S_j ($j \neq i$)) unchanged. See Amit (1989) for detailed discussions on this point.

$$m_{t+1} = \frac{1}{N} \sum_i \xi_i^1 \mathrm{sgn}(h_i^t) = \frac{1}{N} \sum_i \mathrm{sgn}(m_t + \xi_i^1 N_i^t). \qquad (7.61)$$

If the noise term $\xi_i^1 N_i^t$ is negligibly small compared to m_t, then the time development of m_t is described by $m_{t+1} = \mathrm{sgn}(m_t)$, which immediately leads to $m_1 = 1$ or $m_1 = -1$ at the next time step $t = 1$ depending on the sign of the initial condition $m_{t=0}$. Noise cannot be ignored actually, and we have to consider its effects as follows.

Since N_i^t is composed of the sum of many terms involving stochastic variables ξ_i^μ, this N_i^t would follow a Gaussian distribution according to the central limit theorem if the terms in the sum were independent of each other. This independence actually does not hold because S_j^t in N_i^t has been affected by all other $\{\xi_i^\mu\}$ during the past update of states. In the Amari–Maginu dynamics, one nevertheless accepts the approximate treatment that N_i^t follows a Gaussian distribution with vanishing mean and variance σ_t^2, denoted by $N(0, \sigma_t^2)$. The appropriateness of this approximation is judged by the results it leads to.

If we assume that N_i^t obeys the distribution $N(0, \sigma_t^2)$, the same should hold for $\xi_i^1 N_i^t$. Then the time development of m_t can be derived in the large-N limit from (7.61) as

$$m_{t+1} = \int Du \, \mathrm{sgn}(m_t + \sigma_t u) = F\left(\frac{m_t}{\sigma_t}\right), \qquad (7.62)$$

where $F(x) = 2 \int_0^x Du$.

7.5.3 Time evolution of the variance

Our description of dynamics in terms of macroscopic variables m_t and σ_t becomes a closed one if we know the time evolution of the variance of noise σ_t^2 since we have already derived the equation for m_t in (7.62). We may assume that the first pattern to be retrieved has $\xi_i^1 = 1$ ($\forall i$) without loss of generality as can be checked by the gauge transformation $S_i \to S_i \xi_i^1$, the ferromagnetic gauge.

With the notation $E[\cdots]$ for the average (expectation value), the variance of N_i^t is written as

$$\sigma_t^2 = E[(N_i^t)^2] = \frac{1}{N^2} \sum_{\mu \neq 1} \sum_{\nu \neq 1} \sum_{j \neq i} \sum_{j' \neq i} E[\xi_i^\mu \xi_i^\nu \xi_j^\mu \xi_{j'}^\nu S_j^t S_{j'}^t]. \qquad (7.63)$$

The expectation value in this sum is classified into four types according to the combination of indices:

1. For $\mu = \nu$ and $j = j'$, $E[\cdots] = 1$. The number of such terms is $(p-1)(N-1)$.
2. For $\mu \neq \nu$ and $j = j'$, $E[\cdots] = 0$, which can be neglected.
3. Let us write $v_3 \equiv E[\xi_j^\mu \xi_{j'}^\mu S_j^t S_{j'}^t]$ for the contribution of the case $\mu = \nu$ and $j \neq j'$. The number of terms is $(p-1)(N-1)(N-2)$.
4. When $\mu \neq \nu$ and $j \neq j'$, we have to evaluate $v_4 \equiv E[\xi_i^\mu \xi_i^\nu \xi_j^\mu \xi_{j'}^\nu S_j^t S_{j'}^t]$ explicitly. The number of terms is $(p-1)(p-2)(N-1)(N-2)$.

For evaluation of v_3, it is convenient to write the $\xi_j^\mu, \xi_{j'}^\mu$-dependence of $S_j^t S_{j'}^t$ explicitly. According to (7.60),

$$S_j^t = \text{sgn}(m_{t-1} + Q + \xi_j^\mu R + N^{-1}\xi_j^\mu \xi_{j'}^\mu S_{j'}^{t-1})$$

$$S_{j'}^t = \text{sgn}(m_{t-1} + Q' + \xi_{j'}^\mu R + N^{-1}\xi_{j'}^\mu \xi_j^\mu S_j^{t-1}),$$

(7.64)

where Q, Q', and R are defined by

$$Q = \frac{1}{N}\sum_{\nu\neq 1,\mu}\sum_{k\neq j}\xi_j^\nu \xi_k^\nu S_k^{t-1}, Q' = \frac{1}{N}\sum_{\nu\neq 1,\mu}\sum_{k\neq j}\xi_{j'}^\nu \xi_k^\nu S_k^{t-1}, R = \frac{1}{N}\sum_{k\neq j,j'}\xi_k^\mu S_k^{t-1}.$$

(7.65)

We assume these are independent Gaussian variables with vanishing mean and variance $\sigma_{t-1}^2, \sigma_{t-1}^2$, and σ_{t-1}^2/p, respectively. By writing Y_{11} for the contribution to v_3 when $\xi_j^\mu = \xi_{j'}^\mu = 1$ and similarly for $Y_{1-1}, Y_{-11}, Y_{-1-1}$, we have

$$v_3 = \frac{1}{4}(Y_{11} + Y_{1-1} + Y_{-11} + Y_{-1-1}).$$

(7.66)

The sum of the first two terms is

$$Y_{11} + Y_{1-1} = \int P(Q)P(Q')P(R)\,\mathrm{d}Q\,\mathrm{d}Q'\,\mathrm{d}R$$

$$\cdot \left\{ \text{sgn}\left(m_{t-1} + Q + R + \frac{S_{j'}^{t-1}}{N}\right) \text{sgn}\left(m_{t-1} + Q' + R + \frac{S_j^{t-1}}{N}\right) \right.$$

$$\left. - \text{sgn}\left(m_{t-1} + Q + R - \frac{S_{j'}^{t-1}}{N}\right) \text{sgn}\left(m_{t-1} + Q' - R - \frac{S_j^{t-1}}{N}\right) \right\}. (7.67)$$

The Gaussian integral over Q and Q' gives the function F used in (7.62). The integral over R can be performed by expanding F to first order assuming that $R \pm S_{j,j'}^{t-1}/N$ is much smaller than m_{t-1} in the argument of F. The remaining $Y_{-11} + Y_{-1-1}$ gives the same answer, and the final result is

$$Nv_3 = \frac{2}{\pi\alpha}\exp\left(-\frac{m_{t-1}^2}{\sigma_{t-1}^2}\right) + \frac{4m_{t-1}}{\sqrt{2\pi}\sigma_{t-1}}\exp\left(-\frac{m_{t-1}^2}{2\sigma_{t-1}^2}\right) F\left(\frac{m_{t-1}}{\sigma_{t-1}}\right).$$

(7.68)

A similar calculation yields for v_4

$$N^2 v_4 = \frac{2m_{t-1}^2}{\pi\sigma_{t-1}^2}\exp\left(-\frac{m_{t-1}^2}{\sigma_{t-1}^2}\right).$$

(7.69)

We have thus obtained σ_t^2. To be more accurate in discussing the variance, it is better to subtract the square of the average from the average of the square. The

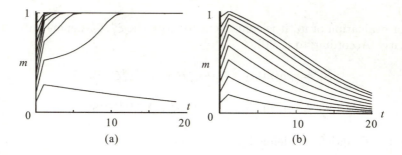

FIG. 7.7. Memory retrieval by synchronous dynamics: (a) $\alpha = 0.08$, and (b) $\alpha = 0.20$

relation $(E[N_i^t])^2 = p^2 v_4$ coming from the definition of v_4 can be conveniently used in this calculation:

$$\sigma_{t+1}^2 = \alpha + \frac{2}{\pi} \exp\left(-\frac{m_{t-1}^2}{\sigma_{t-1}^2}\right) + \frac{4\alpha m_t m_{t+1}}{\sqrt{2\pi}\sigma_t} \exp\left(-\frac{m_{t-1}^2}{2\sigma_{t-1}^2}\right). \tag{7.70}$$

The two equations (7.62) and (7.70) determine the dynamical evolution of the macroscopic parameters m_t and σ_t.

7.5.4   Limit of applicability

We show some explicit solutions of the time evolution equations (7.62) and (7.70). Figure 7.7 depicts m_t for various initial conditions with $\alpha = 0.08$ (a) and 0.20 (b). When $\alpha = 0.20$, the state tends to $m_t \to 0$ as $t \to \infty$ for any initial conditions, a retrieval failure. For $\alpha = 0.08$, m_t approaches 1, a successful retrieval, when the initial condition is larger than a threshold $m_0 > m_{0c}$. These analytical results have been confirmed to agree with simulations, at least qualitatively.

Detailed simulations, however, show that the assumption of Gaussian noise is close to reality when retrieval succeeds, but noise does not obey a Gaussian distribution when retrieval fails (Nishimori and Ozeki 1993). The system moves towards a spin glass state when retrieval fails, in which case the phase space has a very complicated structure and the dynamics cannot be described in terms of only two variables m_t and σ_t.

For α larger than a critical value α_c, the system moves away from the embedded pattern even if the initial condition is exactly at the pattern $m_0 = 1$. Numerical evaluation of (7.62) and (7.70) reveals that this critical value is $\alpha_c = 0.16$. This quantity should in principle be identical to the critical value $\alpha_c = 0.138$ for the existence of the retrieval phase at $T = 0$ in the Hopfield model (see Fig. 7.5). Note that equilibrium replica analysis has been performed also for the system with synchronous dynamics under the Hebb rule (Fontanari and Köberle 1987). The RS result $\alpha_c = 0.138$ is shared by this synchronous case. It is also known that the RSB changes α_c only by a very small amount (Steffan and Kühn 1994).

Considering several previous steps in deriving the time evolution equation is known to improve results (Okada 1995), but this method is still inexact as it uses

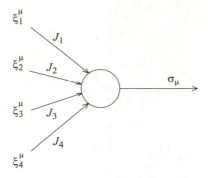

FIG. 7.8. Simple perceptron

a Gaussian assumption of noise distribution. It is in general impossible to solve exactly and explicitly the dynamics of associative memory with an extensive number of patterns embedded (p proportional to N). Closed-form approximation based on a dynamical version of equiparitioning gives a reasonably good description of the dynamical evolution of macrovariables (Coolen and Sherrington 1994). Another approach is to use a dynamical generating functional to derive various relations involving correlation and response functions (Rieger *et al.* 1989; Horner *et al.* 1989), which gives some insight into the dynamics, in particular in relation to the fluctuation–dissipation theorem.

7.6 Perceptron and volume of connections

It is important to investigate the limit of performance of a single neuron. In the present section, we discuss the properties of a simple perceptron with the simplest possible activation function. The point of view employed here is a little different from the previous arguments on associative memory because the synaptic couplings are dynamical variables here.

7.6.1 *Simple perceptron*

A *simple perceptron* is an element that gives an output σ^μ according to the following rule: given N inputs $\xi_1^\mu, \xi_2^\mu, \ldots, \xi_N^\mu$ and synaptic connections $J_1, J_2, \ldots, J_N$ (Fig. 7.8) then

$$\sigma^\mu = \mathrm{sgn}\left(\sum_j J_j \xi_j^\mu - \theta\right). \tag{7.71}$$

Here θ is a threshold and $\mu\,(=1,2,\ldots,p)$ is the index of the input–output pattern to be realized by the perceptron. The perceptron is required to adjust the weights $\boldsymbol{J} = \{J_j\}$ so that the input $\{\xi_i^\mu\}_{i=1,\ldots,N}$ leads to the desired output σ^μ. We assume $\xi_j^\mu, \sigma^\mu = \pm 1$; then the task of the perceptron is to classify p input patterns into two classes, those with $\sigma^\mu = 1$ and those $\sigma^\mu = -1$.

Let us examine an example of the case $N = 2$. By choosing J_1, J_2, and θ appropriately as shown in Fig. 7.9, we have $\sum_j J_j \xi_j^\mu - \theta = 0$ on the dashed line

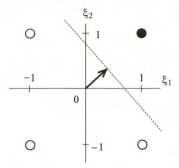

FIG. 7.9. Linearly separable task. The arrow is the vector $\boldsymbol{J} = {}^t(J_1, J_2)$.

perpendicular to the vector $\boldsymbol{J} = {}^t(J_1, J_2)$. This means that the output is $\sigma^1 = 1$ for the input $\xi_1^1 = \xi_2^1 = 1$ denoted by the full circle and $\sigma^\mu = -1$ for the open circles, thus separating the full and open circles by the dashed line. However, if there were two full circles at $(1,1)$ and $(-1,-1)$ and the rest were open, no straight line would separate the circles. It should then be clear that a simple perceptron is capable of realizing classification tasks corresponding to a bisection of the $\{\xi_i^\mu\}$ space by a hypersurface (a line in the case of two dimensions). This condition is called *linear separability*.

7.6.2 *Perceptron learning*

Although it has the constraint of linear separability, a simple perceptron plays very important roles in the theory of learning to be discussed in the next chapter. *Learning* means to change the couplings $\boldsymbol{J}$ gradually, reflecting given examples of input and output. The goal is to adjust the properties of the elements (neurons) or the network so that the correct input–output pairs are realized appropriately. We shall assume $\theta = 0$ in (7.71) for simplicity.

If the initial couplings of a perceptron are given randomly, the correct input–output pairs are not realized. However, under the following *learning rule* (the rule to change the couplings), it is known that the perceptron eventually reproduces correct input–output pairs as long as the examples are all linearly separable and the number of couplings N is finite (the convergence theorem):

$$J_j(t + \Delta t) = J_j(t) + \begin{cases} 0 & \text{(correct output for the } \mu th \text{ pattern)} \\ \eta \sigma^\mu \xi_j^\mu & \text{(otherwise).} \end{cases} \tag{7.72}$$

This is called the *perceptron learning rule*. Here η is a small positive constant. We do not present a formal proof of the theorem here. It is nevertheless to be noted that, by the rule (7.72), the sum of input signals $\sum_j J_j \xi_j^\mu$ increases by $\eta \sigma^\mu N$ when the previous output was incorrect, whereby the output $\text{sgn}(\sum_j J_j \xi_j^\mu)$ is pushed towards the correct value σ^μ.

7.6.3 *Capacity of a perceptron*

An interesting question is how many input–output pairs can be realized by a simple perceptron. The number of input–output pairs (patterns) will be denoted by p. Since we are interested in the limit of large N, the following normalization of the input signal will turn out to be appropriate:

$$\sigma^\mu = \text{sgn}\left(\frac{1}{\sqrt{N}}\sum_j J_j \xi_j^\mu\right). \tag{7.73}$$

The couplings are assumed to be normalized, $\sum_{j=1}^N J_j^2 = N$.

As we present examples (patterns or tasks to be realized) $\mu = 1, 2, 3, \ldots$ to the simple perceptron, the region in the $\boldsymbol{J}$-space where these examples are correctly reproduced by (7.73) shrinks gradually. Beyond a certain limit of the number of examples, the volume of this region vanishes and no $\boldsymbol{J}$ will reproduce some of the tasks. Such a limit is termed the *capacity* of the perceptron and is known to be $2N$. We derive a generalization of this result using techniques imported from spin glass theory (Gardner 1987, 1988).

The condition that the μth example is correctly reproduced is, as is seen by multiplying both sides of (7.73) by σ^μ,

$$\Delta^\mu \equiv \frac{\sigma^\mu}{\sqrt{N}}\sum_j J_j \xi_j^\mu > 0. \tag{7.74}$$

We generalize this inequality by replacing the right hand side by a positive constant κ

$$\Delta^\mu > \kappa \tag{7.75}$$

and calculate the volume of the subspace in the $\boldsymbol{J}$-space satisfying this relation. If we use (7.74), the left hand side can be very close to zero, and, when the input deviates slightly from the correct $\{\xi_j^\mu\}$, Δ^μ might become negative, producing an incorrect output. In (7.75) on the other hand, a small amount of error (or noise) in the input would not affect the sign of Δ^μ so that the output remains intact, implying a larger capability of error correction (or noise removal).

Let us consider the following volume (*Gardner volume*) in the $\boldsymbol{J}$-space satisfying (7.75) with the normalization condition taken into account:

$$V = \frac{1}{V_0}\int \prod_j dJ_j \,\delta\!\left(\sum_j J_j^2 - N\right)\prod_\mu \Theta(\Delta^\mu - \kappa), \quad V_0 = \int \prod_j dJ_j \,\delta\!\left(\sum_j J_j^2 - N\right), \tag{7.76}$$

where $\Theta(x)$ is a step function: $\Theta(x) = 1$ for $x > 0$ and 0 for $x < 0$.

Since we are interested in the typical behaviour of the system for *random* examples of input–output pairs, it is necessary to take the configurational average of the extensive quantity $\log V$ over the randomness of ξ_i^μ and σ^μ, similar to the spin glass theory.[17] For this purpose we take the average of V^n and let n tend to zero, following the prescription of the replica method,

$$[V^n] = \left[\frac{1}{V_0^n} \int \prod_{j,\alpha} \mathrm{d}J_j^\alpha \, \delta \left(\sum_j (J_j^\alpha)^2 - N \right) \prod_{\alpha,\mu} \Theta \left(\frac{\sigma^\mu}{\sqrt{N}} \sum_j J_j^\alpha \xi_j^\mu - \kappa \right) \right].$$

(7.77)

7.6.4 *Replica representation*

To proceed further with the calculation of (7.77), we use the integral representation of the step function

$$\Theta(y - \kappa) = \int_\kappa^\infty \frac{\mathrm{d}\lambda}{2\pi} \int_{-\infty}^\infty \mathrm{d}x \, e^{ix(\lambda - y)}$$

(7.78)

and carry out the average $[\cdots]$ over $\{\xi_i^\mu\}$:

$$\left[\prod_{\alpha,\mu} \Theta \left(\frac{\sigma^\mu}{\sqrt{N}} \sum_j J_j^\alpha \xi_j^\mu - \kappa \right) \right]$$

$$= \int_{-\infty}^\infty \prod_{\alpha,\mu} \mathrm{d}x_\mu^\alpha \int_\kappa^\infty \prod_{\alpha,\mu} \mathrm{d}\lambda_\mu^\alpha \exp \left\{ i \sum_{\alpha,\mu} x_\mu^\alpha \lambda_\mu^\alpha + \sum_{j,\mu} \log \cos \left(\frac{\sigma^\mu}{\sqrt{N}} \sum_\alpha x_\mu^\alpha J_j^\alpha \right) \right\}$$

$$\approx \left\{ \int_{-\infty}^\infty \prod_\alpha \mathrm{d}x^\alpha \int_\kappa^\infty \prod_\alpha \mathrm{d}\lambda^\alpha \right.$$

$$\left. \cdot \exp \left(i \sum_\alpha x^\alpha \lambda^\alpha - \frac{1}{2} \sum_\alpha (x^\alpha)^2 - \sum_{(\alpha\beta)} q_{\alpha\beta} x^\alpha x^\beta \right) \right\}^p.$$

(7.79)

Here we have dropped the trivial factor (a power of 2π), set $q_{\alpha\beta} = \sum_j J_j^\alpha J_j^\beta / N$, and used the approximation $\log \cos(x) \approx -x^2/2$ valid for small x. The notation $(\alpha\beta)$ is for the $n(n-1)/2$ different replica pairs. Apparently, $q_{\alpha\beta}$ is interpreted as the spin glass order parameter in the space of couplings.

Using the normalization condition $\sum_j J_j^2 = N$ and the Fourier representation of the delta function to express the definition of $q_{\alpha\beta}$, we find

$$[V^n] = V_0^{-n} \int \prod_{(\alpha\beta)} \mathrm{d}q_{\alpha\beta} \mathrm{d}F_{\alpha\beta} \prod_\alpha \mathrm{d}E_\alpha \prod_{j,\alpha} \mathrm{d}J_j^\alpha \{(7.79)\} \exp \left\{ -iN \sum_\alpha E_\alpha \right.$$

$$\left. -iN \sum_{(\alpha\beta)} F_{\alpha\beta} q_{\alpha\beta} + i \sum_\alpha E_\alpha \sum_j (J_j^\alpha)^2 + i \sum_{(\alpha\beta)} F_{\alpha\beta} \sum_j J_j^\alpha J_j^\beta \right\}$$

---

[17]In the present problem of capacity, ξ_i^μ and σ^μ are independent random numbers. In the theory of learning to be discussed in the next chapter, σ^μ is a function of $\{\xi_i^\mu\}$.

$$= V_0^{-n} \int \prod_{(\alpha\beta)} dq_{\alpha\beta} \, dF_{\alpha\beta} \prod_\alpha dE_\alpha \, e^{NG}. \tag{7.80}$$

Here

$$G = \alpha G_1(q_{\alpha\beta}) + G_2(F_{\alpha\beta}, E_\alpha) - i\sum_\alpha E_\alpha - i\sum_{(\alpha\beta)} F_{\alpha\beta} q_{\alpha\beta} \tag{7.81}$$

with

$$G_1(q_{\alpha\beta}) = \log \int_{-\infty}^\infty \prod_\alpha dx^\alpha \int_\kappa^\infty \prod_\alpha d\lambda^\alpha$$

$$\cdot \exp\left(i\sum_\alpha x^\alpha \lambda^\alpha - \frac{1}{2}\sum_\alpha (x^\alpha)^2 - \sum_{(\alpha\beta)} q_{\alpha\beta} x^\alpha x^\beta \right) \tag{7.82}$$

$$G_2(F_{\alpha\beta}, E_\alpha) = \log \int_{-\infty}^\infty \prod_\alpha dJ^\alpha$$

$$\cdot \exp\left\{ i\left(\sum_\alpha E_\alpha (J^\alpha)^2 + \sum_{(\alpha\beta)} F_{\alpha\beta} J^\alpha J^\beta \right) \right\}. \tag{7.83}$$

The reader should not confuse the factor $\alpha \, (= p/N)$ in front of $G_1(q_{\alpha\beta})$ in (7.81) with the replica index.

7.6.5 Replica-symmetric solution

The assumption of replica symmetry $q_{\alpha\beta} = q, F_{\alpha\beta} = F, E_\alpha = E$ makes it possible to go further with the evaluation of (7.82) and (7.83). We write the integral in (7.82) as I_1, which is simplified under replica symmetry as

$$I_1 = \int_{-\infty}^\infty \prod_\alpha dx^\alpha \int_\kappa^\infty \prod_\alpha d\lambda^\alpha \exp\left(i\sum_\alpha x^\alpha \lambda^\alpha - \frac{1-q}{2}\sum_\alpha (x^\alpha)^2 - \frac{q}{2}(\sum_\alpha x^\alpha)^2 \right)$$

$$= \int_{-\infty}^\infty \prod_\alpha dx^\alpha \int_\kappa^\infty \prod_\alpha d\lambda^\alpha \int Dy \exp\left(i\sum_\alpha x^\alpha \lambda^\alpha - \frac{1-q}{2}\sum_\alpha (x^\alpha)^2 \right.$$

$$\left. + iy\sqrt{q}\sum_\alpha x^\alpha \right)$$

$$= \int Dy \left\{ \int_{-\infty}^\infty dx \int_\kappa^\infty d\lambda \exp\left(-\frac{1-q}{2}x^2 + ix(\lambda + y\sqrt{q}) \right) \right\}^n. \tag{7.84}$$

The quantity in the final braces $\{\cdots\}$, to be denoted as $L(q)$, is, after integration over x,

$$L(q) = \int_\kappa^\infty d\lambda \frac{1}{\sqrt{1-q}} \exp\left(-\frac{(\lambda + y\sqrt{q})^2}{2(1-q)} \right) = 2 \operatorname{Erfc}\left(\frac{\kappa + y\sqrt{q}}{\sqrt{2(1-q)}} \right), \tag{7.85}$$

where Erfc(x) is the complementary error function $\int_x^\infty e^{-t^2} dt$. Thus $G_1(q_{\alpha\beta})$ in (7.82) is, in the limit $n \to 0$,

$$G_1(q) = n \int Dy \, \log L(q). \tag{7.86}$$

The integral over J in $G_2(F, E)$, on the other hand, can be evaluated by a multidimensional Gaussian integral using the fact that the exponent is a quadratic form in J.[18] The eigenvalues of the quadratic form of J, $E \sum_\alpha (J^\alpha)^2 + F \sum_{(\alpha\beta)} J^\alpha J^\beta$, are $E+(n-1)F/2$ (without degeneracy) and $E-F/2$ (degeneracy $(n-1)$), from which G_2 is, apart from a trivial constant,

$$\begin{aligned} G_2(F, E) &= -\frac{1}{2} \log \left(E + \frac{n-1}{2} F \right) - \frac{n-1}{2} \log \left(E - \frac{F}{2} \right) \\ &\to -\frac{n}{2} \log \left(E - \frac{F}{2} \right) - \frac{nF}{4E - 2F}. \end{aligned} \tag{7.87}$$

Substitution of (7.86) and (7.87) into (7.81) gives

$$\frac{1}{n} G = \alpha \int Dy \, \log L(q) - \frac{1}{2} \log \left(E - \frac{F}{2} \right) - \frac{F}{4E - 2F} - iE + \frac{i}{2} Fq. \tag{7.88}$$

According to the method of steepest descent, E and F can be eliminated in the limit $N \to \infty$ by extremization of G with respect to E and F:

$$E = \frac{i(1 - 2q)}{2(1 - q)^2}, \quad F = -\frac{iq}{(1 - q)^2}. \tag{7.89}$$

Then, (7.88) can be written only in terms of q, and the following expression results:

$$\frac{1}{n} G = \alpha \int Dy \, \log L(q) + \frac{1}{2} \log(1 - q) + \frac{1}{2(1 - q)}. \tag{7.90}$$

Various J are allowed for small p, but as p approaches the capacity limit, the freedom of choice is narrowed in the J-space, and eventually only a single one survives. Then q becomes unity by the definition $N^{-1} \sum_j J_j^\alpha J_j^\beta$, so that the capacity of the perceptron, α_c, is obtained by extremization of (7.90) with respect to q and setting $q \to 1$. Using the limiting form of the complementary error function Erfc(x) $\approx e^{-x^2}/2x$ as $x \to \infty$,

$$\frac{1}{2(1 - q)^2} = \frac{\alpha}{2(1 - q)^2} \int_{-\kappa}^\infty Dy \, (\kappa + y)^2. \tag{7.91}$$

The final result for the capacity is

---

[18]Another idea is to reduce the quadratic form in the exponent involving $F(\sum_\alpha J^\alpha)^2$ to a linear form by Gaussian integration.

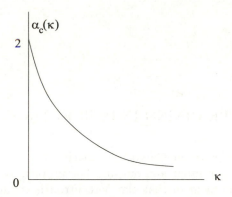

FIG. 7.10. Capacity of a simple perceptron

$$\alpha_c(\kappa) = \left\{ \int_{-\kappa}^{\infty} Dy \, (\kappa + y)^2 \right\}^{-1}. \tag{7.92}$$

In the limit $\kappa \to 0$, $\alpha_c(0) = 2$, which is the capacity $2N$ referred to in the first part of §7.6.3. The function $\alpha_c(\kappa)$ is monotonically decreasing as shown in Fig. 7.10. We mention without proof here that the RS solution is known to be stable as long as $\alpha < 2$ (Gardner 1987, 1988), which means that the J-space does not have a complicated structure below the capacity.

Bibliographical note

Most of the developments in neural network theory from statistical-mechanical points of view are covered in Amit (1989), Hertz et al. (1991), and Domany et al. (1991) as far as references until around 1990 are concerned. More up-to-date expositions are found in Coolen (2001) and Coolen and Sherrington (2001). Activities in recent years are centred mainly around the dynamical aspects of memory retrieval and the problem of learning. The latter will be treated in the next chapter. Reviews concerning the former topic are given in Coolen (2001), Coolen and Sherrington (2001), and Domany et al. (1995).

8

LEARNING IN PERCEPTRON

In the previous chapter we calculated the capacity of a simple perceptron under random combinations of input and output. The problem of learning is different from the capacity problem in that the perceptron is required to simulate the functioning of another perceptron even for new inputs, not to reproduce random signals as in the previous chapter. For this purpose, the couplings are gradually adjusted so that the probability of correct output increases. An important objective of the theory of learning is to estimate the functional relation between the number of examples and the expected error under a given algorithm to change couplings. The argument in this book will be restricted to learning in simple perceptrons.

8.1 Learning and generalization error

We first explain a few basic notions of learning. In particular, we introduce the generalization error, which is the expected error rate for a new input not included in the given examples used to train the perceptron.

8.1.1 *Learning in perceptron*

Let us prepare two perceptrons, one of which is called a *teacher* and the other a *student*. The two perceptrons share a common input $\boldsymbol{\xi} = \{\xi_j\}$ but the outputs are different because of differences in the couplings. The set of couplings of the teacher will be denoted by $\boldsymbol{B} = \{B_j\}$ and that of the student by $\boldsymbol{J} = \{J_j\}$ (Fig. 8.1). All these vectors are of dimensionality N. The student compares its output with that of the teacher and, if necessary, modifies its own couplings so that the output tends to coincide with the teacher output. The teacher couplings do not change. This procedure is termed *supervised learning*. The student couplings $\boldsymbol{J}$ change according to the output of the teacher; the student tries to simulate the teacher, given only the teacher output without explicit knowledge of the teacher couplings. In the case of *unsupervised learning*, by contrast, a perceptron changes its couplings only according to the input signals. There is no teacher which gives an ideal output. The perceptron adjusts itself so that the structure of the input signal (e.g. the distribution function of inputs) is well represented in the couplings. We focus our attention on supervised learning.

Supervised learning is classified into two types. In *batch learning* (or *off-line learning*) the student learns a given set of input–output pairs repeatedly until these examples by the teacher are reproduced sufficiently well. Then the same process is repeated for additional examples. The student learns to reproduce the

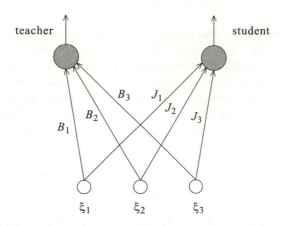

FIG. 8.1. Teacher and student perceptrons with $N = 3$

examples faithfully. A disadvantage is that the time and memory required are large.

The protocol of *on-line learning* is to change the student couplings immediately for a given example that is not necessarily used repeatedly. The student may not be able to answer correctly to previously given examples in contrast to batch learning. One can instead save time and memory in on-line learning. Another advantage is that the student can follow changes of the environment (such as the change of the teacher structure, if any) relatively quickly.

In general, the teacher and student may have more complex structures than a simple perceptron. For example, multilayer networks (where a number of simple elements are connected to form successive layers) are frequently used in practical applications because of their higher capabilities of information processing than simple perceptrons. Nevertheless, we mainly discuss the case of a simple perceptron to elucidate the basic concepts of learning, from which analyses of more complex systems start. References for further study are listed at the end of the chapter.

8.1.2 *Generalization error*

One of the most important quantities in the theory of learning is the *generalization error*, which is, roughly speaking, the expectation value of the error in the student output for a new input. An important goal of the theory of learning is to clarify the behaviour of generalization error as a function of the number of examples.

To define the generalization error more precisely, we denote the input signals to the student and teacher by u and v, respectively,

$$u = \sum_{j=1}^{N} J_j \xi_j, \quad v = \sum_{j=1}^{N} B_j \xi_j. \tag{8.1}$$

Suppose that both student and teacher are simple perceptrons. Then the outputs are $\text{sgn}(u)$ and $\text{sgn}(v)$, respectively. It is also assumed that the components of student and teacher couplings take arbitrary values under normalization $\sum_j J_j^2 = \sum_j B_j^2 = N$. We discuss the case where the components of the input vector are independent stochastic variables satisfying $[\xi_i] = 0$ and $[\xi_i \xi_j] = \delta_{ij}/N$. Other detailed properties of $\boldsymbol{\xi}$ (such as whether ξ_i is discrete or continuous) will be irrelevant to the following argument in the limit of large N. The overlap between the student and teacher couplings is written as R:

$$R = \frac{1}{N} \sum_{j=0}^{N} B_j J_j. \tag{8.2}$$

As the process of learning proceeds, the structure of the student usually approaches that of the teacher ($\boldsymbol{J} \approx \boldsymbol{B}$) and consequently R becomes closer to unity.

We consider the limit of large N since the description of the system then often simplifies significantly. For example, the stochastic variables u and v follow the Gaussian distribution of vanishing mean, variance unity, and covariance (average of uv) R when $N \to \infty$ according to the central limit theorem,

$$P(u,v) = \frac{1}{2\pi\sqrt{1-R^2}} \exp\left(-\frac{u^2 + v^2 - 2Ruv}{2(1-R^2)}\right). \tag{8.3}$$

The average with respect to this probability distribution function corresponds to the average over the input distribution.

To define the generalization error, we introduce the *training energy* or *training cost* for p examples:

$$E = \sum_{\mu=1}^{p} V(\boldsymbol{J}, \sigma_\mu, \boldsymbol{\xi}_\mu), \tag{8.4}$$

where $\sigma_\mu = \text{sgn}(v_\mu)$ is the teacher output to the μth input vector $\boldsymbol{\xi}_\mu$. Note that the training energy depends on the teacher coupling $\boldsymbol{B}$ only through the teacher output σ_μ; the student does not know the detailed structure of the teacher ($\boldsymbol{B}$) but is given only the output σ_μ. The *generalization function* $E(\boldsymbol{J}, \sigma)$ is the average of $V(\boldsymbol{J}, \sigma, \boldsymbol{\xi})$ over the distribution of input $\boldsymbol{\xi}$. A common choice of training energy for the simple perceptron is the number of incorrect outputs

$$E = \sum_{\mu=1}^{p} \Theta(-\sigma_\mu u_\mu) \left(= \sum_{\mu=1}^{p} \Theta(-u_\mu v_\mu)\right), \tag{8.5}$$

where $\Theta(x)$ is the step function. The generalization function corresponding to the energy (8.5) is the probability that $\text{sgn}(u)$ is different from $\text{sgn}(v)$, which is evaluated by integrating $P(u,v)$ over the region of $uv < 0$:

$$E(R) \equiv \int du\, dv\, P(u,v)\Theta(-uv) = \frac{1}{\pi} \cos^{-1} R. \tag{8.6}$$

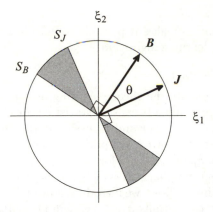

FIG. 8.2. Coupling vectors of the teacher and student, and the input vectors to
give incorrect student output (shaded).

The generalization error ϵ_g of a simple perceptron is the average of the generalization function over the distributions of teacher and student structures. The
generalization function $E(R)$ depends on the teacher and student structures only
through the overlap R. Since the overlap R is a self-averaging quantity for a large
system, the generalization function is effectively equal to the generalization error
in a simple perceptron. The overlap R will be calculated later as a function of the
number of examples divided by the system size $\alpha = p/N$. Thus the generalization
error is represented as a function of α through the generalization function:

$$\epsilon_g(\alpha) = E(R(\alpha)). \tag{8.7}$$

The generalization error as a function of the number of examples $\epsilon_g(\alpha)$ is called
the *learning curve*.

It is possible to understand (8.6) intuitively as follows. The teacher gives
output 1 for input vectors lying above the plane perpendicular to the coupling
vector $\boldsymbol{B}$, $S_B : \sum_j B_j \xi_j = 0$ (i.e. the output is 1 for $\sum_j B_j \xi_j > 0$); the teacher
output is -1 if the input is below the plane $\sum_j B_j \xi_j < 0$. Analogously, the student determines its output according to the position of the input vector relative
to the surface S_J perpendicular to the coupling vector $\boldsymbol{J}$. Hence the probability
of an incorrect output from the student (generalization error) is the ratio of the
subspace bounded by S_B and S_J to the whole space (Fig. 8.2). The generaliza-
tion error is therefore equal to $2\theta/2\pi$, where θ is the angle between $\boldsymbol{B}$ and $\boldsymbol{J}$. In
terms of the inner product R between $\boldsymbol{B}$ and $\boldsymbol{J}$, θ/π is expressed as (8.6).

8.2 Batch learning

Statistical-mechanical formulation is a very powerful tool to analyse batch learning. A complementary approach, PAC (probably almost correct) learning, gives
very general bounds on learning curves (Abu-Mostafa 1989; Hertz *et al.* 1991)

while statistical mechanics makes it possible to calculate the learning curve explicitly for a specified learning algorithm.

8.2.1 Bayesian formulation

It is instructive to formulate the theory learning as a problem of statistical inference using the Bayes formula because a statistical-mechanical point of view is then naturally introduced (Opper and Haussler 1991; Opper and Kinzel 1995). We would like to infer the correct couplings of the teacher $\boldsymbol{B}$, given p examples of teacher output σ_μ ($= \text{sgn}(v_\mu)$) for input $\boldsymbol{\xi}_\mu$ ($\mu = 1, \ldots, p$). The result of inference is the student couplings $\boldsymbol{J}$, which can be derived using the posterior $P(\boldsymbol{J}|\sigma_1, \ldots, \sigma_p)$. For this purpose, we introduce a prior $P(\boldsymbol{J})$, and the conditional probability to produce the outputs $\sigma_1, \ldots, \sigma_p$ will be written as $P(\sigma_1, \ldots, \sigma_p|\boldsymbol{J})$. Then the Bayes formula gives

$$P(\boldsymbol{J}|\sigma_1, \ldots, \sigma_p) = \frac{P(\sigma_1, \ldots, \sigma_p|\boldsymbol{J})P(\boldsymbol{J})}{Z}, \tag{8.8}$$

where Z is the normalization. Although it is possible to proceed without explicitly specifying the form of the conditional probability $P(\sigma_1, \ldots, \sigma_p|\boldsymbol{J})$, it is instructive to illustrate the idea for the case of *output noise* defined as

$$\sigma = \begin{cases} \text{sgn}(v) & \text{prob. } (1 + \text{e}^{-\beta})^{-1} \\ -\text{sgn}(v) & \text{prob. } (1 + \text{e}^{\beta})^{-1}. \end{cases} \tag{8.9}$$

The zero-temperature limit ($\beta = 1/T \to \infty$) has no noise in the output ($\sigma = \text{sgn}(v)$) whereas the high-temperature extreme ($\beta \to 0$) yields perfectly random output. The conditional probability of output σ_μ, given the teacher connections $\boldsymbol{B}$ for a fixed single input $\boldsymbol{\xi}_\mu$, is thus

$$P(\sigma_\mu|\boldsymbol{B}) = \frac{\Theta(\sigma_\mu v_\mu)}{1 + \text{e}^{-\beta}} + \frac{\Theta(-\sigma_\mu v_\mu)}{1 + \text{e}^{\beta}} = \frac{\exp\left\{-\beta\Theta(-\sigma_\mu v_\mu)\right\}}{1 + \text{e}^{-\beta}}. \tag{8.10}$$

The full conditional probability for p independent examples is therefore

$$P(\sigma_1, \ldots, \sigma_p|\boldsymbol{B}) = \prod_{\mu=1}^{p} P(\sigma_\mu|\boldsymbol{B}) = (1 + \text{e}^{-\beta})^{-p} \exp\left\{-\beta \sum_{\mu=1}^{p} \Theta(-\sigma_\mu v_\mu)\right\}. \tag{8.11}$$

The explicit expression of the posterior is, from (8.8),

$$P(\boldsymbol{J}|\sigma_1, \ldots, \sigma_p) = \frac{\exp\left\{-\beta \sum_{\mu=1}^{p} \Theta(-\sigma_\mu u_\mu)\right\} P(\boldsymbol{J})}{Z}, \tag{8.12}$$

which has been derived for a fixed set of inputs $\boldsymbol{\xi}_1, \ldots, \boldsymbol{\xi}_p$. Note that v_μ ($= \boldsymbol{B} \cdot \boldsymbol{\xi}_\mu$) in the argument of the step function Θ in (8.11) has been replaced by u_μ ($= \boldsymbol{J} \cdot \boldsymbol{\xi}_\mu$) in (8.12) because in the latter we consider $P(\sigma_1, \ldots, \sigma_p|\boldsymbol{J})$ instead of $P(\sigma_1, \ldots, \sigma_p | \boldsymbol{B})$ in the former. We shall assume a uniform prior on the sphere, $P(\boldsymbol{J}) \propto \delta(\boldsymbol{J}^2 - N)$ and $P(\boldsymbol{B}) \propto \delta(\boldsymbol{B}^2 - N)$ as mentioned in §8.1.2. The expression (8.12) has the form of a Boltzmann factor with the energy

$$E = \sum_{\mu=1}^{p} \Theta(-\sigma_\mu u_\mu). \tag{8.13}$$

This is the same energy as we introduced before, in (8.5). For large β the system favours students with smaller training errors. Equation (8.12) motivates us to apply statistical mechanics to the theory of learning.

A few comments are in order. First, *input noise*, in contrast with output noise, stands for random deviations of $\boldsymbol{\xi}$ to the student from those to the teacher as well as deviations of the teacher couplings from the true value. Input noise causes various non-trivial complications, which we do not discuss here (Györgyi and Tishby 1990). Second, *training error* is defined as the average of the training energy per example E/p. The average is taken first over the posterior (8.12) and then over the configuration of input $\boldsymbol{\xi}$. Training error measures the average error for examples already given to the student whereas generalization error is the expected error for a new example. We focus our attention on generalization error.

8.2.2 Learning algorithms

It is instructive to list here several popular learning algorithms and their properties. The *Bayesian algorithm* (Opper and Haussler 1991) to predict the output σ for a new input $\boldsymbol{\xi}$ compares the probability of $\sigma = 1$ with that of $\sigma = -1$ and chooses the larger one, in analogy with (5.15) for error-correcting codes,

$$\sigma = \operatorname{sgn}(V^+ - V^-), \tag{8.14}$$

where $V^\pm$ is the probability (or the volume of the relevant phase space) of the output 1 or -1, respectively,

$$V^\pm = \int d\boldsymbol{J} \, \Theta(\pm \boldsymbol{J} \cdot \boldsymbol{\xi}) P(\boldsymbol{J} | \sigma_1, \ldots, \sigma_p). \tag{8.15}$$

This is the best possible (Bayes-optimal) strategy to minimize the generalization error, analogously to the case of error-correcting codes. In fact the generalization error in the limit of large p and N with $\alpha = p/N$ fixed has been evaluated for the Bayesian algorithm as (Opper and Haussler 1991)

$$\epsilon_{\mathrm{g}} \approx \frac{0.44}{\alpha} \tag{8.16}$$

for sufficiently large α, which is the smallest among all learning algorithms (some of which will be explained below). A drawback is that a single student is unable to follow the Bayesian algorithm because one has to explore the whole coupling space to evaluate $V^{\pm}$, which is impossible for a single student with a single set of couplings J. Several methods to circumvent this difficulty have been proposed, one of which is to form a layered network with a number of simple perceptrons in the intermediate layer. All of these perceptrons have couplings generated by the posterior (8.12) and receive a common input. The number of perceptrons in the layer with $\sigma = 1$ is proportional to V^{+} and that with $\sigma = -1$ to V^{-}. The final output is decided by the majority rule of the outputs of these perceptrons (*committee machine*).

An alternative learning strategy is the *Gibbs algorithm* in which one chooses a single coupling vector J following the posterior (8.12). This gives a typical realization of J and can be implemented by a single student perceptron. The performance is slightly worse than the Bayes result. The asymptotic form of the generalization error is, for $T \to 0$,

$$\epsilon_{\mathrm{g}} \approx \frac{0.625}{\alpha} \tag{8.17}$$

in the regime $\alpha = p/N \to \infty$ (Györgyi and Tishby 1990). We shall present the derivation of this result (8.17) later since the calculations are similar to those in the previous chapter and include typical techniques used in many other cases. The noise-free ($T = 0$) Gibbs learning is sometimes called the *minimum error algorithm*.

In the posterior (8.12) we may consider more general forms of energy than the simple error-counting function (8.13). Let us assume that the energy is additive in the number of examples and write

$$E = \sum_{\mu=1}^{p} V(\sigma_{\mu} u_{\mu}). \tag{8.18}$$

The *Hebb algorithm* has $V(x) = -x$; the minimum of this energy with respect to J has $J \propto \sum_{\mu} \sigma_{\mu} \boldsymbol{\xi}_{\mu}$, reminiscent of the Hebb rule (7.4) for associative memory, as can be verified by minimization of E under the constraint $J^2 = N$ using a Lagrange multiplier. This Hebb algorithm is simple but not necessarily efficient in the sense that the generalization error falls rather slowly (Vallet 1989):

$$\epsilon_{\mathrm{g}} \approx \frac{0.40}{\sqrt{\alpha}}. \tag{8.19}$$

The parameter $x = \sigma J \cdot \boldsymbol{\xi} = \sigma u$ may be interpreted as the level of confidence of the student about its output: positive x means a correct answer, and as x increases, small fluctuations in J or $\boldsymbol{\xi}$ become less likely to lead to a wrong answer. In the space of couplings, the subspace with $x > 0$ for all examples is called the *version space*. It seems reasonable to choose a coupling vector J that gives the largest possible x in the version space. More precisely, we first choose those J that have the stability parameter x larger than a given threshold κ.

Then we increase the value of κ until such $\boldsymbol{J}$ cease to exist. The vector $\boldsymbol{J}$ chosen at this border of existence (vanishing version space) is the one with the largest stability. This method is called the *maximum stability algorithm*. The maximum stability algorithm can be formulated in terms of the energy

$$V(x) = \begin{cases} \infty & x < \kappa \\ 0 & x \geq \kappa. \end{cases} \tag{8.20}$$

One increases κ until the volume of subspace under the energy (8.20) vanishes in the $\boldsymbol{J}$-space. The asymptotic form of the generalization error for the maximum stability algorithm has been evaluated as (Opper *et al.* 1990)

$$\epsilon_{\mathrm{g}} \approx \frac{0.50}{\alpha}, \tag{8.21}$$

which lies between the Bayes result (8.19) and the minimum error (Gibbs) algorithm (8.17). The inverse-α law of the generalization error has also been derived from arguments based on mathematical statistics (Haussler *et al.* 1991).

8.2.3 *High-temperature and annealed approximations*

The problem simplifies significantly in the high-temperature limit and yet non-trivial behaviour results. In particular, we show that the generalization error can be calculated relatively easily using the example of the Gibbs algorithm (Seung *et al.* 1992).

The partition function in the high-temperature limit ($\beta \to 0$) is

$$Z = \int \mathrm{d}\boldsymbol{J}\, P(\boldsymbol{J}) \left(1 - \beta \sum_\mu \Theta(-\sigma_\mu u_\mu) \right) + \mathcal{O}(\beta^2) \equiv \int \mathrm{d}R\, Z(R) + \mathcal{O}(\beta^2) \tag{8.22}$$

according to (8.12). The partition function with R specified is rewritten as

$$Z(R) = V_R \left(1 - \frac{\beta \int_R \mathrm{d}\boldsymbol{J}\, P(\boldsymbol{J}) \sum_\mu \Theta(-\sigma_\mu u_\mu)}{V_R} \right), \quad V_R = \int_R \mathrm{d}\boldsymbol{J}\, P(\boldsymbol{J}). \tag{8.23}$$

The integral with suffix R runs over the subspace with a fixed value of the overlap R. The integral over R in (8.22) is dominated by the value of R that maximizes $Z(R)$ or minimizes the corresponding free energy. The configurational average of the free energy for fixed R is

$$\beta F(R) = -[\log Z(R)] = -\log V_R + \beta p E(R), \tag{8.24}$$

where the relation (8.6) has been used. The volume V_R is evaluated using the angle θ between $\boldsymbol{B}$ and $\boldsymbol{J}$ as

$$V_R \propto \sin^{N-2}\theta = (1 - R^2)^{(N-2)/2}. \tag{8.25}$$

The reason is that the student couplings $\boldsymbol{J}$ with a fixed angle θ to $\boldsymbol{B}$ lie on a hypersphere of radius $N\sin\theta$ and dimensionality $N - 2$ in the coupling space.

Suppose that Fig. 8.2 is for $N = 3$; that is, J and B both lie on a sphere of radius $N (= 3)$. Then the J with a fixed angle θ to B all lie on a circle (dimensionality $N - 2 = 1$) with radius $N \sin \theta$ on the surface of the sphere.

We have to minimize the free energy $F(R)$ in the large-N limit. From the explicit expression of the free energy per degree of freedom

$$\beta f(R) = \frac{\tilde{\alpha}}{\pi} \cos^{-1} R - \frac{1}{2} \log(1 - R^2), \qquad (8.26)$$

where $\tilde{\alpha} = \beta \alpha$, we find

$$\frac{R}{\sqrt{1 - R^2}} = \frac{\tilde{\alpha}}{\pi}. \qquad (8.27)$$

The generalization error $\epsilon_{\mathrm{g}} = E(R)$ is thus

$$\epsilon_{\mathrm{g}} = \frac{1}{\pi} \cos^{-1} \frac{\tilde{\alpha}}{\sqrt{\pi^2 + \tilde{\alpha}^2}} \approx \frac{1}{\tilde{\alpha}}, \qquad (8.28)$$

where the last expression is valid in the large-$\tilde{\alpha}$ limit. The combination $\tilde{\alpha} = \beta \alpha$ is the natural parameter in the high-temperature limit (small β) because one should present more and more examples (large α) to overwhelm the effects of noise. The result (8.28) shows that the inverse-α law holds for the learning curve in the high-temperature limit.

Annealed approximation is another useful technique to simplify calculations (Seung *et al.* 1992). One performs the configurational average of Z, instead of $\log Z$, in the annealed approximation. This approximation can be implemented relatively easily in many problems for which the full quenched analysis is difficult. One should, however, appreciate that this is an uncontrolled approximation and the result is sometimes qualitatively unreliable.

8.2.4 *Gibbs algorithm*

We now give an example of derivation of the learning curve for the case of the noise-free Gibbs algorithm (minimum error algorithm) in the limit of large N and p with the ratio $\alpha = p/N$ fixed (Györgyi and Tishby 1990). The techniques used here are analogous to those in the previous chapter and are useful in many other cases of batch learning.

According to (8.6), the generalization error $\epsilon_{\mathrm{g}}(\alpha)$ is determined by the relation between R and α. It is useful for this purpose to calculate the volume of the space of student couplings J under the posterior (8.12):

$$V = \frac{1}{Z} \int \mathrm{d}J \, \delta(\sum_j J_j^2 - N) \, \mathrm{e}^{-\beta E} \qquad (8.29)$$

with $E = \sum_\mu \Theta(-\sigma_\mu u_\mu)$. In the noise-free Gibbs algorithm the student chooses J with a uniform probability in the version space satisfying $E = 0$. Correspondingly, the limit $\beta \to \infty$ will be taken afterwards.

We investigate the typical macroscopic behaviour of the system by taking the configurational average over the input vectors, which are regarded as quenched random variables. It is particularly instructive to derive the average of $\log V$, corresponding to the entropy (logarithm of the volume of the relevant space). We use the replica method and express the average of V^n in terms of the order parameter $q_{\alpha\beta}$. The normalization Z is omitted since it does not play a positive role. The following equation results if we note that Θ is either 0 or 1:

$$[V^n] = \int \prod_\alpha dR_\alpha \int \prod_{(\alpha\beta)} dq_{\alpha\beta} \int \prod_{\alpha,j} dJ_j^\alpha$$

$$\cdot \prod_\alpha \delta(\sum_j (J_j^\alpha)^2 - N) \prod_\alpha \delta(\sum_j B_j J_j^\alpha - NR^\alpha) \prod_{(\alpha\beta)} \delta(\sum_j J_j^\alpha J_j^\beta - Nq_{\alpha\beta})$$

$$\cdot \left[\prod_{\alpha,\mu} \{ e^{-\beta} + (1 - e^{-\beta})\Theta(u_\mu^\alpha v_\mu) \} \right]. \tag{8.30}$$

8.2.5   *Replica calculations*

Evaluation of (8.30) proceeds by separating it into two parts: the first half independent of the input (the part including the delta functions), and the second half including the effects of input signals (in the brackets $[\cdots]$). The former is denoted by I_1^N, and we write the three delta functions in terms of Fourier transforms to obtain, under the RS ansatz,

$$I_1^N = \int \prod_{\alpha,j} dJ_j^\alpha \exp \left[i \left\{ E \sum_{\alpha,j} (J_j^\alpha)^2 + F \sum_{(\alpha\beta),j} J_j^\alpha J_j^\beta + G \sum_{\alpha,j} J_j^\alpha B_j \right. \right.$$

$$\left. \left. - N \left(nE + \frac{n(n-1)}{2} qF + nRG \right) \right\} \right], \tag{8.31}$$

where the integrals over the parameters R, q, E, F, and G have been omitted in anticipation of the use of the method of steepest descent.

It is possible to carry out the above integral for each j independently. The difference between this and (7.83) is only in G, and the basic idea of calculation is the same. One diagonalizes the quadratic form and performs Gaussian integration independently for each eigenmode.[19] It is to be noted that, for a given j, the term involving G has the form $GB_j \sum_\alpha J_j^\alpha$ (uniform sum over α) and correspondingly the uniform mode in the diagonalized form (with the eigenvalue $E + (n-1)F/2$) has a linear term. In other words, for a given j, the contribution of the uniform mode $u (= \sum_\alpha J_j^\alpha)$ to the quadratic form is

$$i \left(E + \frac{n-1}{2} F \right) u^2 + iGB_j u. \tag{8.32}$$

---

[19]One may instead decompose the term involving the double sum $(\alpha\beta)$ to a linear form by Gaussian integration.

By integrating over u and summing the result over j, we obtain $-iG^2N/[4\{E + (n-1)F/2\}]$ in the exponent. Thus in the limit $n \to 0$

$$g_1(E, F, G) \equiv \frac{1}{nN} \log I_1^N = -\frac{1}{2} \log \left(E - \frac{F}{2} \right)$$

$$-\frac{F}{4E - 2F} - \frac{iG^2}{4E - 2F} - iE - iGR + i\frac{qF}{2}. \tag{8.33}$$

The extremization condition of $g_1(E, F, G)$ with respect to E, F, and G yields

$$2E - F = \frac{i}{1 - q}, \quad \frac{F + iG^2}{-2E + F} = \frac{q}{1 - q}, \quad iG = \frac{R}{1 - q}. \tag{8.34}$$

We can now eliminate E, F, and G from (8.33) to obtain

$$g_1 = \frac{1}{2} \log(1 - q) + \frac{1 - R^2}{2(1 - q)}. \tag{8.35}$$

Next, the second half I_2^N of (8.30) is decomposed into a product over μ, and its single factor is

$$(I_2^N)^{1/p} = \left[2\Theta(v) \prod_\alpha \{e^{-\beta} + (1 - e^{-\beta})\Theta(u^\alpha)\} \right]. \tag{8.36}$$

Here, since the contribution from the region $u > 0, v > 0$ is the same as that from $u < 0, v < 0$, we have written only the former and have multiplied the whole expression by 2. It is convenient to note here that u and v are Gaussian variables with the following correlations:

$$[u^\alpha u^\beta] = (1 - q)\delta_{\alpha,\beta} + q, \quad [vu^\alpha] = R, \quad [v^2] = 1. \tag{8.37}$$

These quantities can be expressed in terms of $n + 2$ uncorrelated Gaussian variables t and z^α ($\alpha = 0, \ldots, n$), both of which have vanishing mean and variance unity:

$$v = \sqrt{1 - \frac{R^2}{q}} \, z^0 + \frac{R}{\sqrt{q}} t, \quad u^\alpha = \sqrt{1 - q} \, z^\alpha + \sqrt{q} t \quad (\alpha = 1, \ldots, n). \tag{8.38}$$

Equation (8.36) is rewritten using these variables as

$$(I_2^N)^{1/p} = 2 \int Dt \int Dz^0 \Theta \left(\sqrt{1 - \frac{R^2}{q}} \, z^0 + \frac{R}{\sqrt{q}} t \right)$$

$$\cdot \prod_{\alpha=1}^n \int Dz^\alpha \{e^{-\beta} + (1 - e^{-\beta})\Theta(\sqrt{1 - q} \, z^\alpha + \sqrt{q} t)\}$$

$$= 2 \int \mathrm{D}t \int_{-Rt/\sqrt{q-R^2}}^{\infty} \mathrm{D}z^0$$

$$\cdot \left\{ \int \mathrm{D}z \{ e^{-\beta} + (1 - e^{-\beta}) \Theta(\sqrt{1-q}\, z + \sqrt{q}t) \} \right\}^n . \qquad (8.39)$$

We collect the two factors (8.35) and (8.39) and take the limit $n \to 0$ to find

$$f \equiv \lim_{n \to 0} \frac{1}{nN} \log[V^n] = 2\alpha \int \mathrm{D}t \int_{-Rt/\sqrt{q-R^2}}^{\infty} \mathrm{D}z^0 \log \left\{ e^{-\beta} \right.$$

$$\left. + (1 - e^{-\beta}) \int_{-\sqrt{q/(1-q)}\, t}^{\infty} \mathrm{D}z \right\} + \frac{1}{2} \log(1 - q) + \frac{1 - R^2}{2(1 - q)}. \qquad (8.40)$$

The extremization condition of (8.40) determines R and q.

8.2.6 Generalization error at $T = 0$

It is possible to carry out the calculations for finite temperatures (Györgyi and Tishby 1990). However, formulae simplify significantly for the noise-free case of $T = 0$, which also serves as a prototype of the statistical-mechanical evaluation of the learning curve. We therefore continue the argument by restricting ourselves to $T = 0$. Let us then take the limit $\beta \to \infty$ in (8.40) and change the variables as

$$u = \frac{R}{\sqrt{q}} z^0 - \sqrt{\frac{q - R^2}{q}}\, t, \quad v = \sqrt{\frac{q - R^2}{q}}\, z^0 + \frac{R}{\sqrt{q}} t \qquad (8.41)$$

to obtain

$$f = 2\alpha \int_0^{\infty} \mathrm{D}v \int_{-\infty}^{\infty} \mathrm{D}u \log \int_w^{\infty} \mathrm{D}z + \frac{1}{2} \log(1 - q) + \frac{1 - R^2}{2(1 - q)} \qquad (8.42)$$

$$w = \frac{\sqrt{q - R^2}\, u - Rv}{\sqrt{1 - q}}. \qquad (8.43)$$

The extremization condition of (8.42) with respect to R and q leads to the following equations:

$$\alpha \sqrt{\frac{2}{\pi}} \int_0^{\infty} \mathrm{D}v \int_{-\infty}^{\infty} \mathrm{D}u \, \frac{u e^{-w^2/2}}{\int_w^{\infty} \mathrm{D}z} = \frac{q - 2R^2}{1 - 2q} \sqrt{\frac{q - R^2}{1 - q}} \qquad (8.44)$$

$$\alpha \sqrt{\frac{2}{\pi}} \int_0^{\infty} \mathrm{D}v \int_{-\infty}^{\infty} \mathrm{D}u \, \frac{v e^{-w^2/2}}{\int_w^{\infty} \mathrm{D}z} = \frac{1 - 3q + 2R^2}{(1 - 2q)\sqrt{1 - q}}. \qquad (8.45)$$

The solution of these equations satisfies the relation $q = R$. Indeed, by setting $q = R$ and rewriting the above two equations, we obtain the same formula

$$\frac{\alpha}{\pi} \int_{-\infty}^{\infty} \mathrm{D}x \, \frac{e^{-qx^2/2}}{\int_{\sqrt{q}x}^{\infty} \mathrm{D}z} = \frac{q}{\sqrt{1 - q}}. \qquad (8.46)$$

We have changed the variables $u = x + \sqrt{q/(1-q)}\, y, v = y$ in deriving this equation. The relation $q = R$ implies that the student and teacher are in a certain sense equivalent because of the definitions $q_{\alpha\beta} = N^{-1} \sum J_j^\alpha J_j^\beta$ and $R = N^{-1} \sum B_j J_j^\alpha$.

The behaviour of the learning curve in the limit of a large number of examples ($\alpha \to \infty$) can be derived by setting $q = 1 - \epsilon$ ($|\epsilon| \ll 1$) since $\boldsymbol{J}$ is expected to be close to $\boldsymbol{B}$ in this limit. Then (8.46) reduces to

$$\epsilon \approx \frac{\pi^2}{c^2 \alpha^2}, \quad c = \int_{-\infty}^{\infty} \mathrm{D}x \frac{e^{-x^2/2}}{\int_x^\infty \mathrm{D}z}. \tag{8.47}$$

Substitution of this into (8.6) gives the asymptotic form of the learning curve as

$$\epsilon_{\mathrm{g}} \approx \frac{\sqrt{2}}{c\alpha} = \frac{0.625}{\alpha}, \tag{8.48}$$

as already mentioned in (8.17). This formula shows that the generalization error of a simple perceptron decreases in inverse proportion to the number of examples in batch learning. It is also known that the present RS solution is stable (Györgyi and Tishby 1990).

8.2.7 *Noise and unlearnable rules*

We have discussed the case where the teacher and student are both simple perceptrons with continuous weights and their binary outputs coincide if the student coupling agrees with the teacher coupling. There are many other possible scenarios of learning. For example, the output of the teacher may follow a stochastic process (output noise) as mentioned before, or the input signal may include noise (input noise). In both of these cases, the student cannot reproduce the teacher output perfectly even when the couplings agree. In the simple case we have been treating, output noise deteriorates the generalization error just by a factor mildly dependent on the noise strength (temperature) (Opper and Haussler 1991; Opper and Kinzel 1995). Input noise causes more complications like RSB (Györgyi and Tishby 1990). More complex structures of the unit processing elements than a simple perceptron lead to such effects even for output noise (Watkin and Rau 1993).

Another important possibility we have not discussed so far is that the structure of the student is different from that of the teacher, so that the student cannot reproduce the teacher output in principle (*unlearnable* or *unrealizable rules*). If the teacher is composed of layers of perceptrons for instance, the teacher may be able to perform linearly non-separable rules. A simple perceptron as the student cannot follow such teacher outputs faithfully. Or, if the threshold θ appearing in the output $\mathrm{sgn}(u - \theta)$ is different between the teacher and student, the student cannot follow the teacher output even if $\boldsymbol{J}$ coincides with $\boldsymbol{B}$. A more naïve instance is the weight mismatch where the components of the student vector $\boldsymbol{J}$ may take only discrete values whereas the teacher vector $\boldsymbol{B}$ is continuous. These and more examples of unlearnable rules have been investigated extensively (Seung *et al.* 1992; Watkin and Rau 1992, 1993; Domany *et al.* 1995; Wong *et al.* 1997).

A general observation is that RSB and spin glass states emerge in certain regions of the phase diagram in these unlearnable cases. Non-monotonic behaviour of the generalization error as a function of temperature is also an interesting phenomenon in some of these instances.

8.3   On-line learning

The next topic is on-line learning in which the student changes its couplings immediately after the teacher output is given for an input. In practical applications, on-line learning is often more important than batch learning because the former requires less memory and time to implement and, furthermore, adjustment to changing environments is easier. Similar to the batch case, we mainly investigate the learning curve for simple perceptrons under various learning algorithms. The formula for the generalization error (8.6) remains valid for on-line learning as well.

8.3.1   *Learning algorithms*

Suppose that both student and teacher are simple perceptrons, and the student couplings change according to the perceptron learning algorithm (7.72) as soon as an input is given. Expressing the student coupling vector after m steps of learning by $\boldsymbol{J}^m$, we can write the *on-line perceptron learning algorithm* as

$$\boldsymbol{J}^{m+1} = \boldsymbol{J}^m + \Theta(-\text{sgn}(u)\text{sgn}(v))\,\text{sgn}(v)\boldsymbol{x} = \boldsymbol{J}^m + \Theta(-uv)\,\text{sgn}(v)\boldsymbol{x}, \qquad (8.49)$$

where $u = \sqrt{N}\boldsymbol{J}^m \cdot \boldsymbol{x}/|\boldsymbol{J}^m|$ and $v = \sqrt{N}\boldsymbol{B} \cdot \boldsymbol{x}/|\boldsymbol{B}|$ are input signals to the student and teacher, respectively, and $\boldsymbol{x}$ is the normalized input vector, $|\boldsymbol{x}| = 1$. The σ^μ in (7.72) corresponds to $\text{sgn}(v)$ and $\eta\boldsymbol{\xi}$ to $\boldsymbol{x}$. We normalize $\boldsymbol{x}$ to unity since, in contrast to §7.6.2, the dynamics of learning is investigated here in the limit of large N and we should take care that various quantities do not diverge. The components of the input vector $\boldsymbol{x}$ are assumed to have no correlations with each other, and independent $\boldsymbol{x}$ is drawn at each learning step; in other words, infinitely many independent examples are available. This assumption may not be practically acceptable, especially when the number of examples is limited. The effects of removing this condition have been discussed as the problem of the *restricted training set* (Sollich and Barber 1997; Barber and Sollich 1998; Heskes and Wiegerinck 1998; Coolen and Saad 2000).

Other typical learning algorithms include the *Hebb algorithm*

$$\boldsymbol{J}^{m+1} = \boldsymbol{J}^m + \text{sgn}(v)\boldsymbol{x} \qquad (8.50)$$

and the *Adatron algorithm*

$$\boldsymbol{J}^{m+1} = \boldsymbol{J}^m - u\Theta(-uv)\boldsymbol{x}. \qquad (8.51)$$

The Hebb algorithm changes the student couplings by $\text{sgn}(v)\boldsymbol{x}$ irrespective of the student output (correct or not) so that the inner product $\boldsymbol{J} \cdot \boldsymbol{x}$ tends to yield the correct output $\text{sgn}(v)$ for the next input. The Adatron algorithm is a little similar to the perceptron algorithm (8.49), the difference being that the amount of correction for incorrect output is proportional to u.

8.3.2 *Dynamics of learning*

To develop a general argument, we write the learning algorithm in the form

$$\boldsymbol{J}^{m+1} = \boldsymbol{J}^m + f(\mathrm{sgn}(v), u)\boldsymbol{x}. \tag{8.52}$$

The choice of arguments of the function f, $\mathrm{sgn}(v)$ and u, shows that the student knows only the teacher output $\mathrm{sgn}(v)$, not the input signal v to the teacher, the latter involving the information on the teacher couplings.

Some algorithms may be interpreted as the dynamics along the gradient of an energy (or cost function) with respect to $\boldsymbol{J}$. For example, the Hebb algorithm (8.50) can be expressed as

$$\boldsymbol{J}^{m+1} = \boldsymbol{J}^m - \eta_m \frac{\partial V(\sigma \boldsymbol{x} \cdot \boldsymbol{J}_m)}{\partial \boldsymbol{J}^m} \tag{8.53}$$

with $\sigma = \mathrm{sgn}(v), \eta_m = 1$, and $V(y) = -y$. If we identify V with the cost function analogously to batch learning (§8.2.1), (8.53) represents a process to reduce V by gradient descent with constant learning rate $\eta_m = 1$.[20] Such a viewpoint will be useful later in §8.3.5 where adaptive change of η_m is discussed. We use the expression (8.52) here, which does not assume the existence of energy, to develop a general argument for the simple perceptron.

The learning dynamics (8.52) determines each component J_i of the coupling vector $\boldsymbol{J}$ precisely. However, we are mainly interested in the macroscopic properties of the system in the limit $N \gg 1$, such as the coupling length $l^m = |\boldsymbol{J}^m|/\sqrt{N}$ and the overlap $R^m = (\boldsymbol{J}^m \cdot \boldsymbol{B})/|\boldsymbol{J}^m||\boldsymbol{B}|$ to represent the teacher–student proximity. We shall assume that $|\boldsymbol{B}|$ is normalized to $\sqrt{N}$. Our goal in the present section is to derive the equations to describe the time development of these macroscopic quantities R and l from the microscopic relation (8.52).

To this end, we first square both sides of (8.52) and write $u_m = (\boldsymbol{J}^m \cdot \boldsymbol{x})/l^m$ to obtain

$$N\{(l^{m+1})^2 - (l^m)^2\} = 2[fu_m]l^m + [f^2], \tag{8.54}$$

where fu and f^2 have been replaced with their averages by the distribution function (8.3) of u and v assuming the self-averaging property. By writing $l^{m+1} - l^m = \mathrm{d}l$ and $1/N = \mathrm{d}t$, we can reduce (8.54) to a differential equation

$$\frac{\mathrm{d}l}{\mathrm{d}t} = [fu] + \frac{[f^2]}{2l}. \tag{8.55}$$

Here t is the number of examples in units of N and can be regarded as the time of learning.

---

[20]The error *back propagation algorithm* for multilayer networks, used quite frequently in practical applications, is formulated as a gradient descent process with the training error as the energy or cost function. Note that this algorithm is essentially of batch type, not on-line (Hertz *et al.* 1991). A more sophisticated approach employs natural gradient in information geometry (Amari 1997).

The equation for R is derived by taking the inner product of both sides of (8.52) and $\boldsymbol{B}$ and using the relations $\boldsymbol{B} \cdot \boldsymbol{J}^m = N l^m R^m$ and $v = \boldsymbol{B} \cdot \boldsymbol{x}$. The result is

$$\frac{dR}{dt} = \frac{[fv] - [fu]R}{l} - \frac{R}{2l^2}[f^2].$$
(8.56)

The learning curve $\epsilon_g = E(R(t))$ is obtained by solving (8.55) and (8.56) and substituting the solution $R(t)$ into (8.6).

8.3.3 Generalization errors for specific algorithms

We next present explicit solutions for various learning algorithms. The first one is the perceptron algorithm

$$f = \Theta(-uv)\operatorname{sgn}(v).$$
(8.57)

The averages in the equations of learning (8.55) and (8.56) reduce to the following expressions after the integrals are evaluated:

$$[fu] = -[fv] = \int du\, dv\, P(u,v)\Theta(-uv)u\operatorname{sgn}(v) = \frac{R-1}{\sqrt{2\pi}}$$
(8.58)

$$[f^2] = \int du\, dv\, P(u,v)\Theta(-uv) = E(R) = \frac{1}{\pi}\cos^{-1}R.$$
(8.59)

Insertion of these equations into (8.55) and (8.56) will give the time dependence of the macroscopic parameters $R(t)$ and $l(t)$, from which we obtain the learning curve $\epsilon_g(t)$.

It is interesting to check the explicit asymptotic form of the learning curve as l grows and R approaches unity. Setting $R = 1 - \epsilon$ and $l = 1/\delta$ with $\epsilon, \delta \ll 1$, we have from (8.55) and (8.56)

$$\frac{d\delta}{dt} = -\frac{\sqrt{2\epsilon}}{2\pi}\delta^3 + \frac{\epsilon\delta^2}{\sqrt{2\pi}}, \quad \frac{d\epsilon}{dt} = \frac{\sqrt{2\epsilon}}{2\pi}\delta^2 - \sqrt{\frac{2}{\pi}}\epsilon\delta,$$
(8.60)

where use has been made of (8.58) and (8.59). The solution is

$$\epsilon = \left(\frac{1}{3\sqrt{2}}\right)^{2/3} t^{-2/3}, \quad \delta = \frac{2\sqrt{\pi}}{(3\sqrt{2})^{1/3}} t^{-1/3}.$$
(8.61)

The final asymptotic form of the generalization error is therefore derived as (Barkai *et al.* 1995)

$$\epsilon_g = E(R) \approx \frac{\sqrt{2}}{\pi(3\sqrt{2})^{1/3}} t^{-1/3} = 0.28\, t^{-1/3}.$$
(8.62)

Comparison of this result with the learning curve of batch learning (8.48) for the Gibbs algorithm, $\epsilon_g \propto \alpha^{-1}$, reveals that the latter converges much more rapidly to the ideal state $\epsilon_g = 0$ if the numbers of examples in both cases are assumed to be equal ($\alpha = t$). It should, however, be noted that the cost (time and memory) to learn the given number of examples is much larger in the batch case than in on-line learning.

A similar analysis applies to the Hebb algorithm

$$f(\text{sgn}(v), u) = \text{sgn}(v).\tag{8.63}$$

The averages in (8.55) and (8.56) are then

$$[fu] = \frac{2R}{\sqrt{2\pi}}, \quad [f^2] = 1, \quad [fv] = \sqrt{\frac{2}{\pi}}.\tag{8.64}$$

The asymptotic solution of the dynamical equations is derived by setting $R = 1 - \epsilon$ and $l = 1/\delta$ with $\epsilon, \delta \ll 1$:

$$\frac{d\delta}{dt} = -\sqrt{\frac{2}{\pi}}\delta^2, \quad \frac{d\epsilon}{dt} = \frac{\delta^2}{2} - \frac{4}{\sqrt{2\pi}}\epsilon\delta.\tag{8.65}$$

The solution is

$$\epsilon = \frac{\pi}{4t}, \quad \delta = \sqrt{\frac{\pi}{2}}\frac{1}{t},\tag{8.66}$$

and the corresponding generalization error is (Biehl *et al.* 1995)

$$\epsilon_{\text{g}} \approx \frac{1}{\sqrt{2\pi}}t^{-1/2} = 0.40\,t^{-1/2}.\tag{8.67}$$

It is seen that the learning curve due to the Hebb algorithm, $\epsilon_{\text{g}} \propto t^{-1/2}$, approaches zero faster than in the case of the perceptron algorithm, $\epsilon_{\text{g}} \propto t^{-1/3}$. It is also noticed that the learning curve (8.67) shows a relaxation of comparable speed to the corresponding batch result (8.19) if we identify t with α.

The Adatron algorithm of on-line learning has

$$f(\text{sgn}(v), u) = -u\Theta(-uv),\tag{8.68}$$

and integrals in the dynamical equations (8.55) and (8.56) are

$$[fu] = -\sqrt{2}\int_0^\infty Du\, u^2\text{Erfc}\left(\frac{Ru}{\sqrt{2(1-R^2)}}\right), \quad [fv] = \frac{(1-R)^{3/2}}{\pi} + R[fu].\tag{8.69}$$

With (8.68), $[f^2]$ is equal to $-[fu]$. The asymptotic form for $\epsilon = 1 - R \ll 1$ is, writing $c = 8/(3\sqrt{2\pi})$,

$$[fu] \approx -\frac{4(2\epsilon)^{3/2}}{\pi}\int_0^\infty y^2\text{d}y\,\text{Erfc}(y) = -c\epsilon^{3/2}, \quad [fv] = \left(-c + \frac{2\sqrt{2}}{\pi}\right)\epsilon^{3/2}.\tag{8.70}$$

To solve the dynamical equations, we note that (8.55) has the explicit form

$$\frac{dl}{dt} = \left(-1 + \frac{1}{2l}\right) c\epsilon^{3/2}, \tag{8.71}$$

which implies that l does not change if $l = 1/2$. It is thus convenient to restrict ourselves to $l = 1/2$, and then we find that (8.56) is

$$\frac{d\epsilon}{dt} = 2\left(c - \frac{2\sqrt{2}}{\pi}\right)\epsilon^{3/2}. \tag{8.72}$$

This equation is solved as $\epsilon = 4/(kt)^2$, where $k = 4\sqrt{2}/\pi - 2c$. The generalization error is therefore (Biehl and Riegler 1994)

$$\epsilon_g \approx \frac{2\sqrt{2}}{\pi k} \cdot \frac{1}{t} = \frac{3}{2t}. \tag{8.73}$$

It is remarkable that *on-line* Adatron learning leads to a very fast convergence comparable to the batch case discussed in §8.2.2 after identification of α and t.

8.3.4   *Optimization of learning rate*

Let us go back to the dynamical equation of learning (8.52) with the learning rate η written explicitly:

$$J^{m+1} = J^m + \eta_m f(\text{sgn}(v), u)x. \tag{8.74}$$

The constant learning rate $\eta_m = 1$ as in (8.52) keeps each component of J fluctuating even after sufficient convergence to the desired result. Apparently, it seems desirable to change the coupling vector J rapidly (large η) at the initial state of learning and, in the later stage, adjust J carefully with small η. Such an adjustment of learning rate would lead to an acceleration of convergence. We discuss this topic in the present and next subsections. We first formulate the problem as an optimization of the learning rate, taking the case of the perceptron algorithm as an example (Inoue *et al.* 1997). Then a more general framework is developed in the next subsection without explicitly specifying the algorithm.

The discrete learning dynamics (8.74) is rewritten in terms of differential equations in the limit of large N as in (8.55) and (8.56) with f multiplied by η. For the perceptron algorithm with simple perceptrons as the teacher and student, the resulting equations are

$$\frac{dl}{dt} = \frac{\eta(R - 1)}{\sqrt{2\pi}} + \frac{\eta^2 \cos^{-1} R}{2\pi l} \tag{8.75}$$

$$\frac{dR}{dt} = -\frac{\eta(R^2 - 1)}{\sqrt{2\pi}l} - \frac{\eta^2 R \cos^{-1} R}{2\pi l^2}, \tag{8.76}$$

where we have used (8.58) and (8.59).

Since $\epsilon_g = E(R) = 0$ for $R = 1$, the best strategy to adjust η is to accelerate the increase of R towards $R = 1$ by maximizing the right hand side of (8.76)

with respect to η at each value of R. We thus differentiate the right hand side of
(8.76) by η and set the result to zero to obtain

$$\eta = -\frac{\sqrt{2\pi}\, l(R^2 - 1)}{2R \cos^{-1} R}. \tag{8.77}$$

Then the dynamical equations (8.75) and (8.76) simplify to

$$\frac{dl}{dt} = -\frac{l(R-1)^3(R+1)}{4R^2 \cos^{-1} R} \tag{8.78}$$

$$\frac{dR}{dt} = \frac{(R^2 - 1)^2}{4R \cos^{-1} R}. \tag{8.79}$$

Taking the ratio of both sides of these equations, we can derive the exact relation
between R and l as

$$l = \frac{cR}{(R+1)^2}. \tag{8.80}$$

The constant c is determined by the initial condition. This solution indicates
that l approaches a constant as $R \to 1$.

The asymptotic solution of $R(t)$ for $R = 1 - \epsilon$ ($|\epsilon| \ll 1$) can easily be derived
from (8.79) as $\epsilon \approx 8/t^2$. Therefore the generalization error is asymptotically

$$\epsilon_{\mathrm{g}} = \frac{\cos^{-1} R}{\pi} \approx \frac{4}{\pi t}, \tag{8.81}$$

which is much faster than the solution with constant learning rate (8.62). The
learning rate decreases asymptotically as $\eta \propto 1/t$ as we expected before.

A weakness of this method is that the learning rate depends explicitly on
R, as seen in (8.77), which is unavailable to the student. This difficulty can be
avoided by using the asymptotic form $\eta \propto 1/t$ for the whole period of learning although the optimization at intermediate values of t may not be achieved. Another
problem is that the simple optimization of the learning rate does not necessarily
lead to an improved convergence property in other learning algorithms such as
the Hebb algorithm. Optimization of the learning algorithm itself (to change the
functional form of f adaptively) is a powerful method to overcome this point
(Kinouchi and Caticha 1992).

8.3.5   *Adaptive learning rate for smooth cost function*

The idea of adaptive learning rate can be developed for a very general class of
learning algorithms including both learnable and unlearnable rules (Müller *et
al.* 1998). Let us assume that the cost function for a single input $V(\boldsymbol{x}, \boldsymbol{\sigma}; \boldsymbol{J})$
is differentiable by $\boldsymbol{J}$.[21] The output may be vector-valued $\boldsymbol{\sigma}$. The goal is to
minimize the total energy

$$E(\boldsymbol{J}) = [V(\boldsymbol{x}, \boldsymbol{\sigma}; \boldsymbol{J})], \tag{8.82}$$

---

[21]Note that this condition excludes the perceptron and Adatron algorithms.

which is identified with the generalization error ϵ_g. The energy is assumed to be differentiable around the optimal state $J = B$:

$$E(J) = E(B) + \frac{1}{2}{}^t(J - B)K(B)(J - B). \tag{8.83}$$

Here $K(B)$ is the value of the second-derivative matrix (Hessian) of the energy $E(J)$ at $J = B$. The on-line dynamics to be discussed here is specified by

$$J^{m+1} = J^m - \eta_m K^{-1}(J)\frac{\partial V(x, \sigma; J)}{\partial J} \tag{8.84}$$

$$\eta_{m+1} = \eta_m + a\{b(V(x, \sigma; J) - E_t) - \eta_m\}, \tag{8.85}$$

where $E_t = \sum_\mu V(x_\mu, \sigma_\mu; J)/p$ is the training error, and a and b are positive constants. Equation (8.84) is a gradient descent with the direction modified by $K^{-1}(J)$: the eigendirection of $K(J)$ with the smallest eigenvalue has the fastest rate of change. According to (8.85), η increases if the current error V exceeds the cumulative training error E_t. The final term with the negative sign in (8.85) has been added to suppress uncontrolled increase of the learning rate.

The corresponding differential form of the learning dynamics is

$$\frac{dJ}{dt} = -\eta K^{-1}(J)\left[\frac{\partial V}{\partial J}\right]$$
$$\frac{d\eta}{dt} = a\eta\{b([V] - E_t) - \eta\}. \tag{8.86}$$

We now expand V on the right hand sides of these equations around $J = B$:

$$\left[\frac{\partial V}{\partial J}\right] \approx K(B)(J - B)$$
$$[V] - E_t \approx E(B) - E_t + \frac{1}{2}{}^t(J - B)K(B)(J - B). \tag{8.87}$$

Then the dynamical equations reduce to

$$\frac{dJ}{dt} = -\eta(J - B)$$
$$\frac{d\eta}{dt} = a\eta\left\{\frac{b}{2}{}^t(J - B)K(B)(J - B) - \eta\right\}. \tag{8.88}$$

These equations can be rewritten in terms of the energy as

$$\frac{dE(J)}{dt} = -2\eta\{E(J) - E(B)\}$$
$$\frac{d\eta}{dt} = ab\eta\{E(J) - E(B)\} - a\eta^2. \tag{8.89}$$

The solution is easily found to be

$$\epsilon_g = E(J) = E(B) + \frac{1}{b}\left(\frac{1}{2} - \frac{1}{a}\right)\frac{1}{t} \tag{8.90}$$

$$\eta = \frac{1}{2t}.$$

(8.91)

Choosing $a > 2$, we have thus shown that the learning rate with the $1/t$ law (8.91) leads to rapid convergence (8.90) just as in the previous subsection but under a very general condition. It is possible to generalize the present framework to the cases without a cost function or without explicit knowledge of the Hessian K (Müller *et al.* 1998).

8.3.6   *Learning with query*

Inputs satisfying the condition $u = 0$ lie on the border of the student output $\mathrm{sgn}(u)$ (*decision boundary*). The student is not sure what output to produce for such inputs. It thus makes sense to teach the student the correct outputs for inputs satisfying $u = 0$ for efficient learning. This idea of restricting examples to a subspace is called *learning with query* (Kinzel and Ruján 1990). One uses the distribution function

$$P(u, v) = \frac{\delta(u)}{\sqrt{2\pi(1 - R^2)}} \exp\left(-\frac{u^2 + v^2 - 2Ruv}{2(1 - R^2)}\right)$$

(8.92)

instead of (8.3). This method works only for the Hebb algorithm among the three algorithms discussed so far because the perceptron algorithm has $f \propto \Theta(-uv)$ which is indefinite at $u = 0$, and the f for the Adatron algorithm is proportional to u and vanishes under the present condition $\delta(u)$.

   The dynamical equations (8.55) and (8.56) have the following forms for the Hebb algorithm with query:

$$\frac{dl}{dt} = \frac{1}{2l}$$

(8.93)

$$\frac{dR}{dt} = \frac{1}{l}\sqrt{\frac{2(1 - R^2)}{\pi}} - \frac{R}{2l^2}.$$

(8.94)

The first equation for l immediately gives $l = \sqrt{t}$, and the second has the asymptotic solution for small ϵ ($R = 1 - \epsilon$) as $\epsilon \approx \pi/16t$. The generalization error then behaves asymptotically as

$$\epsilon_{\mathrm{g}} \approx \frac{1}{2\sqrt{2\pi}\sqrt{t}}.$$

(8.95)

Comparison with the previous result (8.67) reveals that asking queries reduces the prefactor by a half in the generalization error.

   It is possible to improve the performance further by combining query and optimization of the learning rate. Straightforward application of the ideas of the present section and §8.3.4 leads to the optimized learning rate

$$\eta = \frac{1}{R}\sqrt{\frac{2(1 - R^2)}{\pi}}$$

(8.96)

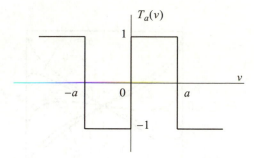

FIG. 8.3. Output $T_a(v)$ of the reversed-wedge-type perceptron

for the Hebb algorithm with query. The overlap is solved as $R = \sqrt{1 - c\mathrm{e}^{-2\alpha/\pi}}$ and the generalization error decays asymptotically as

$$\epsilon_{\mathrm{g}} \approx \frac{c}{\pi}\mathrm{e}^{-\alpha/\pi}, \qquad (8.97)$$

a very fast exponential convergence.

We have so far discussed learning with query based on a heuristic argument to restrict the training set to the decision boundary. It is possible to formulate this problem in a more systematic way using an appropriate cost function to be extremized (such as information gain); one can then construct the best possible algorithm to ask queries depending upon various factors. See Sollich (1994) and references cited therein for this and related topics.

8.3.7 *On-line learning of unlearnable rule*

It is relatively straightforward, compared to the batch case, to analyse on-line learning for unlearnable rules. Among various types of unlearnable rules (some of which were mentioned in §8.2.7), we discuss here the case of the reversed-wedge-type non-monotonic perceptron as the teacher (Inoue *et al.* 1997; Inoue and Nishimori 1997). The student is the usual simple perceptron. The input signal $\boldsymbol{\xi}$ is shared by the student and teacher. After going through the synaptic couplings, the input signal becomes u for the student and v for the teacher as defined in (8.1). The student output is $\mathrm{sgn}(u)$ and the teacher output is assumed to be $T_a(v) = \mathrm{sgn}\{v(a - v)(a + v)\}$, see Fig. 8.3.

The generalization error is obtained by integration of the distribution function of u and v, (8.3), over the region where the student output is different from that of the teacher:

$$\epsilon_{\mathrm{g}} = E(R) \equiv \int \mathrm{d}u\,\mathrm{d}v\,P(u,v)\Theta(-T_a(v)\mathrm{sgn}(u)) = 2\int_{-\infty}^{0} \mathrm{D}t\,\Omega(R,t), \qquad (8.98)$$

where

$$\Omega(R,t) = \int \mathrm{D}z\,\{\Theta(-z\sqrt{1-R^2} - Rt - a) + \Theta(z\sqrt{1-R^2} + Rt)$$

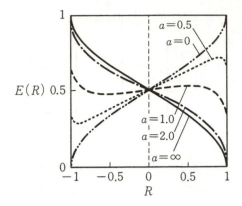

FIG. 8.4. Generalization error of the non-monotonic perceptron

$$- \Theta(z\sqrt{1 - R^2} + Rt - a)\}. \tag{8.99}$$

We have used in the derivation of (8.98) that (1) $\int dv P(u, v) \Theta(-T_a(v)\text{sgn}(u))$ is an even function of u (and thus the integral over $u < 0$ is sufficient if we multiply the result by 2) and that (2) z and t can be written in terms of two independent Gaussian variables with vanishing mean and variance unity as $u = t$ and $v = z\sqrt{1 - R^2} + Rt$. In Fig. 8.4 we have drawn this generalization error as a function of R. This non-monotonic perceptron reduces to the simple perceptron in the limit $a \to \infty$; then $E(R)$ is a monotonically decreasing function of R and has the minimum value $E(R) = 0$ at $R = 1$. When $a = 0$, the student output is just the opposite of the teacher output for the same input, and we have $E(R) = 0$ at $R = -1$.

For intermediate values $0 < a < \infty$, the generalization error does not vanish at any R, and the student cannot simulate the teacher irrespective of the learning algorithm. As one can see in Fig. 8.4, when $0 < a < a_{c1} = \sqrt{2\log 2} = 1.18$, there is a minimum of $E(R)$ in the range $-1 < R < 0$. This minimum is the global minimum when $0 < a < a_{c2} \approx 0.08$. We have shown in Fig. 8.5 the minimum value of the generalization error (a) and the R that gives this minimum as functions of a (b). In the range $a > a_{c2}$ the minimum of the generalization error is achieved when the student has the same couplings as the teacher, $R = 1$, but the minimum value does not vanish because the structures are different. In the case of $0 < a < a_{c2}$, the minimum of the generalization error lies in the range $-1 < R < 0$.

The generalization error as a function of R, (8.98), does not depend on the type of learning. The general forms of the dynamical equations (8.55) and (8.56) also remain intact, but the function f is different. For instance, the Hebb algorithm in the case of a non-monotonic teacher has $f = \text{sgn}\{v(a - v)(a + v)\}$. Then

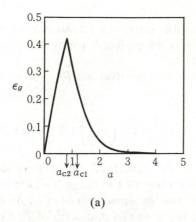
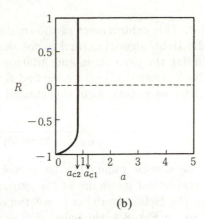

(a)   (b)

FIG. 8.5. Minimum value of the generalization error (a) and overlap R that
gives the minimum of the generalization error (b).

$$[fu] = \sqrt{\frac{2}{\pi}}R(1 - 2e^{-a^2/2}), \quad [fv] = \sqrt{\frac{2}{\pi}}(1 - 2e^{-a^2/2}), \quad [f^2] = 1. \quad (8.100)$$

Accordingly, the dynamical equations read

$$\frac{dl}{dt} = \frac{1}{2l} + \sqrt{\frac{2}{\pi}}R(1 - 2e^{-a^2/2}) \quad (8.101)$$

$$\frac{dR}{dt} = -\frac{R}{2l^2} + \frac{1}{l}\sqrt{\frac{2}{\pi}}(1 - 2e^{-a^2/2})(1 - R^2). \quad (8.102)$$

Solutions of (8.101) and (8.102) show different behaviour according to the value
of a. To see this, we set $R = 0$ on the right hand side of (8.102):

$$\frac{dR}{dt} \approx \frac{1}{l}\sqrt{\frac{2}{\pi}}(1 - 2e^{-a^2/2}). \quad (8.103)$$

This shows that R increases from zero when $a > a_{c1} = \sqrt{2\log 2}$ and decreases if
$0 < a < a_{c1}$. Since (8.102) has a fixed point at $R = 1$ and $l \to \infty$, R approaches
one as $t \to \infty$ when $a > a_{c1}$, and the learning curve is determined asymptotically
from (8.101) and (8.102) using $R = 1 - \epsilon$ and $l = 1/\delta$. It is not difficult to check
that $\epsilon \approx (2k^2t)^{-1}, \delta \approx (kt)^{-1}$ holds as $\epsilon, \delta \ll 1$ with $k = \sqrt{2}(1 - 2e^{-a^2/2})/\sqrt{\pi}$.
Substituting this into (8.98), we find the final result

$$\epsilon_g \approx \frac{\sqrt{2\epsilon}}{\pi} + \frac{2}{\sqrt{\pi}}\text{Erfc}\left(\frac{a}{\sqrt{2}}\right)$$

$$= \frac{1}{\sqrt{2\pi}(1 - 2e^{-a^2/2})\sqrt{t}} + \frac{2}{\sqrt{\pi}}\text{Erfc}\left(\frac{a}{\sqrt{2}}\right). \quad (8.104)$$

The second term on the right hand side is the asymptotic value as $t \to \infty$ and
this coincides with the theoretical minimum of generalization error shown in Fig.

8.5(a). This achievement of the smallest possible error is a remarkable feature of the Hebb algorithm, and is not shared by other on-line learning algorithms including the perceptron and Adatron algorithms.

Next, when $0 < a < a_{c1}$, we find $R \to -1$ as $t \to \infty$. If we set $R = -1 + \epsilon, l = 1/\delta$, the asymptotic form is evaluated as

$$\epsilon_g \approx \frac{1}{\sqrt{6\pi}(1 - 2e^{-a^2/2})} \frac{1}{\sqrt{t}} + 1 - \frac{2}{\sqrt{\pi}} \mathrm{Erfc}\left(\frac{a}{\sqrt{2}}\right). \qquad (8.105)$$

The asymptotic value after the second term on the right hand side is larger than the theoretical minimum of the generalization error in the range $0 < a < a_{c1}$. Thus the Hebb algorithm is not the optimal one for small values of a. As one can see in Fig. 8.4, the value $R = -1$ does not give the minimum of $E(R)$ in the case of $0 < a < a_{c1}$. This is the reason why the Hebb algorithm, which gives $R \to -1$ as $t \to \infty$, does not lead to convergence to the optimal state.

Bibliographical note

We have elucidated the basic ideas of learning in very simple cases. There are many interesting problems not discussed here, including the effects of noise, perceptron with continuous output, multilayer networks, on-line Bayesian learning, path-integral formalism, support vector machines, restricted training set, information-theoretical approaches, and unsupervised learning. These and other topics are discussed in various review articles (Watkin and Rau 1993; Domany *et al.* 1995; Wong *et al.* 1997; Saad 1998).

9

OPTIMIZATION PROBLEMS

A decision-making problem is often formulated as minimization or maximization of a multivariable function, an optimization problem. In the present chapter, after a brief introduction, we show that methods of statistical mechanics are useful to study some optimization problems. Then we discuss mathematical properties of simulated annealing, an approximate numerical method for generic optimization problems. In particular we analyse the method of generalized transition probability, which is attracting considerable attention recently because of its rapid convergence properties.

9.1 Combinatorial optimization and statistical mechanics

The goal of an *optimization problem* is to find the variables to minimize (or maximize) a multivariable function. When the variables take only discrete values under some combinatorial constraints, the problem is called a *combinatorial optimization problem*. The function to be minimized (or maximized) $f(x_1, x_2, \ldots, x_n)$ is termed the *cost function* or the *objective function*. It is sufficient to discuss minimization because maximization of f is equivalent to minimization of $-f$.

An example of a combinatorial optimization problem familiar to physicists is to find the ground state of an Ising model. The variables are the set of spins $\{S_1, S_2, \ldots, S_N\}$ and the Hamiltonian is the cost function. The ground state is easily determined if all the interactions are ferromagnetic, but this is not the case in spin glasses. The possible number of spin configurations is 2^N, and we would find the correct ground state of a spin glass system if we checked all of these states explicitly. However, the number 2^N grows quite rapidly with the increase of N, and the ground-state search by such a naïve method quickly runs into the practical difficulty of explodingly large computation time. Researchers have been trying to find an algorithm to identify the ground state of a spin glass by which one has to check less than an exponential number of states (i.e. power of N). These efforts have so far been unsuccessful except for a few special cases.

The ground-state determination is an example of an *NP* (non-deterministic polynomial) *complete* problem. Generally, the algorithm to solve an NP complete problem can be transformed into another NP complete problem by a polynomial-time algorithm; but no polynomial-time algorithm has been found so far for any of the NP complete problems. These statements define the class 'NP complete'. It is indeed expected that we will need exponentially large computational efforts to solve any NP complete problem by any algorithm.

An exponential function of N grows quite rapidly as N increases, and it is virtually impossible to solve an NP complete problem for any reasonable value of N. There are many examples of NP complete problems including the spin glass ground state, travelling salesman, number partitioning, graph partitioning, and knapsack problems. We shall discuss the latter three problems in detail later. A description of the satisfiability problem will also be given. Here in this section, a few words are mentioned on the travelling salesman problem.

In the *travelling salesman problem*, one is given N cities and distances between all pairs of cities. One is then asked to find the shortest path to return to the original city after visiting all the cities. The cost function is the length of the path. There are about $N!$ possible paths; starting from a city, one can choose the next out of $N - 1$, and the next out of $N - 2$, and so on. The precise number of paths is $(N-1)!/2$ because, first, the equivalence of all cities as the starting point gives the dividing factor of N, and, second, any path has an equivalent one with the reversed direction, accounting for the factor 2 in the denominator. Since the factorial increases more rapidly than the exponential function, identification of the shortest path is obviously very difficult. The travelling salesman problem is a typical NP complete problem and has been studied quite extensively. It also has some practical importance such as the efficient routing of merchandise delivery. A statistical-mechanical analysis of the travelling salesman problem is found in Mézard *et al.* (1987).

Before elucidating statistical mechanical approaches to combinatorial optimization problems, we comment on a difference in viewpoints between statistical mechanics and optimization problems. In statistical mechanics, the target of primary interest is the behaviour of macroscopic quantities, whereas details of microscopic variables in the optimized state play more important roles in optimization problems. For example, in the travelling salesman problem, the shortest path itself is usually much more important than the path length. In the situation of spin glasses, this corresponds to clarification of the state of each spin in the ground state. Such a point of view is somewhat different from the statistical-mechanical standpoint in which the properties of macroscopic order parameters are of paramount importance. These distinctions are not always very clear cut, however, as is exemplified by the important role of the TAP equation that is designed to determine *local* magnetizations.

9.2   Number partitioning problem

9.2.1   *Definition*

In the *number partitioning problem*, one is given a set of positive numbers $A = \{a_1, a_2, \ldots, a_N\}$ and asked to choose a subset $B \subset A$ such that the *partition difference*

$$\tilde{E} = \left| \sum_{i \in B} a_i - \sum_{i \in A \setminus B} a_i \right| \tag{9.1}$$

is minimized. The partition difference is the cost function. The problem is *unconstrained* if one can choose any subset B. In the *constrained* number partitioning problem, the size of the set $|B|$ is fixed. In particular, when $|B|$ is half the total size, $|B| = N/2$ (N even) or $|B| = (N \pm 1)/2$ (N odd), the problem is called the *balanced* number partitioning problem.

The number partitioning problem is known to belong to the class NP complete (Garey and Johnson 1979). Below we develop a detailed analysis of the unconstrained number partitioning problem as it is simpler than the constrained problem (Sasamoto *et al.* 2001; Ferreira and Fontanari 1998). We will be interested mainly in the number of partitions for a given value of the partition difference.

9.2.2 *Subset sum*

For simplicity, we assume that N is even and the a_i are positive integers, not exceeding an integer L, with the gcd unity. The number partitioning problem is closely related to the problem of *subset sum* in which we count the number of configurations, $C(E)$, with a given value E of the Hamiltonian

$$H = \sum_{i=1}^{N} a_i n_i. \tag{9.2}$$

Here n_i is a dynamical variable with the value of 0 or 1. The cost function of the number partitioning problem (9.1), to be denoted as $\tilde{H}$, is related to this Hamiltonian (9.2) using the relation $S_i = 2n_i - 1$ as

$$\tilde{H} = \left| \sum_{i=1}^{N} a_i S_i \right| = \left| 2H - \sum_{i=1}^{N} a_i \right|. \tag{9.3}$$

The spin configuration $S_i = 1$ indicates that $a_i \in B$ and $S_i = -1$ otherwise. From (9.3) we have $2E - \sum_i a_i = \pm\tilde{E}$, and the number of configurations, $C(E)$, for the subset sum (9.2) is translated into that for the number partitioning problem (9.3), $\tilde{C}(\tilde{E})$, by the relation

$$\tilde{C}(\tilde{E}) = \begin{cases} C\left(\dfrac{\tilde{E} + \sum_i a_i}{2}\right) + C\left(\dfrac{-\tilde{E} + \sum_i a_i}{2}\right) & (\tilde{E} \neq 0) \\[4mm] C\left(\dfrac{\sum_i a_i}{2}\right) & (\tilde{E} = 0). \end{cases} \tag{9.4}$$

The following analysis will be developed for a given set $\{a_i\}$; no configurational average will be taken.

9.2.3 *Number of configurations for subset sum*

It is straightforward to write the partition function of the subset sum:

$$Z = \sum_{\{n_i\}} e^{-\beta H} = \prod_{i=1}^{N} (1 + e^{-\beta a_i}). \tag{9.5}$$

By expanding the right hand side in powers of $e^{-\beta} \equiv w$, we can express the above equation as

$$Z = \sum_{E=0}^{E_{\max}} C(E) w^E, \tag{9.6}$$

where $E_{\max} = a_1 + \cdots + a_N$. Since E is a positive integer, the coefficient $C(E)$ of the polynomial (9.6) is written in terms of a contour integral:

$$C(E) = \frac{1}{2\pi i} \oint \frac{dw}{w^{E+1}} Z = \frac{1}{2\pi i} \oint \frac{e^{\log Z}}{w^{E+1}} dw, \tag{9.7}$$

where the integral is over a closed contour around the origin of the complex-w plane. Since $\log Z$ is proportional to the system size $N \, (\gg 1)$, we can evaluate (9.7) by steepest descent.

It is useful to note that the value of the energy E has a one-to-one correspondence with the inverse temperature β through the expression of the expectation value

$$E = \sum_{i=1}^{N} \frac{a_i}{1 + e^{\beta a_i}} \tag{9.8}$$

when the fluctuations around this thermal expectation value are negligible, which is the case in the large-N limit. Let us therefore write the inverse temperature as β_0 which yields the given value of the energy, E_0. It is then convenient to rewrite (9.7) by specifying the integration contour as the circle of radius $e^{-\beta_0}$ and using the phase variable defined by $w = e^{-\beta_0 + i\theta}$:

$$C(E_0) = \frac{1}{2\pi} \int_{-\pi}^{\pi} d\theta \, e^{\log Z + \beta_0 E_0 - iE_0\theta}. \tag{9.9}$$

It is easy to verify that the saddle point of the integrand is at $\theta = 0$.[22] We therefore expand $\log Z$ in powers of $\theta \, (= i\beta - i\beta_0)$ to second order and find

$$C(E_0) = e^{\log Z]_{\beta=\beta_0} + \beta_0 E_0} \cdot \frac{1}{2\pi} \int_{-\pi}^{\pi} d\theta \exp\left(-\frac{\theta^2}{2} \frac{\partial^2}{\partial \beta^2} \log Z \Big]_{\beta=\beta_0} \right). \tag{9.10}$$

Since the exponent in the integrand is proportional to N, we may extend the integration range to $\pm\infty$. The result is

$$C(E_0) = \frac{e^{\beta_0 E_0} \prod_i (1 + e^{-\beta_0 a_i})}{\sqrt{2\pi \sum_i a_i^2 / (1 + e^{\beta_0 a_i})(1 + e^{-\beta_0 a_i})}}. \tag{9.11}$$

[22] There exist other saddle points if the gcd of $\{a_i\}$ is not unity.

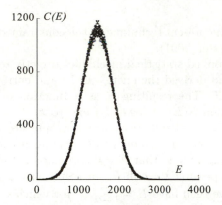

F<small>IG</small>. 9.1. The number of configurations as a function of the energy for the subset sum with $N = 20$, $L = 256$, and $\{a_i\} = \{218, 13, 227, 193, 70, 134, 89, 198, 205, 147, 227, 190, 64, 168, 4, 209, 27, 239, 192, 131\}$. The theoretical prediction (9.11) is indistinguishable from the numerical results plotted in dots.

This formula gives the number of configurations of the subset sum as a function of E_0 through

$$E_0 = \sum_{i=1}^{N} \frac{a_i}{1 + e^{\beta_0 a_i}}. \tag{9.12}$$

Figure 9.1 depicts the result of numerical verification of (9.11).

9.2.4 *Number partitioning problem*

The number of partitions for a given value of $\tilde{E}$ of the number partitioning problem, $\tilde{C}(\tilde{E})$, can be derived from (9.11) using (9.4). For example, the optimal configuration with $\tilde{E} = 0$ has $\beta_0 = 0$ according to the relation $2E - \sum_i a_i = \pm \tilde{E}$ and (9.12). Then, (9.11) with $\beta_0 = 0$ gives the result for the number of optimal partitions as

$$\tilde{C}(\tilde{E} = 0) = \frac{2^{N+1}}{\sqrt{2\pi \sum_i a_i^2}}. \tag{9.13}$$

An important restriction on the applicability of the present argument is that L should not exceed 2^N for the following reason. If we choose $\{a_i\}$ from $\{1, \ldots, L\}$ using a well-behaved distribution function, then $\sum_i a_i^2 = \mathcal{O}(L^2)$, and (9.13) gives $\tilde{C}(E = 0) = \mathcal{O}(2^N/L)$. Since $\tilde{C}$ is the number of partitions, $2^N/L$ must not be smaller than one, which leads to the condition $2^N > L$. This condition may correspond to the strong crossover in the behaviour of the probability to find a perfect partition ($\tilde{E} = 0$) found numerically (Gent and Walsh 1996). The other case of $2^N < L$ is hard to analyse by the present method although it is quite interesting from the viewpoint of information science (Gent and Walsh 1996). It is possible to apply the present technique to the constrained number partitioning problem (Sasamoto *et al.* 2001).

Mertens (1998) applied statistical mechanics directly to the system with the Hamiltonian (9.3) and derived the energy $\tilde{E}(T)$ and entropy $\tilde{S}(T)$ as functions of the temperature T. The resulting $\tilde{S}$ as a function of $\tilde{E}$ differs from ours under the identification $\tilde{S}(\tilde{E}) = \log \tilde{C}(\tilde{E})$ except at $\tilde{E} = 0$. In particular, his result gives $\partial \tilde{S}/\partial \tilde{E} > 0$ whereas we have the opposite inequality $\partial \tilde{S}/\partial \tilde{E} < 0$ as can be verified from (9.11) and (9.12). Figure 9.1 shows that the latter possibility is realized if we note that $\tilde{E} = 0$ corresponds to the peak of the curve at $E = \sum_i a_i/2$. It is possible to confirm our result also in the solvable case with $a_1 = \cdots = a_N = 1$: we then have $\tilde{H} = |\sum_i S_i|$ for which

$$\tilde{C}(\tilde{E}) = \binom{N}{(N + \tilde{E})/2}. \tag{9.14}$$

This is a monotonically decreasing function of $\tilde{E}$. Equations (9.11) and (9.12) with $a_1 = \cdots = a_N = 1$ reproduce (9.14) for sufficiently large N.

The system described by the Hamiltonian (9.3) is anomalous because the number of configurations (partitions) decreases with increasing energy. We should be careful in applying statistical mechanics to such a system; the calculated entropy may not necessarily give the logarithm of the number of configurations as exemplified above.

9.3   Graph partitioning problem

The next example of statistical-mechanical analysis of an optimization problem is the *graph partitioning problem*. It will be shown that the graph partitioning problem is equivalent to the SK model in a certain limit.

9.3.1 *Definition*

Suppose we are given N nodes $V = \{v_1, v_2, \ldots, v_N\}$ and the set of edges between them $E = \{(v_i, v_j)\}$. Here N is an even integer. A *graph* is a set of such nodes and edges. In the graph partitioning problem, we should divide the set V into two subsets V_1 and V_2 with the same size $N/2$ and, at the same time, minimize the number of edges connecting nodes in V_1 with those in V_2. The cost function is the number of edges between V_1 and V_2. Let us consider an example of the graph specified by $N = 6, E = \{(1,2),(1,3),(2,3),(2,4),(4,5)\}$. The cost function has the value $f = 1$ for the partitioning $V_1 = \{1,2,3\}, V_2 = \{4,5,6\}$ and $f = 3$ for $V_1 = \{1,2,4\}, V_2 = \{3,5,6\}$ (Fig. 9.2).

It is known that the graph partitioning problem is NP complete. The problem has direct instances in real-life applications such as the configuration of components on a computer chip to minimize wiring lengths.

The problem of a *random graph*, in which each pair of nodes (v_i, v_j) has an edge between them with probability p, is conveniently treated by statistical mechanics (Fu and Anderson 1986). We assume in this book that p is of order

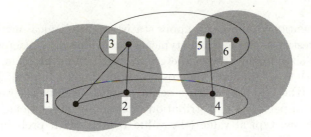

FIG. 9.2. A graph partitioning with $N = 6$

1 and independent of N. Thus the number of edges emanating from each node is pN on average, which is a very large number for large N. The methods of mean-field theory are effectively applied in such a case.

9.3.2 Cost function

We start our argument by expressing the cost function $f(p)$ in terms of the Ising spin Hamiltonian. Let us write $S_i = 1$ when the node v_i belongs to the set V_1 and $S_i = -1$ otherwise. The existence of an edge between v_i and v_j is represented by the coupling $J_{ij} = J$, and $J_{ij} = 0$ otherwise. The Hamiltonian is written as

$$H = -\sum_{i<j} J_{ij}S_iS_j = -\frac{1}{2}\left(\sum_{i\in V_1,j\in V_1} + \sum_{i\in V_2,j\in V_2} + \sum_{i\in V_1,j\in V_2} + \sum_{i\in V_2,j\in V_1}\right)J_{ij}$$

$$+ \left(\sum_{i\in V_1,j\in V_2} + \sum_{i\in V_2,j\in V_1}\right)J_{ij} = -\frac{J}{2}\cdot\frac{2N(N-1)p}{2} + 2f(p)J, \qquad (9.15)$$

where $N(N-1)p/2$ is the total number of edges. The cost function and the Hamiltonian are therefore related to each other as

$$f(p) = \frac{H}{2J} + \frac{1}{4}N(N-1)p. \qquad (9.16)$$

This equation shows that the cost function is directly related to the ground state of the Hamiltonian (9.15) under the condition that the set V is divided into two subsets of exactly equal size

$$\sum_{i=1}^{N} S_i = 0. \qquad (9.17)$$

Equation (9.17) does not change if squared. Expansion of the squared expression may be interpreted as the Hamiltonian of a system with uniform antiferromagnetic interactions between all pairs. Thus the partitioning problem of a random graph has been reduced to the ferromagnetic Ising model with diluted interactions (i.e. some J_{ij} are vanishing) and with the additional constraint of uniform infinite-range antiferromagnetic interactions.

As was mentioned above, we apply statistical-mechanical methods that are suited to investigate the average (typical) behaviour of a macroscopic system of very large size. In the case of the graph partitioning problem, the principal objective is then to evaluate the cost function in the limit of large N. The Hamiltonian (the cost function) is self-averaging, and hence we calculate its average with respect to the distribution of random interactions, which should coincide with the typical value (the value realized with probability 1) in the thermodynamic limit $N \to \infty$.

9.3.3 *Replica expression*

Let us calculate the configurational average by the replica method. The replica average of the partition function of the system (9.15) is

$$[Z^n] = (1-p)^{N(N-1)/2}\mathrm{Tr}\prod_{i<j}\left\{1 + p_0 \exp\left(\beta J \sum_{\alpha=1}^{n} S_i^\alpha S_j^\alpha\right)\right\}, \qquad (9.18)$$

where $p_0 = p/(1-p)$, and Tr denotes the sum over spin variables under the condition (9.17). We shall show below that (9.18) can be transformed into the expression

$$[Z^n] = (1-p)^{N(N-1)/2}\exp\left\{\frac{N(N-1)}{2}\log(1+p_0) - \frac{N}{2}(\beta J c_1 n + \beta^2 J^2 c_2 n^2)\right\}$$

$$\cdot \mathrm{Tr}\exp\left\{\frac{(\beta J)^2}{2}c_2 \sum_{\alpha,\beta}\left(\sum_i S_i^\alpha S_i^\beta\right)^2 + \mathcal{O}\left(\beta^3 J^3 \sum_{i<j}(\sum_\alpha S_i^\alpha S_j^\alpha)^3\right)\right\}, \qquad (9.19)$$

where

$$c_j = \frac{1}{j!}\sum_{l=1}^{\infty}\frac{(-1)^{l-1}}{l}p_0^l l^j. \qquad (9.20)$$

To derive (9.19) we expand the logarithmic and exponential functions as follows:

$$\sum_{i<j}\log\{1 + p_0 \exp(\beta J \sum_\alpha S_i^\alpha S_j^\alpha)\}$$

$$= \sum_{l=1}^{\infty}\frac{(-1)^{l-1}}{l}p_0^l \sum_{k_1=0}^{\infty}\cdots\sum_{k_l=0}^{\infty}\frac{(\beta J)^{k_1+\cdots+k_l}}{k_1!\ldots k_l!}\sum_{i<j}(\sum_\alpha S_i^\alpha S_j^\alpha)^{k_1+\cdots+k_l}.$$

It is convenient to rewrite this formula as a power series in βJ. The constant term corresponds to $k_1 = \cdots = k_l = 0$, for which the sum over l gives $\{N(N-1)/2\}\log(1+p_0)$. The coefficient of the term linear in βJ is, according to (9.17), a constant

$$\sum_{l=1}^{\infty}\frac{(-1)^{l-1}}{l}p_0^l \cdot l \cdot \frac{1}{2}\sum_\alpha\{(\sum_i S_i^\alpha)^2 - N\} = -\frac{Nn}{2}c_1. \qquad (9.21)$$

Next, the coefficient of the quadratic term is, after consideration of all possibilities corresponding to $k_1 + \cdots + k_l = 2$ (such as $(k_1 = k_2 = 1), (k_1 = 2, k_2 = 0)$, and so on),

$$
\sum_{l=1}^{\infty} \frac{(-1)^{l-1}}{l} p_0^l \sum_{i<j} (\sum_{\alpha} S_i^{\alpha} S_j^{\alpha})^2 \left(\binom{l}{2} + \frac{l}{2!} \right)
$$

$$
= \sum_{l=1}^{\infty} \frac{(-1)^{l-1}}{l} p_0^l \cdot \frac{l^2}{2} \sum_{\alpha,\beta} \{ (\sum_i S_i^{\alpha} S_i^{\beta})^2 - N \}
$$

$$
= \frac{c_2}{2} \sum_{\alpha,\beta} (\sum_i S_i^{\alpha} S_i^{\beta})^2 - \frac{Nn^2}{2} c_2. \tag{9.22}
$$

Equation (9.19) results from (9.21) and (9.22).

9.3.4   *Minimum of the cost function*

Recalling (9.16), we write the lowest value of the cost function as

$$
f(p) = \frac{N^2}{4} p + \frac{1}{2J} E_{\mathrm{g}}
$$

$$
E_{\mathrm{g}} = \lim_{\beta \to \infty} \lim_{n \to 0} \left(-\frac{1}{n\beta} \right) \left[\mathrm{Tr} \exp \left\{ \frac{(\beta J)^2}{2} c_2 \sum_{\alpha,\beta} (\sum_i S_i^{\alpha} S_i^{\beta})^2 \right. \right.
$$

$$
\left. \left. + \mathcal{O} \left(\beta^3 J^3 \sum_{i<j} (\sum_{\alpha} S_i^{\alpha} S_j^{\alpha})^3 \right) \right\} - 1 \right], \tag{9.23}
$$

where we have used the fact that the contribution of the term linear in βJ to $f(p)$ is $Np/4$ from $c_1 = p_0/(1 + p_0) = p$. Equation (9.23) is similar to the SK model expression (2.12) with $J_0 = 0$ and $h = 0$. The additional constraint (9.17) is satisfied automatically because there is no spontaneous magnetization in the SK model when the centre of distribution J_0 and the external field h are both vanishing.

Since J has been introduced as an arbitrarily controllable parameter, we may choose $J = \tilde{J}/\sqrt{N}$ and reduce (9.23) to the same form as (2.12). Then, if $N \gg 1$, the term proportional to $\beta^3 J^3$ is negligibly smaller than the term of $\mathcal{O}(\beta^2 J^2)$, and we can apply the results for the SK model directly. Substituting $c_2 = p(1-p)/2$, we finally have

$$
f(p) = \frac{N^2}{4} p + \frac{\sqrt{N}}{2\tilde{J}} U_0 \sqrt{c_2} \tilde{J} N = \frac{N^2}{4} p + \frac{1}{2} U_0 N^{3/2} \sqrt{p(1-p)}. \tag{9.24}
$$

Here U_0 is the ground-state energy per spin of the SK model and is approximately $U_0 = -0.38$ according to numerical studies. The first term on the right hand side of (9.24) is interpreted as the number of edges between V_1 and V_2,

$N^2/4$, multiplied by the ratio of actually existing edges p. The second term is the correction to this leading contribution, which is the non-trivial contribution derived from the present argument.

9.4 Knapsack problem

The third topic is a maximization problem of a cost function under constraints expressed by inequalities. The replica method is again useful to clarify certain aspects of this problem (Korutcheva *et al.* 1994; Fontanari 1995; Inoue 1997).

9.4.1 *Knapsack problem and linear programming*

Suppose that there are N items, each of which has the weight a_j and the value c_j. The *knapsack problem* is to maximize the total value by choosing appropriate items when the total weight is constrained not to exceed b. This may be seen as the maximization of the total value of items to be carried in a knapsack within a weight limit when one climbs a mountain.

With the notation $S_j = 1$ when the jth item is chosen to be carried and $S_j = -1$ otherwise, the cost function (the value to be maximized) U and the constraint are written as follows:

$$U = \sum_{j=1}^{N} c_j \cdot \frac{S_j + 1}{2}, \quad Y = \sum_{j=1}^{N} a_j \cdot \frac{S_j + 1}{2} \le b. \tag{9.25}$$

A generalization of (9.25) is to require many (K) constraints:

$$Y_k = \frac{1}{2} \sum_j a_{kj}(S_j + 1) \le b_k, \quad (k = 1, \ldots, K). \tag{9.26}$$

When S_j is continuous, the minimization (or maximization) problem of a linear cost function under linear constraints is called *linear programming*.

In this section we exemplify a statistical-mechanical approach to such a class of problems by simplifying the situation so that the c_j are constant c and so are the b_k ($= b$). It is also assumed that a_{kj} is a Gaussian random variable with mean $\frac{1}{2}$ and variance σ^2,

$$a_{kj} = \frac{1}{2} + \xi_{kj}, \quad P(\xi_{kj}) = \frac{1}{\sqrt{2\pi}\sigma} \exp\left(-\frac{\xi_{kj}^2}{2\sigma^2}\right). \tag{9.27}$$

The constraint (9.26) is then written as

$$Y_k - b = \frac{1}{2} \sum_j (1 + S_j)\xi_{kj} + \frac{1}{4} \sum_j S_j + \frac{N}{4} - b \le 0. \tag{9.28}$$

The first and second terms in (9.28), where sums over j appear, are of $\mathcal{O}(N)$ at most, so that (9.28) is satisfied by any $\boldsymbol{S} = \{S_j\}$ if $b \gg N/4$. Then one can carry virtually all items in the knapsack ($S_j = 1$). In the other extreme case $N/4 \gg b$, one should leave almost all items behind ($S_j = -1$). The system shows the most interesting behaviour in the intermediate case $b = N/4$, to which we restrict ourselves.

9.4.2   *Relaxation method*

It is sometimes convenient to relax the condition of discreteness of variables to solve a combinatorial optimization problem, the *relaxation method*, because computations are sometimes easier and faster with continuous numbers. If discrete values are necessary as the final answer, it often suffices to accept the discrete values nearest to the continuous solution.

It is possible to discuss the large-scale knapsack problem, in which the number of constraints K is of the same order as N, with S_j kept discrete (Korutcheva *et al.* 1994). However, in the present section we follow the idea of the relaxation method and use continuous variables satisfying $\sum_j S_j^2 = N$ because the problem is then formulated in a very similar form as that of the perceptron capacity discussed in §7.6 (Inoue 1997). Hence the S_j are assumed to be continuous real numbers satisfying $\sum_j S_j^2 = N$. We are to maximize the cost function

$$U = \frac{cN}{2} + \frac{c\sqrt{N}}{2}M \tag{9.29}$$

under the constraint

$$Y_k = \frac{1}{2}\sum_j (1 + S_j)\xi_{kj} + \frac{\sqrt{N}}{4}M \le 0, \quad M = \frac{1}{\sqrt{N}}\sum_j S_j. \tag{9.30}$$

The normalization in the second half of (9.30) implies that $\sum_j S_j$ is of order $\sqrt{N}$ when $b = N/4$, and consequently about half of the items are carried in the knapsack. Consistency of this assumption is confirmed if M is found to be of order unity after calculations using this assumption. This will be shown to be indeed the case. M is the coefficient of the deviation of order $\sqrt{N}$ from the average number of items $N/2$ to be carried in the knapsack.

9.4.3   *Replica calculations*

Let V be the volume of subspace satisfying the condition (9.30) in the space of the variables $\boldsymbol{S}$. The typical behaviour of the system is determined by the configurational average of the logarithm of V over the random variables $[\zeta_{kj}]$. According to the replica method, the configurational average of V^n is, similarly to (7.77),

$$[V^n] = \left[V_0^{-n}\int\prod_{\alpha,i}\mathrm{d}S_i^\alpha\prod_\alpha \delta\left(\sum_j S_j^\alpha - \sqrt{N}M\right)\delta\left(\sum_j(S_j^\alpha)^2 - N\right)\right.$$

$$\cdot \prod_{\alpha,k} \Theta\left(-\frac{1}{\sqrt{N}}\sum_j (1+S_j^\alpha)\xi_{kj} - \frac{M}{2}\right)\Bigg], \tag{9.31}$$

where V_0 is defined as the quantity with only the part of $\sum_j (S_j^\alpha)^2 = N$ kept in the above integrand.

Equation (9.31) has almost the same form as (7.77), so that we can evaluate the former very similarly to §7.6. We therefore write only the result here. Under the assumption of replica symmetry, we should extremize the following G:

$$[V^n] = \exp\{nNG(q, E, F, \tilde{M})\}$$

$$G = \alpha G_1(q) + G_2(E, F, \tilde{M}) - \frac{i}{2}qF + iE$$

$$G_1(q) = \log \int_{M/2}^\infty \prod_\alpha \frac{d\lambda^\alpha}{2\pi} \int_{-\infty}^\infty \prod_\alpha dx^\alpha$$

$$\cdot \exp\left(i\sum_\alpha x^\alpha \lambda^\alpha - \sigma^2 \sum_\alpha (x^\alpha)^2 - \sigma^2(1+q)\sum_{(\alpha\beta)} x^\alpha x^\beta\right)$$

$$G_2(E, F, \tilde{M}) = \log \int_{-\infty}^\infty \prod_\alpha dS^\alpha$$

$$\cdot \exp\left(-i\tilde{M}\sum_\alpha S^\alpha - iF\sum_{(\alpha\beta)} S^\alpha S^\beta - iE\sum_\alpha (S^\alpha)^2\right),$$

where q is the RS value of $q_{\alpha\beta} = N^{-1}\sum_i S_i^\alpha S_i^\beta$, and $\alpha = K/N$ (not to be confused with the replica index α). Extremization by $\tilde{M}$ readily shows that $\tilde{M} = 0$. We also perform the integration and eliminate E and F by extremization with respect to these variables, as in §7.6, to find

$$G = \alpha \int Dy \log L(q) + \frac{1}{2}\log(1-q) + \frac{1}{2(1-q)} \tag{9.32}$$

$$L(q) = 2\sqrt{\pi}\, \mathrm{Erfc}\left(\frac{M/2 + y\sigma\sqrt{1+q}}{\sqrt{2(1-q)}\,\sigma}\right). \tag{9.33}$$

As the number of items increases with the ratio $\alpha\,(=K/N)$ fixed, the system reaches a limit beyond which one cannot carry items. We write M_{opt} for the value of M at this limit. To evaluate the limit explicitly we note that there is only one way to choose items to carry at the limit, which implies $q = 1$. We thus extremize (9.32) with respect to q and take the limit $q \to 1$ to obtain M_{opt} as a function of α as

$$\alpha(M_{\mathrm{opt}}) = \left\{\frac{1}{4}\int_{-M_{\mathrm{opt}}/(2\sqrt{2}\sigma)}^\infty Dy \left(\frac{M_{\mathrm{opt}}}{\sigma} + 2\sqrt{2}y\right)^2\right\}^{-1}. \tag{9.34}$$

Figure 9.3 shows M_{opt} as a function of α when $\sigma = 1/12$. Stability analysis of

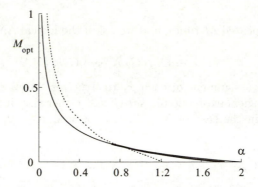

FIG. 9.3. M_{opt} as a function of α (full line) and the AT line (dashed)

the RS solution leads to the AT line:

$$\alpha \left\{ \int \mathrm{D}t (1 - L_2 + L_1^2) \right\}^2 = 1, \quad L_k = \frac{\int_{I_z} \mathrm{D}z\, z^k}{\int_{I_z} \mathrm{D}z}, \quad \int_{I_z} = \int_{(M/2\sigma + t\sqrt{1+q})/\sqrt{1-q}}^{\infty}, \tag{9.35}$$

which is shown dashed in Fig. 9.3. The RS solution is stable in the range $\alpha <$ 0.846 but not beyond. The 1RSB solution for M_{opt} is also drawn in Fig. 9.3, but is hard to distinguish from the RS solution at this scale of the figure. We may thus expect that further RSB solutions would give qualitatively similar results.

9.5   Satisfiability problem

Another interesting instance of the optimization problem is the satisfiability problem. One forms a logical expression out of many logical variables in a certain special way. The problem is to determine whether or not an assignment of each logical variable to 'true' or 'false' exists so that the whole expression is 'true'. It is possible to analyse some aspects of this typical NP complete problem (Garey and Johnson 1979) by statistical-mechanical methods. Since the manipulations are rather complicated, we describe only important ideas and some of the steps of calculations below. The reader is referred to the original papers for more details (Kirkpatrick and Selman 1994; Monasson and Zecchina 1996, 1997, 1998; Monasson *et al.* 1999, 2000).

9.5.1   *Random satisfiability problem*

Let us define a class of satisfiability problems, a *random K-satisfiability problem* (*K*-SAT). Suppose that there are N logical variables $x_1, \ldots, x_N$, each of which is either 'true' or 'false'. Then we choose $K (< N)$ of the variables $x_{i_1}, \ldots, x_{i_K}$ and negate each with probability $\frac{1}{2}$: $x_{i_k} \to \overline{x}_{i_k} (= \mathrm{NOT}(x_{i_k}))$. A *clause* C_l is formed by logical OR ($\vee$) of these K variables; for example,

$$C_l = x_{i_1} \vee \overline{x}_{i_2} \vee \overline{x}_{i_3} \vee x_{i_4} \vee \cdots \vee x_{i_K}. \tag{9.36}$$

This process is repeated M times, and we ask if the logical AND ($\wedge$) of these M clauses

$$F \equiv C_1 \wedge C_2 \wedge \cdots \wedge C_M \qquad (9.37)$$

gives 'true'. If an assignment of each x_i to 'true' or 'false' exists so that F is 'true', then this logical expression is *satisfiable*. Otherwise, it is *unsatisfiable*.

For instance, in the case of $N = 3, M = 3$, and $K = 2$, we may form the following clauses:

$$C_1 = x_1 \vee \overline{x}_2, \quad C_2 = x_1 \vee x_3, \quad C_3 = \overline{x}_2 \vee x_3, \qquad (9.38)$$

and $F = C_1 \wedge C_2 \wedge C_3$. This F is satisfied by $x_1 =$ 'true', $x_2 =$ 'false', and $x_3 =$ 'true'.

It is expected that the problem is difficult (likely to be unsatisfiable) if M is large since the number of conditions in (9.37) is large. The other limit of large N should be easy (likely to be satisfiable) because the number of possible choices of $x_{i_1}, \ldots, x_{i_K}$ out of $x_1, \ldots, x_N$ is large and hence C_l is less likely to share the same x_i with other clauses. It can indeed be shown that the problem undergoes a phase transition between easy and difficult regions as the ratio $\alpha \equiv M/N$ crosses a critical value α_c in the limit $N, M \to \infty$ with α and K fixed. Evidence for this behaviour comes from numerical experiments (Kirkpatrick and Selman 1994; Hogg *et al.* 1996), rigorous arguments (Goerdt 1996), and replica calculations (below): for $\alpha < \alpha_c$ (easy region), one finds an exponentially large number of solutions to satisfy a K-SAT, a finite entropy. The number of solutions vanishes above α_c (difficult region) and the best one can do is to minimize the number of unsatisfied clauses (MAX K-SAT). The critical values are $\alpha_c = 1$ (exact) for $K = 2$, 4.17 for $K = 3$ (numerical), and $2^K \log 2$ (asymptotic) for sufficiently large K. The order parameter changes continuously at α_c for $K = 2$ but discontinuously for $K \geq 3$. The RS solution gives the correct answer for $\alpha < \alpha_c$ and one should consider RSB when $\alpha > \alpha_c$.

It is also known that the K-SAT is NP complete for $K \geq 3$ (Garey and Johnson 1979) in the sense that there is no generic polynomial-time algorithm to find a solution when we know that a solution exists. By contrast, a linear-time algorithm exists to find a solution for $K = 1$ and 2 (Aspvall *et al.* 1979). The qualitatively different behaviour of $K = 2$ and $K \geq 3$ mentioned above may be related to this fact.

9.5.2 *Statistical-mechanical formulation*

Let us follow Monasson and Zecchina (1997, 1998) and Monasson *et al.* (2000) and formulate the random K-SAT. The energy to be minimized in the K-SAT is the number of unsatisfied clauses. We introduce an Ising variable S_i which is 1 if the logical variable x_i is 'true' and -1 for x_i 'false'. The variable with quenched randomness C_{li} is equal to 1 if the clause C_l includes x_i, $C_{li} = -1$ if $\overline{x}_i$ is included, and $C_{li} = 0$ when x_i does not appear in C_l. Thus one should choose K non-vanishing C_{li} from $\{C_{l1}, \ldots, C_{lN}\}$ and assign ± 1 randomly for those non-vanishing components, $\sum_{i=1}^{N} C_{li}^2 = K$. Satisfiability is judged by the value of $\sum_{i=1}^{N} C_{li}S_i$: if it is larger than $-K$ for all l, the problem is satisfiable because at least one of the $C_{li}S_i$ is 1 (satisfied) in the clause. The energy or the Hamiltonian is thus formulated using Kronecker's delta as

$$E(\boldsymbol{S}) = \sum_{l=1}^{M} \delta \left(\sum_{i=1}^{N} C_{li}S_i, -K \right). \tag{9.39}$$

Vanishing energy means that the problem is satisfiable. Macroscopic properties of the problem such as the expectation values of the energy and entropy can be calculated from this Hamiltonian by statistical-mechanical techniques. One can observe in (9.39) that Kronecker's delta imposes a relation between K spins and thus represents an interaction between K spins. The qualitative difference between $K = 2$ and $K \geq 3$ mentioned in the previous subsection may be related to this fact; in the infinite-range r-spin interacting model discussed in Chapter 5, the transition is of second order for $r = 2$ but is of first order for $r \geq 3$.

Averaging over quenched randomness in C_{li} can be performed by the replica method. Since M clauses are independent of each other, we find

$$[Z^n] = \text{Tr}\, \zeta_K(\boldsymbol{S})^M \tag{9.40}$$

$$\zeta_K(\boldsymbol{S}) = \left[\exp \left\{ -\frac{1}{T} \sum_{\alpha=1}^{n} \delta \left(\sum_{i=1}^{N} C_i S_i^{\alpha}, -K \right) \right\} \right], \tag{9.41}$$

where Tr denotes the sum over $\boldsymbol{S}$. The configurational average $[\cdots]$ is taken over the random choice of the C_i. We have introduced the temperature T to control the average energy. It is useful to note in (9.41) that

$$\delta \left(\sum_{i=1}^{N} C_i S_i^{\alpha}, -K \right) = \prod_{i=1; C_i \neq 0}^{N} \delta(S_i^{\alpha}, -C_i), \tag{9.42}$$

where the product runs over all the i for which C_i is not vanishing. Then $\zeta_K(\boldsymbol{S})$ of (9.41) is

$$\zeta_K(\boldsymbol{S}) = \frac{1}{2^K} \sum_{C_1=\pm1} \cdots \sum_{C_K=\pm1} N^{-K} \sum_{i_1=1}^{N} \cdots \sum_{i_K=1}^{N} \exp \left\{ -\frac{1}{T} \sum_{\alpha=1}^{n} \prod_{k=1}^{K} \delta(S_{i_k}^{\alpha}, -C_k) \right\}, \tag{9.43}$$

where we have neglected corrections of $\mathcal{O}(N^{-1})$. This formula for $\zeta_K(\boldsymbol{S})$ is conveniently rewritten in terms of $\{c(\boldsymbol{\sigma})\}_{\boldsymbol{\sigma}}$, the set of the number of sites with a specified spin pattern in the replica space $\boldsymbol{\sigma} = \{\sigma^1, \ldots, \sigma^n\}$:

$$Nc(\boldsymbol{\sigma}) = \sum_{i=1}^{N} \prod_{\alpha=1}^{n} \delta(S_i^{\alpha}, \sigma^{\alpha}). \tag{9.44}$$

Equation (9.43) depends on the spin configuration $\boldsymbol{S}$ only through $\{c(\boldsymbol{\sigma})\}$. Indeed, if we choose $\sigma_k^{\alpha} = -C_k S_{i_k}^{\alpha}$, Kronecker's delta in the exponent is $\delta(\sigma_k^{\alpha}, 1)$.

All such terms (there are $Nc(-C_k\boldsymbol{\sigma}_k)$ of them) in the sum over i_1 to i_K give the same contribution. Hence (9.43) is written as

$$\zeta_K(\boldsymbol{S}) = \zeta_K(\{c\}) \equiv \frac{1}{2^K} \sum_{C_1=\pm 1} \cdots \sum_{C_K=\pm 1} \sum_{\boldsymbol{\sigma}_1} \cdots \sum_{\boldsymbol{\sigma}_K} c(-C_1\boldsymbol{\sigma}_1)\dots c(-C_K\boldsymbol{\sigma}_K)$$

$$\cdot \exp\left\{-\frac{1}{T}\sum_{\alpha=1}^{n}\prod_{k=1}^{K}\delta(\sigma_k^\alpha,1)\right\}. \tag{9.45}$$

We may drop the C_i-dependence of $c(-C_i\boldsymbol{\sigma}_i)$ in (9.45) due to the relation $c(\boldsymbol{\sigma}) = c(-\boldsymbol{\sigma})$. This last equation is equivalent to the assumption that the overlap of odd numbers of replica spins vanishes. To see this, we expand the right hand side of (9.44) as

$$c(\boldsymbol{\sigma}) = \frac{1}{N}\sum_i\prod_\alpha \frac{1+S_i^\alpha\sigma_i^\alpha}{2} = \frac{1}{2^n}\left(1 + \sum_\alpha Q^\alpha\sigma^\alpha\right.$$

$$\left. + \sum_{\alpha<\beta} Q^{\alpha\beta}\sigma^\alpha\sigma^\beta + \sum_{\alpha<\beta<\gamma} Q^{\alpha\beta\gamma}\sigma^\alpha\sigma^\beta\sigma^\gamma + \dots\right), \tag{9.46}$$

where

$$Q^{\alpha\beta\gamma\cdots} = \frac{1}{N}\sum_i S_i^\alpha S_i^\beta S_i^\gamma \dots. \tag{9.47}$$

The symmetry $c(\boldsymbol{\sigma}) = c(-\boldsymbol{\sigma})$ follows from the relation $Q^\alpha = Q^{\alpha\beta\gamma} = \dots = 0$ for odd numbers of replica indices, which is natural if there is no symmetry breaking of the ferromagnetic type. We assume this to be the case here.

The partition function (9.41) now has a compact form

$$[Z^n] = \int\prod_{\boldsymbol{\sigma}} dc(\boldsymbol{\sigma})\, e^{-NE_0(\{c\})}$$

$$\cdot \mathrm{Tr}\prod_{\boldsymbol{\sigma}}\delta\left\{c(\boldsymbol{\sigma}) - N^{-1}\sum_{i=1}^{N}\prod_{\alpha=1}^{n}\delta(S_i^\alpha,\sigma^\alpha)\right\} \tag{9.48}$$

$$E_0(\{c\}) = -\alpha\log\left\{\sum_{\boldsymbol{\sigma}_1,\dots,\boldsymbol{\sigma}_K} c(\boldsymbol{\sigma}_1)\dots c(\boldsymbol{\sigma}_K)\right.$$

$$\left. \cdot\prod_{\alpha=1}^{n}\left(1 + (e^{-\beta}-1)\prod_{k=1}^{K}\delta(\sigma_k^\alpha,1)\right)\right\}, \tag{9.49}$$

where one should not confuse $\alpha = M/N$ in front of the logarithm with the replica index. The trace operation over the spin variables $\boldsymbol{S}$ (after the Tr symbol in (9.48)) gives the entropy as

$$\frac{N!}{\prod_\sigma (Nc(\sigma))!} = \exp\left(-N\sum_\sigma c(\sigma)\log c(\sigma)\right) \tag{9.50}$$

by Stirling's formula. If we apply the steepest descent method to the integral in (9.48), the free energy is given as, in the thermodynamic limit $N, M \to \infty$ with α fixed,

$$-\frac{\beta F}{N} = -E_0(\{c\}) - \sum_\sigma c(\sigma)\log c(\sigma) \tag{9.51}$$

with the condition $\sum_\sigma c(\sigma) = 1$.

9.5.3 *Replica-symmetric solution and its interpretation*

The free energy (9.51) is to be extremized with respect to $c(\sigma)$. The simple RS solution amounts to assuming $c(\sigma)$ which is symmetric under permutation of $\sigma^1, \ldots, \sigma^n$. It is convenient to express the function $c(\sigma)$ in terms of the distribution function of local magnetization $P(m)$ as

$$c(\sigma) = \int_{-1}^{1} dm\, P(m) \prod_{\alpha=1}^{n} \frac{1 + m\sigma^\alpha}{2}. \tag{9.52}$$

It is clear that this $c(\sigma)$ is RS. The extremization condition of the free energy (9.51) leads to the following self-consistent equation of $P(m)$ as shown in Appendix D:

$$P(m) = \frac{1}{2\pi(1-m^2)} \int_{-\infty}^{\infty} du \cos\left(\frac{u}{2}\log\frac{1+m}{1-m}\right)$$
$$\cdot \exp\left\{-\alpha K + \alpha K \int_{-1}^{1} \prod_{k=1}^{K-1} dm_k\, P(m_k) \cos\left(\frac{u}{2}\log A_{K-1}\right)\right\} \tag{9.53}$$

$$A_{K-1} = 1 + (e^{-\beta} - 1) \prod_{k=1}^{K-1} \frac{1 + m_k}{2}. \tag{9.54}$$

The free energy is written in terms of the solution of the above equation as

$$-\frac{\beta F}{N} = \log 2 + \alpha(1-K) \int_{-1}^{1} \prod_{k=1}^{K} dm_k\, P(m_k) \log A_K$$
$$+ \frac{\alpha K}{2} \int_{-1}^{1} \prod_{k=1}^{K-1} dm_k P(m_k) \log A_{K-1} - \frac{1}{2}\int_{-1}^{1} dm P(m)\log(1-m^2). \tag{9.55}$$

It is instructive to investigate the simplest case of $K = 1$ using the above result. When $K = 1$, A_0 is $e^{-\beta}$ and (9.53) gives

$$P(m) = \frac{1}{2\pi(1-m^2)} \int_{-\infty}^{\infty} du \cos\left(\frac{u}{2}\log\frac{1+m}{1-m}\right) e^{-\alpha+\alpha\cos(u\beta/2)}. \tag{9.56}$$

Expressing the exponential cosine in terms of modified Bessel functions (using $e^{z \cos \theta} = \sum_k I_k(z) e^{ik\theta}$) and integrating the result over u, we find

$$P(m) = e^{-\alpha} \sum_{k=-\infty}^{\infty} I_k(\alpha) \delta \left(m - \tanh \frac{\beta k}{2} \right). \qquad (9.57)$$

In the interesting case of the zero-temperature limit, this equation reduces to

$$P(m) = e^{-\alpha} I_0(\alpha) \delta(m) + \frac{1}{2} \left(1 - e^{-\alpha} I_0(\alpha) \right) \{ \delta(m-1) + \delta(m+1) \}. \qquad (9.58)$$

Inserting the distribution function (9.57) into (9.55), we obtain the free energy

$$-\frac{\beta F}{N} = \log 2 - \frac{\alpha \beta}{2} + e^{-\alpha} \sum_{k=-\infty}^{\infty} I_k(\alpha) \log \cosh \frac{\beta k}{2} \qquad (9.59)$$

which gives, in the limit $\beta \to \infty$,

$$\frac{E(\alpha)}{N} = \frac{\alpha}{2} - \frac{\alpha}{2} e^{-\alpha} \left(I_0(\alpha) + I_1(\alpha) \right). \qquad (9.60)$$

This ground-state energy is positive for all positive α. It means that the $K = 1$ SAT is always unsatisfiable for $\alpha > 0$. The positive weight of $\delta(m \pm 1)$ in (9.58) is the origin of this behaviour: since a spin is fixed to 1 (or -1) with finite probability $(1 - e^{-\alpha} I_0(\alpha))/2$, the addition of a clause to the already existing M clauses gives a finite probability of yielding a 'false' formula because one may choose a spin fixed to the wrong value as the $(M+1)$th clause. If $P(m)$ were to consist only of a delta function at the origin, we might be able to adjust the spin chosen as the $(M+1)$th clause to give the value 'true' for the whole formula; this is not the case, however.

The coefficient $e^{-\alpha} I_0(\alpha)$ of $\delta(m)$ in (9.58) is the probability that a spin is free to flip in the ground state. Since a single free spin has the entropy $\log 2$, the total entropy might seem to be $N e^{-\alpha} I_0(\alpha) \log 2$, which can also be derived directly from (9.59). A more careful inspection suggests subtraction of $N e^{-\alpha} \log 2$ from the above expression to yield the correct ground-state entropy

$$\frac{S}{N} = e^{-\alpha} (I_0(\alpha) - 1) \log 2. \qquad (9.61)$$

The reason for the subtraction of $N e^{-\alpha} \log 2$ is as follows (Sasamoto, private communication).

Let us consider the simplest case of $M = 1$, $K = 1$, and N arbitrary. Since only a single spin is chosen as the clause, the values of all the other spins do not affect the energy. In this sense the ground-state degeneracy is 2^{N-1} and the corresponding entropy is $(N-1) \log 2$. However, such a redundancy is clearly a trivial one, and we should count only the degree(s) of freedom found in the spins which actually exist in the clause, disregarding those of the redundant spins not chosen in the clauses. Accordingly, the real degeneracy of the above example is unity and the entropy is zero instead of $(N-1)\log 2$.

For general M (and $K = 1$), the probability of a spin not to be chosen in a clause is $1 - 1/N$, so that the probability that a spin is not chosen in any M clauses is $(1-1/N)^M$, which reduces to $e^{-\alpha}$ as $M, N \to \infty$ with $\alpha = M/N$ fixed. Thus the contribution from the redundant spins to the entropy is $Ne^{-\alpha}\log 2$, which is to be subtracted as in (9.61). A little more careful argument on the probability for a spin not to be chosen gives the same answer.

The entropy (9.61) is a non-monotonic function of α, starting from zero at $\alpha = 0$ and vanishing again in the limit $\alpha \to \infty$. Thus it is positive for any positive α, implying a macroscopic degeneracy of the ground state, MAX 1-SAT.

Analysis of the case $K \geq 2$ is much more difficult and the solutions are known only partially. We summarize the results below and refer the interested reader to Monasson and Zecchina (1997, 1998) and Monasson $et~al.$ (2000). If $P(m)$ does not have delta peaks at $m = \pm 1$, the spin states are flexible, and the condition $F = C_1 \wedge \cdots \wedge C_M = $ 'true' can be satisfied. This is indeed the case for small α (easy region). The ground-state energy vanishes in this region, and the problem is satisfiable. There are an exponentially large number of solutions, a finite entropy. When α exceeds a threshold α_c, $P(m)$ starts to have delta peaks at $m = \pm 1$ continuously ($K = 2$) or discontinuously ($K \geq 3$) across α_c. The delta peaks at $m = \pm 1$ imply that a finite fraction of spins are completely frozen so that it is difficult to satisfy the condition $F = $ 'true'; there is a finite probability that all the x_i in a clause are frozen to the wrong values. The K-SAT is unsatisfiable in this region $\alpha > \alpha_c$. The critical point is $\alpha_c = 1$ for $K = 2$ and $\alpha_c = 4.17$ for the discontinuous case of $K = 3$. The latter value of 4.17 is from numerical simulations. It is hard to locate the transition point α_c from the RS analysis for $K \geq 3$ because one should compare the RS and RSB free energies, the latter taking place for $\alpha > \alpha_c$. Comparison with numerical simulations indicates that, for any $K \geq 2$, the RS theory is correct for the easy region $\alpha < \alpha_c$ but not for the difficult region $\alpha > \alpha_c$. In the former region an extensive number of solutions exist as mentioned above, but they suddenly disappear at α_c.

9.6   Simulated annealing

Let us next discuss the convergence problem of simulated annealing. Simulated annealing is widely used as a generic numerical method to solve combinatorial optimization problems. This section is not about a direct application of the spin glass theory but is nonetheless closely related to it; the ground-state search of a spin glass system is a very interesting example of combinatorial optimization, and simulated annealing emerged historically through efforts to identify the correct ground state in the complex energy landscape found typically in spin glasses (Kirkpatrick $et~al.$ 1983). We investigate, in particular, some mathematical aspects of simulated annealing with the generalized transition probability, which

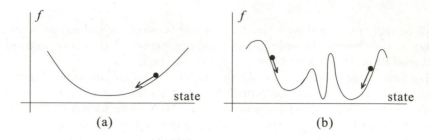

FIG. 9.4. Phase space with a simple structure (a) and a complicated structure (b)

is recently attracting attention for its fast convergence properties and various other reasons (Abe and Okamoto 2001).

9.6.1 *Simulated annealing*

Suppose that we wish to find the optimal state (that minimizes the cost function) by starting from a random initial state and changing the state gradually. If the value of the cost function decreases by a small change of state, we accept the new state as the one that is actually realized at the next step, and we reject the new state if the cost function increases. We generate new states consecutively by this process until no new states are actually accepted, in which case we understand that the optimal state has been reached (Fig. 9.4). This idea is called the *gradient descent method*. If the phase space has a simple structure as in Fig. 9.4(a), the gradient descent always leads to the optimal state. However, this is not necessarily the case if there are local minima which are not true minima as in Fig. 9.4(b), because the system would be trapped in a local minimum for some initial conditions.

It is then useful to introduce transitions induced by thermal fluctuations since they allow processes to increase the value of the cost function with a certain probability. We thus introduce the concept of temperature T as an externally controllable parameter. If the cost function decreases by a small change of state, then we accept the new state just as in the simple gradient descent method. If, on the other hand, the cost function increases, we accept the new state with probability $e^{-\Delta f/T}$ that is determined by the increase of the cost function $\Delta f \, (> 0)$ and the temperature T. In the initial stage of simulated annealing, we keep the temperature high, which stimulates transitions to increase the cost function with relatively high probability because $e^{-\Delta f/T}$ is close to unity. The system searches the global structure of the phase space by such processes that allow the system to stay around states with relatively high values of the cost function. Then, a gradual decrease of the temperature forces the system to have larger probabilities to stay near the optimal state with low f, which implies that more and more local structures are taken into account. We finally let $T \to 0$ to stop state changes and, if successful, the optimal state will be reached. *Simulated annealing* is the idea to realize the above process numerically to obtain an approximate solution of a combinatorial optimization problem. One can clearly reach the true optimal state if the temperature is lowered infinitesimally slowly. In practical numerical calculations one decreases the temperature at a finite speed and terminates the process before the temperature becomes exactly zero if a certain criterion is satisfied. Simulated annealing is an approximate numerical method for this reason.

9.6.2 *Annealing schedule and generalized transition probability*

An important practical issue in simulated annealing is the *annealing schedule*, the rate of temperature decrease. If the temperature were decreased too rapidly, the system would be trapped in a local minimum and lose a chance to escape there because a quick decrease of the temperature soon inhibits processes to increase the cost function. If the temperature is changed sufficiently slowly, on the other hand, the system may be regarded to be approximately in an equilibrium state at each T and the system therefore reaches the true optimal state in the limit $T \to 0$. It is, however, impracticable to decrease the temperature infinitesimally slowly. Thus the problem of the annealing schedule arises in which we ask ourselves how fast we can decrease the temperature without being trapped in local (not global) minima.

Fortunately, this problem has already been solved in the following sense (Geman and Geman 1984; Aarts and Korst 1989). When we allow the system to increase the cost function with probability $e^{-\Delta f/T}$ as mentioned before, the system reaches the true optimal state in the infinite-time limit $t \to \infty$ as long as the temperature is decreased so that it satisfies the inequality $T(t) \geq c/\log(t+2)$. Here c is a constant of the order of the system size N. It should, however, be noted that the logarithm $\log(t+2)$ is only mildly dependent on t and the lower bound of the temperature does not approach zero very quickly. Thus the above bound is not practically useful although it is theoretically important.

An inspection of the proof of the result mentioned above reveals that this logarithmic dependence on time t has its origin in the exponential form of the transition probability $e^{-\Delta f/T}$. It further turns out that this exponential function comes from the Gibbs–Boltzmann distribution $P(x) = e^{-f(x)/T}/Z$ of the equilibrium state at temperature T.

However, it is not necessary to use the exponential transition probability, which comes from the equilibrium distribution function, because we are interested only in the limit $T \to 0$ in combinatorial optimization problems. The only requirement is to reach the optimal state in the end, regardless of intermediate steps. Indeed, numerical investigations have shown that the *generalized transition probability* to be explained below has the property of very rapid convergence to the optimal state and is now used actively in many situations (Abe and Okamoto 2001). We show in the following that simulated annealing using the generalized transition probability converges in the sense of weak ergodicity under an appropriate annealing schedule (Nishimori and Inoue 1998). The proof given here includes the convergence proof of conventional simulated annealing with the exponential transition probability as a special case.

9.6.3 *Inhomogeneous Markov chain*

Suppose that we generate states one after another sequentially by a stochastic process starting from an initial state. We consider a *Markov process* in which the next state is determined only by the present state, so that we call it a *Markov chain*. Our goal in the present section is to investigate conditions of convergence of the Markov chain generated by the generalized transition probability explained below. We first list various definitions and notations.

The cost function is denoted by f that is defined on the set of states (phase space) S. The temperature T is a function of time, and accordingly the transition probability G from a state $x\,(\in S)$ to y is also a function of time $t\,(=0,1,2,\ldots)$. This defines an *inhomogeneous Markov chain* in which the transition probability depends on time. This G is written as follows:

$$G(x,y;t) = \begin{cases} P(x,y)A(x,y;T(t)) & (x \neq y) \\ 1 - \sum_{z(\neq x)} P(x,z)A(x,z;T(t)) & (x = y). \end{cases} \qquad (9.62)$$

Here $P(x,y)$ is the *generation probability* (probability to generate a new state)

$$P(x,y) \begin{cases} > 0 & (y \in S_x) \\ = 0 & (\text{otherwise}), \end{cases} \qquad (9.63)$$

where S_x is the *neighbour* of x, the set of states that can be reached by a single step from x. In (9.62), $A(x,y;T)$ is the *acceptance probability* (probability by which the system actually makes a transition to the new state) with the form

$$A(x,y;T) = \min\{1, u(x,y;T)\}$$
$$u(x,y;T) = \left(1 + (q-1)\frac{f(y) - f(x)}{T}\right)^{1/(1-q)}. \qquad (9.64)$$

Here q is a real parameter and we assume $q > 1$ for the moment. Equation (9.62) shows that one generates a new trial state y with probability $P(x,y)$, and the system actually makes a transition to it with probability $A(x,y;T)$. According to (9.64), when the change of the cost function $\Delta f = f(y) - f(x)$ is zero or negative, we have $u \geq 1$ and thus $A(x,y;T) = 1$ and the state certainly changes to the new one. If Δf is positive, on the other hand, $u < 1$ and the transition to y is determined with probability $u(x,y;T)$. If $\Delta f < 0$ and the quantity in the large parentheses in (9.64) vanishes or is negative, then we understand that $u \to \infty$ or $A = 1$. It is assumed that the generation probability $P(x,y)$ is *irreducible*: it is possible to move from an arbitrary state in S to another arbitrary state by successive transitions between pairs of states x and y satisfying $P(x,y) > 0$. It should be remarked here that the acceptance probability (9.64) reduces to the previously mentioned exponential form $e^{-\Delta f/T}$ in the limit $q \to 1$. Equation (9.64) is a generalization of the Gibbs–Boltzmann framework in this sense.

Now we choose the annealing schedule as follows:

$$T(t) = \frac{b}{(t+2)^c} \quad (b, c > 0, t = 0, 1, 2, \ldots). \tag{9.65}$$

In the convergence proof of conventional simulated annealing corresponding to $q \to 1$, a logarithm of t appears in the denominator of (9.65). Here in the case of the generalized transition probability (9.62)–(9.64), it will turn out to be appropriate to decrease T as a power of t.

It is convenient to regard $G(x, y; t)$ as a matrix element of a transition matrix $G(t)$:

$$[G(t)]_{x,y} = G(x, y; t). \tag{9.66}$$

We denote the set of probability distributions on $\mathcal{S}$ by $\mathcal{P}$ and regard a probability distribution p as a row vector with element $p(x)$ ($x \in \mathcal{S}$). If at time s the system is in the state described by the probability distribution p_0 ($\in \mathcal{P}$), the probability distribution at time t is given as

$$p(s, t) = p_0 G^{s,t} \equiv p_0 G(s) G(s + 1) \ldots G(t - 1). \tag{9.67}$$

We also introduce the *coefficient of ergodicity* as follows, which is a measure of the state change in a single step:

$$\alpha(G^{s,t}) = 1 - \min \left\{ \sum_{z \in \mathcal{S}} \min\{G^{s,t}(x, z), G^{s,t}(y, z)\} | x, y \in \mathcal{S} \right\}. \tag{9.68}$$

In the next section we prove *weak ergodicity* of the system, which means that the probability distribution is asymptotically independent of the initial condition:

$$\forall s \geq 0: \lim_{t \to \infty} \sup\{\|p_1(s, t) - p_2(s, t)\| \mid p_{01}, p_{02} \in \mathcal{P}\} = 0, \tag{9.69}$$

where $p_1(s, t)$ and $p_2(s, t)$ are distribution functions with different initial conditions,

$$p_1(s, t) = p_{01} G^{s,t}, \quad p_2(s, t) = p_{02} G^{s,t}. \tag{9.70}$$

The norm of the difference of probability distributions in (9.69) is defined by

$$\|p_1 - p_2\| = \sum_{x \in \mathcal{S}} |p_1(x) - p_2(x)|. \tag{9.71}$$

Strong ergodicity to be contrasted with weak ergodicity means that the probability distribution approaches a fixed distribution irrespective of the initial condition:

$$\exists r \in \mathcal{P}, \forall s \geq 0: \lim_{t \to \infty} \sup\{\|p(s, t) - r\| \mid p_0 \in \mathcal{P}\} = 0. \tag{9.72}$$

The conditions for ergodicity are summarized in the following theorems (Aarts and Korst 1989).

Theorem 9.1. (Weak ergodicity) *An inhomogeneous Markov chain is weakly ergodic if and only if there exists a monotonically increasing sequence of integers*

$$0 < t_0 < t_1 < \cdots < t_i < t_{i+1} < \ldots$$

and the coefficient of ergodicity satisfies

$$\sum_{i=0}^{\infty}(1 - \alpha(G^{t_i,t_{i+1}})) = \infty. \tag{9.73}$$

Theorem 9.2. (Strong ergodicity) *An inhomogeneous Markov chain is strongly ergodic if*

- *it is weakly ergodic,*
- *there exists a stationary state $p_t = p_t G(t)$ at each t, and*
- *the above p_t satisfies the condition*

$$\sum_{t=0}^{\infty} \|p_t - p_{t+1}\| < \infty. \tag{9.74}$$

9.6.4   Weak ergodicity

We prove in this section that the Markov chain generated by the generalized transition probability (9.62)–(9.64) is weakly ergodic. The following lemma is useful for this purpose.

Lemma 9.3. (Lower bound to the transition probability) *The transition probability of the inhomogeneous Markov chain defined in §9.6.3 satisfies the following inequality. Off-diagonal elements of G for transitions between different states satisfy*

$$P(x,y) > 0 \Rightarrow \forall t \geq 0 : G(x,y;t) \geq w \left(1 + \frac{(q-1)L}{T(t)}\right)^{1/(1-q)}. \tag{9.75}$$

For diagonal elements we have

$$\forall x \in \mathcal{S} \setminus \mathcal{S}^M, \exists t_1 > 0, \forall t \geq t_1 : G(x,x;t) \geq w \left(1 + \frac{(q-1)L}{T(t)}\right)^{1/(1-q)}. \tag{9.76}$$

Here $\mathcal{S}^M$ is the set of states to locally maximize the cost function

$$\mathcal{S}^M = \{x | x \in \mathcal{S}, \forall y \in \mathcal{S}_x : f(y) \leq f(x)\}, \tag{9.77}$$

L is the largest value of the change of the cost function by a single step

$$L = \max\{|f(x) - f(y)| \mid P(x,y) > 0\}, \tag{9.78}$$

and w is the minimum value of the non-vanishing generation probability

$$w = \min\{P(x,y) \mid P(x,y) > 0, x,y \in \mathcal{S}\}. \tag{9.79}$$

Proof We first prove (9.75) for off-diagonal elements. If $f(y) - f(x) > 0$, then $u(x, y; T(t)) \leq 1$ and therefore

$$G(x, y; t) = P(x, y)A(x, y; T(t)) \geq w \min\{1, u(x, y; T(t))\}$$
$$= w\, u(x, y; T(t)) \geq w \left(1 + \frac{(q-1)L}{T(t)}\right)^{1/(1-q)}. \qquad (9.80)$$

When $f(x) - f(y) \leq 0$, $u(x, y; T(t)) \geq 1$ holds, leading to

$$G(x, y; t) \geq w \min\{1, u(x, y; T(t))\} = w \geq w \left(1 + \frac{(q-1)L}{T(t)}\right)^{1/(1-q)}. \qquad (9.81)$$

To prove the diagonal part (9.76), we note that there exists a state $y_+ \in S_x$ to increase the cost function $f(y_+) - f(x) > 0$ because of $x \in S \setminus S^M$. Then

$$\lim_{t \to \infty} u(x, y; T(t)) = 0, \qquad (9.82)$$

and therefore

$$\lim_{t \to \infty} \min\{1, u(x, y; T(t))\} = 0. \qquad (9.83)$$

For sufficiently large t, $\min\{1, u(x, y; T(t))\}$ can be made arbitrarily small, and hence there exists some t_1 for an arbitrary $\epsilon > 0$ which satisfies

$$\forall t \geq t_1 : \min\{1, u(x, y; T(t))\} < \epsilon. \qquad (9.84)$$

Therefore

$$\sum_{z \in S} P(x, z)A(x, z; T(t))$$
$$= \sum_{\{y_+\}} P(x, y_+) \min\{1, u(x, y_+; T(t))\} + \sum_{z \in S \setminus \{y_+\}} P(x, z) \min\{1, u(x, z; T(t))\}$$
$$< \sum_{\{y_+\}} P(x, y_+)\epsilon + \sum_{z \in S \setminus \{y_+\}} P(x, z) = -(1 - \epsilon) \sum_{\{y_+\}} P(x, y_+) + 1. \qquad (9.85)$$

From this inequality and (9.62) the diagonal element satisfies

$$G(x, x; t) \geq (1 - \epsilon) \sum_{\{y_+\}} P(x, y_+) \geq w \left(1 + \frac{(q-1)L}{T(t)}\right)^{1/(1-q)}. \qquad (9.86)$$

In the final inequality we have used the fact that the quantity in the large parentheses can be chosen arbitrarily small for sufficiently large t. $\square$

It is convenient to define some notation to prove weak ergodicity. Let us write $d(x, y)$ for the minimum number of steps to make a transition from x to y. The maximum value of $d(x, y)$ as a function of y will be denoted as $k(x)$:

$$k(x) = \max\{d(x, y)|y \in \mathcal{S}\}. \tag{9.87}$$

Thus one can reach an arbitrary state within $k(x)$ steps starting from x. The minimum value of $k(x)$ for x such that $x \in \mathcal{S} \setminus \mathcal{S}^M$ is written as R, and the x to give this minimum value will be x^*:

$$R = \min\{k(x)|x \in \mathcal{S} \setminus \mathcal{S}^M\}, \quad x^* = \arg\min\{k(x)|x \in \mathcal{S} \setminus \mathcal{S}^M\}. \tag{9.88}$$

Theorem 9.4. (Weak ergodicity : generalized transition probability) *The inhomogeneous Markov chain defined in §9.6.3 is weakly ergodic if* $0 < c \leq (q-1)/R$.

Proof Let us consider a transition from x to x^*. From (9.67),

$$G^{t-R,t}(x, x^*) = \sum_{x_1,\ldots,x_{R-1}} G(x, x_1; t-R)G(x_1, x_2; t-R+1)\ldots G(x_{R-1}, x^*; t-1). \tag{9.89}$$

There exists a sequence of transitions to reach x^* from x within R steps according to the definitions of x^* and R,

$$x \neq x_1 \neq x_2 \neq \cdots \neq x_k = x_{k+1} = \cdots = x_R = x^*. \tag{9.90}$$

We keep this sequence only in the sum (9.89) and use Lemma 9.3 to obtain

$$G^{t-R,t}(x, x^*) \geq G(x, x_1; t-R)G(x_1, x_2; t-R+1)\ldots G(x_{R-1}, x_R; t-1)$$

$$\geq \prod_{k=1}^{R} w\left(1 + \frac{(q-1)L}{T(t-R+k-1)}\right)^{1/(1-q)}$$

$$\geq w^R\left(1 + \frac{(q-1)L}{T(t-1)}\right)^{R/(1-q)}. \tag{9.91}$$

It therefore follows that the coefficient of ergodicity satisfies the inequality

$$\alpha(G^{t-R,t}) = 1 - \min\left\{\sum_{z \in \mathcal{S}} \min\{G^{t-R,t}(x, z), G^{t-R,t}(y, z)\}|x, y \in \mathcal{S}\right\}$$

$$\leq 1 - \min\{\min\{G^{t-R,t}(x, x^*), G^{t-R,t}(y, x^*)\}|x, y \in \mathcal{S}\}$$

$$\leq 1 - w^R\left(1 + \frac{(q-1)L}{T(t-1)}\right)^{R/(1-q)}. \tag{9.92}$$

We now use the annealing schedule (9.65). According to (9.92) there exists a positive integer k_0 such that the following inequality holds for any integer k satisfying $k \geq k_0$:

$$1 - \alpha(G^{kR-R,kR}) \geq w^R\left(1 + \frac{(q-1)L(kR+1)^c}{b}\right)^{R/(1-q)}$$

$$\geq w^R \left\{ \frac{2(q-1)LR^c}{b} \left(k + \frac{1}{R} \right)^c \right\}^{R/(1-q)}. \qquad (9.93)$$

It is then clear that the following quantity diverges when $0 < c \leq (q-1)/R$:

$$\sum_{k=0}^{\infty} (1 - \alpha(G^{kR-R,kR})) = \sum_{k=0}^{k_0-1} (1 - \alpha(G^{kR-R,kR})) + \sum_{k=k_0}^{\infty} (1 - \alpha(G^{kR-R,kR})). \qquad (9.94)$$

This implies weak ergodicity from Theorem 9.1.   □

This proof breaks down if $q < 1$ since the quantity in the large parentheses of (9.64) may be negative even when $\Delta f > 0$. In numerical calculations, such cases are treated as $u = 0$, no transition. Fast relaxations to the optimal solutions are observed often for $q < 1$ in actual numerical investigations. It is, however, difficult to formulate a rigorous proof for this case.

It is also hard to prove a stronger result of strong ergodicity and approach to the optimal distribution (distribution uniform over the optimal states) for the Markov chain defined in §9.6.3 for general q. Nevertheless, weak ergodicity itself has physically sufficient significance because the asymptotic probability distribution does not depend on the initial condition; it is usually inconceivable that such an asymptotic state independent of the initial condition is not the optimal one or changes with time periodically.

9.6.5 *Relaxation of the cost function*

It is not easy to prove the third condition of strong ergodicity in Theorem 9.2, (9.74), for a generalized transition probability with $q \neq 1$. However, if we restrict ourselves to the conventional transition probability $e^{-\Delta f/T}$ corresponding to $q \to 1$, the following theorem can be proved (Geman and Geman 1984).

Theorem 9.5. (Strong ergodicity : conventional transition probability)
If we replace (9.64) and (9.65) in §9.6.3 by

$$u(x, y; T) = \exp\{-(f(y) - f(x))/T\} \qquad (9.95)$$

$$T(t) \geq \frac{RL}{\log(t+2)}, \qquad (9.96)$$

then this Markov chain is strongly ergodic. The probability distribution in the limit $t \to \infty$ converges to the optimal distribution. Here R and L are defined as in §9.6.4.

To prove this theorem, we set $q \to 1$ in (9.92) in the proof of Theorem 9.4. By using the annealing schedule of (9.96), we find that (9.94) diverges, implying weak ergodicity. It is also well known that the stationary distribution at temperature $T(t)$ for any given t is the Gibbs–Boltzmann distribution, so that the second condition of Theorem 9.2 is satisfied. Some manipulations are necessary to prove the third convergence condition (9.74), and we only point out two facts essential for the proof: the first is that the probability of the optimal state monotonically increases with decreasing temperature due to the explicit form of the Gibbs–Boltzmann distribution. The second one is that the probabilities of non-optimal states monotonically decrease with decreasing temperature at sufficiently low temperature.

A comment is in order on the annealing schedule. As one can see in Theorem 9.4, the constant c in the annealing schedule (9.65) is bounded by $(q-1)/R$, but R is of the order of the system size N by the definition (9.88) of R.[23] Then, as N increases, c decreases and the change of $T(t)$ becomes very mild. The same is true for (9.96). In practice, one often controls the temperature irrespective of the mathematically rigorous result as in (9.65) and (9.96); for example, an exponential decrease of temperature is commonly adopted. If the goal is to obtain an approximate estimation within a given, limited time, it is natural to try fast annealing schedules even when there is no guarantee of asymptotic convergence.

It is instructive to investigate in more detail which is actually faster between the power decay of temperature (9.65) and the logarithmic law (9.96). The time t_1 necessary to reach a very low temperature δ by (9.65) is, using $b/t_1^c \approx \delta$ (where $c = (q-1)/R$),

$$ t_1 \approx \exp\left(\frac{k_1 N}{q-1} \log \frac{b}{\delta}\right). \tag{9.97} $$

Here we have set $R = k_1 N$. For the case of (9.96), on the other hand, from $k_2 N/\log t_2 \approx \delta$,

$$ t_2 \approx \exp\left(\frac{k_2 N}{\delta}\right). \tag{9.98} $$

Both of these are of exponential form in N, which is reasonable because we are discussing generic optimization problems including the class NP complete. An improvement in the case of the generalized transition probability (9.97) is that δ appears as $\log \delta$ whereas it has $1/\delta$-dependence in t_2, which means a smaller coefficient of N for small δ in the former case.

It should also be remarked that smaller temperature does not immediately mean a smaller value of the cost function. To understand this, note that the acceptance probability (9.64) for $q \neq 1$ is, when $T = \delta \ll 1$,

$$ u_1(T = \delta) \approx \left(\frac{\delta}{(q-1)\Delta f}\right)^{1/(q-1)}, \tag{9.99} $$

while for $q = 1$, according to (9.95),

$$ u_2(T = \delta) \approx e^{-\Delta f/\delta}. \tag{9.100} $$

For transitions satisfying $\Delta f/\delta \gg 1$, we have $u_1(\delta) \gg u_2(\delta)$. This implies that transitions to states with high cost function values take place easier for $q \neq 1$ than for $q = 1$ if the temperature is the same. Transition probabilities for $q \neq 1$ cause

[23]The number of steps to reach arbitrary states is at least of the order of the system size.

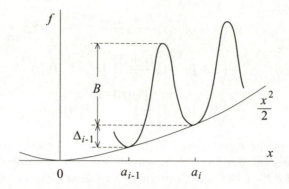

FIG. 9.5. Potential and barriers in one dimension

state searches among wide regions in the phase space even at low temperatures. It would follow that the equilibrium expectation value of the cost function after a sufficiently long time at a fixed temperature is likely to be higher in the case $q \neq 1$ than in $q = 1$. In numerical experiments, however, it is observed in many cases that $q \neq 1$ transition probabilities lead to faster convergence to lower values of the cost function. The reason may be that the relaxation time for $q \neq 1$ is shorter than for $q = 1$, which helps the system escape local minima to relax quickly towards the real minimum.

9.7 Diffusion in one dimension

The argument in the previous section does not directly show that we can reach the optimal state faster by the generalized transition probability. It should also be kept in mind that we have no proof of convergence in the case of $q < 1$. In the present section we fill this gap by the example of diffusion in one dimension due to Shinomoto and Kabashima (1991). It will be shown that the generalized transition probability with $q < 1$ indeed leads to a much faster relaxation to the optimal state than the conventional one $q = 1$ (Nishimori and Inoue 1998).

9.7.1 *Diffusion and relaxation in one dimension*

Suppose that a particle is located at one of the discrete points $x = ai$ (with i integer and $a > 0$) in one dimension and is under the potential $f(x) = x^2/2$. There are barriers between the present and the neighbouring locations $i \pm 1$. The height is B to the left ($i \to i - 1$) and $B + \Delta_i$ to the right ($i \to i + 1$) as shown in Fig. 9.5. Here Δ_i is the potential difference between two neighbouring points, $\Delta_i = f(a(i+1)) - f(ai) = ax + a^2/2$.

The probability $P_t(i)$ that the particle is located at $x = ai$ at time t follows the master equation using the generalized transition probability:

$$\frac{\mathrm{d}P_t(i)}{\mathrm{d}t} = \left(1 + (q-1)\frac{B}{T}\right)^{1/(1-q)} P_t(i+1)$$

$$+ \left(1 + (q-1)\frac{B + \Delta_{i-1}}{T}\right)^{1/(1-q)} P_t(i-1)$$

$$- \left(1 + (q-1)\frac{B + \Delta_i}{T}\right)^{1/(1-q)} P_t(i)$$

$$- \left(1 + (q-1)\frac{B}{T}\right)^{1/(1-q)} P_t(i). \tag{9.101}$$

The first term on the right hand side represents the process that the particle at $i+1$ goes over the barrier B to i, which increases the probability at i. The second term is for $i-1 \to i$, the third for $i \to i+1$, and the fourth for $i \to i-1$. It is required that the transition probability in (9.101) should be positive semi-definite. This condition is satisfied if we restrict q to $q = 1 - (2n)^{-1}$ ($n = 1, 2, \ldots$) since the power $1/(1-q)$ then equals $2n$, which we therefore accept here.

It is useful to take the continuum limit $a \to 0$ to facilitate the analysis. Let us define $\gamma(T)$ and $D(T)$ by

$$\gamma(T) = \frac{1}{T}\left(1 + (q-1)\frac{B}{T}\right)^{q/(1-q)}, \quad D(T) = \left(1 + (q-1)\frac{B}{T}\right)^{1/(1-q)}, \tag{9.102}$$

and expand the transition probability of (9.101) to first order in Δ_i and Δ_{i-1} ($\propto a$) to derive

$$\frac{\mathrm{d}P_t(i)}{\mathrm{d}t} = D(T)\left\{P_t(i+1) - 2P_t(i) + P_t(i-1)\right\}$$

$$+ a\gamma(T)\left\{xP_t(i) + \frac{a}{2}P_t(i) - xP_t(i-1) + \frac{a}{2}P_t(i-1)\right\}. \tag{9.103}$$

We rescale the time step as $a^2 t \to t$ and take the limit $a \to 0$, which reduces the above equation to the form of the *Fokker–Planck equation*

$$\frac{\partial P}{\partial t} = \gamma(T)\frac{\partial}{\partial x}(xP) + D(T)\frac{\partial^2 P}{\partial x^2}. \tag{9.104}$$

We are now ready to study the time evolution of the expectation value of the cost function y:

$$y(t) = \int \mathrm{d}x\, f(x)P(x,t). \tag{9.105}$$

The goal is to find an appropriate annealing schedule $T(t)$ to reduce y to the optimal value 0 as quickly as possible. The time evolution equation for y can be derived from the Fokker–Planck equation (9.104) and (9.105),

$$\frac{\mathrm{d}y}{\mathrm{d}t} = -2\gamma(T)y + D(T). \tag{9.106}$$

Maximization of the rate of decrease of y is achieved by minimization of the right hand side of (9.106) as a function of T at each time (or maximization of the absolute value of this negative quantity). We hence differentiate the right hand side of (9.106) with respect to T using the definitions of $\gamma(T)$ and $D(T)$, (9.102):

$$T_{\text{opt}} = \frac{2yB + (1-q)B^2}{2y + B} = (1-q)B + 2qy + \mathcal{O}(y^2). \tag{9.107}$$

Thus (9.106) is asymptotically, as $y \approx 0$,

$$\frac{dy}{dt} = -2B^{1/(q-1)} \left(\frac{2q}{1-q}\right)^{q/(1-q)} y^{1/(1-q)}, \tag{9.108}$$

which is solved as

$$y = B^{1/q} \left(\frac{1-q}{2q}\right)^{1/q} t^{-(1-q)/q}. \tag{9.109}$$

Substitution into (9.107) reveals the optimal annealing schedule as

$$T_{\text{opt}} \approx (1-q)B + \text{const} \cdot t^{-(1-q)/q}. \tag{9.110}$$

Equation (9.109) indicates that the relaxation of y is fastest when $q = 1/2$ ($n = 1$), which leads to

$$y \approx \frac{B^2}{4} t^{-1}, \quad T_{\text{opt}} \approx \frac{B}{2} + \frac{B^2}{4} t^{-1}. \tag{9.111}$$

The same analysis for the conventional exponential transition probability with $q \to 1$ in the master equation (9.101) leads to the logarithmic form of relaxation (Shinomoto and Kabashima 1991)

$$y \approx \frac{B}{\log t}, \quad T_{\text{opt}} \approx \frac{B}{\log t}. \tag{9.112}$$

Comparison of (9.111) and (9.112) clearly shows a faster relaxation to $y = 0$ in the case of $q = 1/2$.

A remark on the significance of temperature is in order. T_{opt} in (9.107) approaches a finite value $(1-q)B$ in the limit $t \to \infty$, which may seem unsatisfactory. However, the transition probability in (9.101) vanishes at $T = (1-q)B$, not at $T = 0$, if $q \neq 1, a \to 0$, and therefore $T = (1-q)B$ effectively plays the role of absolute zero temperature.

Bibliographical note

The application of statistical mechanics to optimization problems started with simulated annealing (Kirkpatrick et al. 1983). Review articles on optimization problems, not just simulated annealing but including travelling salesman, graph partitioning, matching, and related problems, from a physics point of view are found in Mézard et al. (1987) and van Hemmen and Morgenstern (1987), which cover most materials until the mid 1980s. A more complete account of simulated annealing with emphasis on mathematical aspects is given in Aarts and Korst (1989). Many optimization problems can be formulated in terms of neural networks. Detailed accounts are found in Hertz et al. (1991) and Bishop (1995).

APPENDIX A

EIGENVALUES OF THE HESSIAN

In this appendix we derive the eigenvalues and eigenvectors of the Hessian discussed in Chapter 3. Let us note that the dimensionality of the matrix G is equal to the sum of the spatial dimension of ϵ^α and that of $\eta^{\alpha\beta}$, $n + n(n-1)/2 = n(n+1)/2$. We write the eigenvalue equation as

$$G\boldsymbol{\mu} = \lambda\boldsymbol{\mu}, \ \boldsymbol{\mu} = \begin{pmatrix} \{\epsilon^\alpha\} \\ \{\eta^{\alpha\beta}\} \end{pmatrix}. \tag{A.1}$$

The symbol $\{\epsilon^\alpha\}$ denotes a column from ϵ^1 at the top to ϵ^n at the bottom and $\{\eta^{\alpha\beta}\}$ is for η^{12} to $\eta^{n-1,n}$.

A.1 Eigenvalue 1

There are three types of eigenvectors. The first one $\boldsymbol{\mu}_1$ treated in the present section has the form $\epsilon^\alpha = a, \eta^{\alpha\beta} = b$. The first row of G is written as

$$(A, B, \ldots, B, C, \ldots, C, D \ldots, D), \tag{A.2}$$

so that the first row of the eigenvalue equation $G\boldsymbol{\mu}_1 = \lambda\boldsymbol{\mu}_1$ is

$$Aa + (n-1)Ba + (n-1)Cb + \frac{1}{2}(n-1)(n-2)Db = \lambda_1 a. \tag{A.3}$$

The lower half of the same eigenvalue equation (corresponding to $\{\eta^{\alpha\beta}\}$) is, using the form of the corresponding row of G, $(C, C, D, \ldots, D, P, Q, \ldots, Q, R, \ldots, R)$,

$$2Ca + (n-2)Da + Pb + 2(n-2)Qb + \frac{1}{2}(n-2)(n-3)Rb = \lambda_1 b. \tag{A.4}$$

The factor of 2 in front of the first C comes from the observation that both $G_{\alpha(\alpha\beta)}$ and $G_{\beta(\alpha\beta)}$ are C for fixed $(\alpha\beta)$. The factor $(n-2)$ in front of D reflects the number of replicas γ giving $G_{\gamma(\alpha\beta)} = D$. The $2(n-2)$ in front of Q is the number of choices of replicas satisfying $G_{(\alpha\beta)(\alpha\gamma)} = Q$, and similarly for $(n-2)(n-3)/2$. The condition that both (A.3) and (A.4) have a solution with non-vanishing a and b yields

$$\lambda_1 = \frac{1}{2}(X \pm \sqrt{Y^2 + Z}), \tag{A.5}$$

$$X = A + (n-1)B + P + 2(n-2)Q + \frac{1}{2}(n-2)(n-3)R \tag{A.6}$$

$$Y = A + (n-1)B - P - 2(n-2)Q - \frac{1}{2}(n-2)(n-3)R \qquad (A.7)$$

$$Z = 2(n-1)\{2C + (n-2)D\}^2. \qquad (A.8)$$

This eigenvalue reduces in the limit $n \to 0$ to

$$\lambda_1 = \frac{1}{2}\left\{A - B + P - 4Q + 3R \pm \sqrt{(A-B-P+4Q-3R)^2 - 8(C-D)^2}\right\}. \qquad (A.9)$$

A.2 Eigenvalue 2

The next type of solution μ_2 has $\epsilon^\theta = a$ (for a specific replica θ), $\epsilon^\alpha = b$ (otherwise) and $\eta^{\alpha\beta} = c$ (when α or β is equal to θ), $\eta^{\alpha\beta} = d$ (otherwise). We assume $\theta = 1$ without loss of generality. The first row of the matrix G has the form $(A, B\ldots, B, C, \ldots, C, D, \ldots, D)$. Both B and C appear $n-1$ times, and D exists $(n-1)(n-2)/2$ times. Vector μ_2 is written as $^t(a, b, \ldots, b, c, \ldots, c, d, \ldots, d)$, where there are $n-1$ of b and c, and $(n-1)(n-2)/2$ of d. The first row of the eigenvalue equation $G\mu_2 = \lambda_2\mu_2$ is

$$Aa + (n-1)Bb + Cc(n-1) + \frac{1}{2}Dd(n-1)(n-2) = \lambda_2 a. \qquad (A.10)$$

The present vector μ_2 should be different from the previous μ_1 and these vectors must be orthogonal to each other. A sufficient condition for orthogonality is that the upper halves (with dimensionality n) of μ_1 and μ_2 have a vanishing inner product and similarly for the lower halves. Then, using the notation $\mu_1 = {}^t(x, x, \ldots, x, y, y, \ldots, y)$, we find

$$a + (n-1)b = 0, \quad c + \frac{1}{2}(n-2)d = 0. \qquad (A.11)$$

Equation (A.10) is now rewritten as

$$(A - \lambda_2 - B)a + (n-1)(C-D)c = 0. \qquad (A.12)$$

We next turn to the lower half of the eigenvalue equation corresponding to $\{\eta^{\alpha\beta}\}$. The relevant row of G is $(C, C, D, \ldots, D, P, Q, \ldots, Q, R, \ldots, R)$, where there are $n-2$ of the D, $2(n-2)$ of the Q, and $(n-2)(n-3)/2$ of the R. The eigenvector μ_2 has the form $^t(a, b, \ldots, b, c, \ldots, c, d, \ldots, d)$. Hence we have

$$aC + bC + (n-2)Db + Pc + (n-2)Qc + (n-2)Qd + \frac{1}{2}(n-2)(n-3)Rd = \lambda_2 c. \quad (A.13)$$

This relation can be written as, using (A.11),

$$\frac{n-2}{n-1}(C-D)a + \{P + (n-4)Q - (n-3)R - \lambda_2\}c = 0. \qquad (A.14)$$

The condition that (A.12) and (A.14) have non-vanishing solution yields

$$\lambda_2 = \frac{1}{2}(X \pm \sqrt{Y^2 + Z}), \qquad (A.15)$$

$$X = A - B + P + (n-4)Q - (n-3)R \tag{A.16}$$
$$Y = A - B - P - (n-4)Q + (n-3)R \tag{A.17}$$
$$Z = 4(n-2)(C-D)^2. \tag{A.18}$$

This eigenvalue becomes degenerate with λ_1 in the limit $n \to 0$.

There are n possible choices of the special replica θ, so that we may choose n different eigenvectors μ_2. Dimensionality n corresponding to $\{\epsilon^\alpha\}$ from $n(n+1)/2$ dimensions has thus been exhausted. Within this subspace, the eigenvectors μ_1 and μ_2 cannot all be independent as there are no more than n independent vectors in this space. Therefore we have n independent vectors formed from μ_1 and μ_2. If we recall that λ_1 and λ_2 are both doubly degenerate, the eigenvectors μ_1 and μ_2 are indeed seen to construct a $2n$-dimensional space.

A.3 Eigenvalue 3

The third type of eigenvector μ_3 has $\epsilon^\theta = a, \epsilon^\nu = a$ (for two specific replicas θ, ν) and $\epsilon^\alpha = b$ (otherwise), and $\eta^{\theta\nu} = c, \eta^{\theta\alpha} = \eta^{\nu\alpha} = d$ and $\eta^{\alpha\beta} = e$ otherwise. We may assume $\theta = 1, \nu = 2$ without loss of generality.

A sufficient condition for orthogonality with $\mu_1 = {}^t(x, \ldots, x, y \ldots, y)$ gives

$$2a + (n-2)b = 0, \quad c + 2(n-2)d + \frac{1}{2}(n-2)(n-3)e = 0. \tag{A.19}$$

To check a sufficient condition of orthogonality of μ_3 and μ_2, we write $\mu_2 = {}^t(x, y, \ldots, y, v, \ldots, v, w, \ldots, w)$ and obtain

$$ax + ay + (n-2)by = 0, \quad cv + (n-2)dv = 0, \quad (n-2)dw + \frac{1}{2}(n-2)(n-3)ew = 0. \tag{A.20}$$

From this and the condition $x + (n-1)y = 0$ as derived in (A.11), we obtain

$$a - b = 0, \quad c + (n-2)d = 0, \quad d + \frac{1}{2}(n-3)e = 0. \tag{A.21}$$

From (A.19) and (A.21), $a = b = 0, c = (2-n)d, d = (3-n)e/2$. This relation reduces the upper half of the eigenvalue equation (corresponding to $\{\epsilon^\alpha\}$) to the trivial form $0 = 0$. The relevant row of G is $(\ldots, P, Q, \ldots, Q, R, \ldots, R)$ and $\mu_3 = {}^t(0, \ldots, 0, c, d, \ldots, d, e, \ldots, e)$. Thus the eigenvalue equation is

$$Pc + 2(n-2)Qd + \frac{1}{2}(n-2)(n-3)Re = \lambda_3 c, \tag{A.22}$$

which can be expressed as, using (A.21),

$$\lambda_3 = P - 2Q + R. \tag{A.23}$$

The degeneracy of λ_3 may seem to be $n(n-1)/2$ from the number of choices of θ and ν. However, n vectors have already been used in relation to λ_1, λ_2 and the actual degeneracy (the number of independent vectors) is $n(n-3)/2$. Together with the degeneracy of λ_1 and λ_2, we have $n(n+1)/2$ vectors and exhausted all the eigenvalues.

APPENDIX B

PARISI EQUATION

We derive the free energy in the full RSB scheme for the SK model in this appendix following Duplantier (1981). The necessary work is the evaluation of the term $\mathrm{Tr}\,e^L$ in the free energy (2.17). We set $\beta = J = 1$ during calculations and retrieve these afterwards by dimensionality arguments.

As one can see from the form of the matrix (3.25) in §3.2.1, the diagonal blocks have elements q_K. Thus we may carry out calculations with the diagonal element $q_{\alpha\alpha}$ kept untouched first in the sum of $q_{\alpha\beta}S^\alpha S^\beta$ and add $\beta^2 J^2 q_K/2$ $(q_K \to q(1))$ to βf later to cancel this extra term. We therefore evaluate

$$G = \mathrm{Tr}\,\exp\left(\frac{1}{2}\sum_{\alpha,\beta=1}^n q_{\alpha\beta}S^\alpha S^\beta + h\sum_\alpha^n S^\alpha\right)$$

$$= \exp\left(\frac{1}{2}\sum_{\alpha,\beta} q_{\alpha\beta}\frac{\partial^2}{\partial h_\alpha \partial h_\beta}\right)\prod_\alpha 2\cosh h_\alpha\Bigg]_{h_\alpha=h}. \tag{B.1}$$

If all the $q_{\alpha\beta}$ are equal to q (the RS solution), the manipulation is straightforward and yields

$$G = \exp\left(\frac{q}{2}\frac{\partial^2}{\partial h^2}\right)(2\cosh h)^n, \tag{B.2}$$

where we have used

$$\sum_\alpha \frac{\partial f(h_1,\ldots,h_n)}{\partial h_\alpha}\Bigg]_{h_\alpha=h} = \frac{\partial f(h,\ldots,h)}{\partial h}. \tag{B.3}$$

In the case of 2RSB, we may reach the $n \times n$ matrix $\{q_{\alpha\beta}\}$ in three steps by increasing the matrix dimension as m_2, m_1, n:

(2-1) $(q_2 - q_1)I(m_2)$. Here $I(m_2)$ is an $m_2 \times m_2$ matrix with all elements unity.

(2-2) $(q_2 - q_1)\mathrm{Diag}_{m_1}[I(m_2)] + (q_1 - q_0)I(m_1)$. The matrix $\mathrm{Diag}_{m_1}[I(m_2)]$ has dimensionality $m_1 \times m_1$ with all diagonal blocks equal to $I(m_2)$ and all the other elements zero. The second term $(q_1 - q_0)I(m_1)$ specifies all elements to $q_1 - q_0$, and the first term replaces the block-diagonal part by $q_2 - q_0$.

(2-3) $(q_2 - q_1)\mathrm{Diag}_n[\mathrm{Diag}_{m_1}[I(m_2)]] + (q_1 - q_0)\mathrm{Diag}_n[I(m_1)] + q_0 I(n)$. All the elements are set first to q_0 by the third term. The second term replaces the diagonal block of size $m_1 \times m_1$ by q_1. The elements of the innermost diagonal block of size $m_2 \times m_2$ are changed to q_2 by the first term.

Similarly, for general K-RSB,

(K-1) $(q_K - q_{K-1})I(m_K)$ ($m_K \times m_K$ matrix),

(K-2) $(q_K - q_{K-1})\text{Diag}_{K-1}[I(m_K)] + (q_{K-1} - q_{K-2})I(m_{K-1})$ ($m_{K-1} \times m_{K-1}$ matrix),

and so on. Now, suppose that we have carried out the trace operation for the $m_K \times m_K$ matrix determined by the above procedure. If we denote the result as $g(m_K, h)$, since the elements of the matrix in (K-1) are all $q_K - q_{K-1}$ corresponding to RS, we have from (B.2)

$$g(m_K, h) = \exp\left\{\frac{1}{2}(q_K - q_{K-1})\frac{\partial^2}{\partial h^2}\right\}(2\cosh h)^{m_K}. \qquad (B.4)$$

The next step is (K-2). The matrix in (K-2) is inserted into $q_{\alpha\beta}$ of (B.1), and the sum of terms $(q_K - q_{K-1})\text{Diag}_{K-1}[I(m_K)]$ and $(q_{K-1} - q_{K-2})I(m_{K-1})$ is raised to the exponent. The former can be written by the already-obtained $g(m_K, h)$ in (B.4) and there are m_{K-1}/m_K of this type of contribution. The latter has uniform elements and the RS-type calculation applies. One therefore finds that $g(m_{K-1}, h)$ can be expressed as follows:

$$g(m_{K-1}, h) = \exp\left\{\frac{1}{2}(q_{K-1} - q_{K-2})\frac{\partial^2}{\partial h^2}\right\}[g(m_K, h)]^{m_{K-1}/m_K}. \qquad (B.5)$$

Repeating this procedure, we finally arrive at

$$G = g(n, h) = \exp\left\{\frac{1}{2}q(0)\frac{\partial^2}{\partial h^2}\right\}[g(m_1, h)]^{n/m_1}. \qquad (B.6)$$

In the limit $n \to 0$, the replacement $m_j - m_{j-1} = -\mathrm{d}x$ is appropriate, and (B.5) reduces to the differential relation

$$g(x + \mathrm{d}x, h) = \exp\left\{-\frac{1}{2}\mathrm{d}q(x)\frac{\partial^2}{\partial h^2}\right\}g(x, h)^{1+\mathrm{d}\log x}. \qquad (B.7)$$

In (B.4) we have $m_K \to 1, q_K - q_{K-1} \to 0$ and this equation becomes $g(1, h) = 2\cosh h$. Equation (B.7) is cast into a differential equation

$$\frac{\partial g}{\partial x} = -\frac{1}{2}\frac{\mathrm{d}q}{\mathrm{d}x}\frac{\partial^2 g}{\partial h^2} + \frac{1}{x}g\log g, \qquad (B.8)$$

which may be rewritten using the notation $f_0(x, h) = (1/x)\log g(x, h)$ as

$$\frac{\partial f_0}{\partial x} = -\frac{1}{2}\frac{\mathrm{d}q}{\mathrm{d}x}\left\{\frac{\partial^2 f_0}{\partial h^2} + x\left(\frac{\partial f_0}{\partial h}\right)^2\right\}. \qquad (B.9)$$

By taking the limit $n \to 0$, we find from (B.6)

$$\frac{1}{n}\log\text{Tr}\,e^L = \exp\left(\frac{1}{2}q(0)\frac{\partial^2}{\partial h^2}\right)\frac{1}{x}\log g(x, h)\Bigg]_{x, h \to 0}$$

$$= \exp\left(\frac{1}{2}q(0)\frac{\partial^2}{\partial h^2}\right) f_0(0, h)]_{h \to 0}$$

$$= \int Du\, f_0(0, \sqrt{q(0)}u). \tag{B.10}$$

We have restricted ourselves to the case $h = 0$. The last expression can be confirmed, for example, by expanding $f_0(0, h)$ in powers of h. The final expression of the free energy (2.17) is

$$\beta f = -\frac{\beta^2 J^2}{4}\left\{1 + \int_0^1 q(x)^2 dx - 2q(1)\right\} - \int Du\, f_0(0, \sqrt{q(0)}u). \tag{B.11}$$

Here f_0 satisfies the following Parisi equation:

$$\frac{\partial f_0(x, h)}{\partial x} = -\frac{J^2}{2}\frac{dq}{dx}\left\{\frac{\partial^2 f_0}{\partial h^2} + x\left(\frac{\partial f_0}{\partial h}\right)^2\right\} \tag{B.12}$$

under the initial condition $f_0(1, h) = \log 2 \cosh \beta h$. The parameters β and J have been recovered for correct dimensionality.

APPENDIX   C

CHANNEL CODING THEOREM

In this appendix we give a brief introduction to information theory and sketch the arguments leading to Shannon's channel coding theorem used in Chapter 5.

C.1   Information, uncertainty, and entropy

Suppose that an information source U generates a sequence of *symbols* (or *alphabets*) from the set $\{a_1, a_2, \ldots, a_L\}$ with probabilities $p_1, p_2, \ldots, p_L$, respectively. A single symbol a_i is assumed to be generated one at a time according to this independently, identically distributed probability. The resulting sequence has the form of, for example, $a_2 a_5 a_i a_1 \ldots$.

The *entropy* of this information source is defined by

$$H(U) = -\sum_{i=1}^{L} p_i \log_2 p_i \quad [\text{bit/symbol}]. \tag{C.1}$$

This quantity is a measure of uncertainty about the outcome from the source. For example, if all symbols are generated with equal probability ($p_1 = \cdots = p_L = 1/L$), the entropy assumes the maximum possible value $H = \log_2 L$ as can be verified by extremizing $H(U)$ under the normalization condition $\sum_i p_i = 1$ using the Lagrange multiplier. This result means that the amount of information obtained after observation of the actual outcome (a_1, for instance) is largest in the uniform case. Thus the entropy may also be regarded as the amount of information obtained by observing the actual outcome. The other extreme is the case where one of the symbols is generated with probability 1 and all other symbols with probability 0, resulting in $H = 0$. This vanishing value is also natural since no information is gained by observation of the actual outcome because we know the result from the outset (no uncertainty). The entropy takes intermediate values for other cases with partial uncertainties.

The unit of the entropy is chosen to be [bit/symbol], and correspondingly the base of the logarithm is two. This choice is easy to understand if one considers the case of $L = 2$ and $p_1 = p_2 = 1/2$. The entropy is then $H = \log_2 2 = 1$, which implies that one gains one bit of information by observing the actual outcome of a perfectly randomly generated binary symbol.

A frequently used example is the *binary entropy* $H_2(p)$. A symbol (e.g. 0) is generated with probability p and another symbol (e.g. 1) with $1 - p$. Then the entropy is

$$H_2(p) = -p \log_2 p - (1 - p) \log_2 (1 - p). \tag{C.2}$$

The binary entropy $H_2(p)$ is convex, reaches its maximum $H_2 = 1$ at $p = 1/2$, and is symmetric about $p = 1/2$.

C.2   Channel capacity

To discuss the properties of a transmission channel, it is convenient to introduce a few quantities related to entropy. The first one is the conditional entropy $H(X|Y)$ that is a measure of uncertainty about the set of events X given another event $y \in Y$. For a given conditional probability $P(x|y)$, the following quantity measures the uncertainty about X, given y:

$$H(X|Y = y) = -\sum_x P(x|y) \log_2 P(x|y). \tag{C.3}$$

The *conditional entropy* is defined as the average of $H(X|y)$ over the distribution of y:

$$H(X|Y) = \sum_y P(y) H(X|y)$$

$$= -\sum_y P(y) \sum_x P(x|y) \log_2 P(x|y)$$

$$= -\sum_x \sum_y P(x,y) \log_2 P(x|y), \tag{C.4}$$

where we have used $P(x,y) = P(x|y)P(y)$. Similarly,

$$H(Y|X) = -\sum_x \sum_y P(x,y) \log_2 P(y|x). \tag{C.5}$$

One sometimes uses the *joint entropy* for two sets of events X and Y although it does not appear in the analyses of the present book:

$$H(X,Y) = -\sum_x \sum_y P(x,y) \log_2 P(x,y). \tag{C.6}$$

It is straightforward to verify the following relations

$$H(X,Y) = H(Y) + H(X|Y) = H(X) + H(Y|X) \tag{C.7}$$

from the identity $P(x,y) = P(x|y)P(y) = P(y|x)P(x)$.
    The *mutual information* is defined by

$$I(X,Y) = H(X) - H(X|Y). \tag{C.8}$$

The meaning of this expression is understood relatively easily in the situation of a noisy transmission channel. Suppose that X is the set of inputs to the channel

$H(X)$

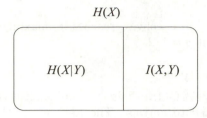

FIG. C.1. Entropy $H(X)$, conditional entropy $H(X|Y)$, and mutual information $I(X,Y)$

and Y is for the output. Then $H(X)$ represents uncertainty about the input without any observation of the output, whereas $H(X|Y)$ corresponds to uncertainty about the input *after* observation of the output. Thus their difference $I(X,Y)$ is the change of uncertainty by learning the channel output, which may be interpreted as the amount of information carried by the channel. Stated otherwise, we arrive at a decreased value of uncertainty $H(X|Y)$ by utilizing the information $I(X,Y)$ carried by the channel (see Fig. C.1). The mutual information is also written as

$$I(X,Y) = H(Y) - H(Y|X). \tag{C.9}$$

The *channel capacity* C is defined as the maximum possible value of the mutual information as a function of the input probability distribution:

$$C = \max_{\{\text{input prob}\}} I(X,Y). \tag{C.10}$$

The channel capacity represents the maximum possible amount of information carried by the channel with a given noise probability.

The concepts of entropy and information can be applied to continuous distributions as well. For a probability distribution density $P(x)$ of a continuous stochastic variable X, the entropy is defined by

$$H(X) = -\int P(x) \log_2 P(x)\, dx. \tag{C.11}$$

The conditional entropy is

$$H(Y|X) = -\int P(x,y) \log_2 P(y|x)\, dx\, dy, \tag{C.12}$$

and the mutual information is given as

$$I(X,Y) = H(X) - H(X|Y) = H(Y) - H(Y|X). \tag{C.13}$$

The channel capacity is the maximum value of mutual information with respect to the input probability distribution function

$$C = \max_{\{\text{input prob}\}} I(X,Y). \tag{C.14}$$

C.3   BSC and Gaussian channel

Let us calculate the channel capacities of the BSC and Gaussian channel using the formulation developed in the previous section.

We first consider the BSC. Suppose that the input symbol of the channel is either 0 or 1 with probabilities r and $1 - r$, respectively:

$$P(x = 0) = r, \quad P(x = 1) = 1 - r. \tag{C.15}$$

The channel has a binary symmetric noise:

$$\begin{aligned} P(y = 0|x = 0) = P(y = 1|x = 1) = 1 - p \\ P(y = 1|x = 0) = P(y = 0|x = 1) = p. \end{aligned} \tag{C.16}$$

Then the probability of the output is easily calculated as

$$P(y = 0) = r(1 - p) + (1 - r)p = r + p - 2rp, \quad P(y = 1) = 1 - P(y = 0). \tag{C.17}$$

The relevant entropies are

$$\begin{aligned} H(Y) = &-(r + p - 2rp)\log_2(r + p - 2rp) \\ &- (1 - r - p + 2rp)\log_2(1 - r - p + 2rp) \\ H(Y|X) = &-p\log_2 p - (1 - p)\log_2(1 - p) = H_2(p) \\ I(X,Y) = &H(Y) - H(Y|X). \end{aligned} \tag{C.18}$$

The channel capacity is the maximum of $I(X,Y)$ with respect to r. This is achieved when $r = 1/2$ (perfectly random input):

$$C = \max_r I(X,Y) = 1 + p\log_2 p + (1 - p)\log_2(1 - p) = 1 - H_2(p). \tag{C.19}$$

Let us next investigate the capacity of the Gaussian channel. Suppose that the input sequence is generated according to a probability distribution $P(x)$. The typical strength (*power*) of an input signal will be denoted by J_0^2:

$$\int P(x)x^2 \, \mathrm{d}x = J_0^2. \tag{C.20}$$

The output Y of the Gaussian channel with noise power J^2 is described by the probability density

$$P(y|x) = \frac{1}{\sqrt{2\pi}J} \exp\left\{-\frac{(y - x)^2}{2J^2}\right\}. \tag{C.21}$$

To evaluate the mutual information using the second expression of (C.13), we express the entropy of the output using

$$P(y) = \int P(y|x)P(x) \, \mathrm{d}x = \frac{1}{\sqrt{2\pi}J} \int \exp\left\{-\frac{(y - x)^2}{2J^2}\right\} P(x) \, \mathrm{d}x \tag{C.22}$$

as

$$H(Y) = -\int P(y) \log_2 P(y)\, dy. \tag{C.23}$$

The conditional entropy is derived from (C.12) and $P(x,y) = P(y|x)P(x)$ as

$$H(Y|X) = \log_2(\sqrt{2\pi}J) + \frac{\log_2 e}{2}. \tag{C.24}$$

Thus the mutual information is

$$I(X,Y) = -\int P(y) \log_2 P(y)\, dy - \log_2(\sqrt{2\pi}J) - \frac{\log_2 e}{2}. \tag{C.25}$$

To evaluate the channel capacity, this mutual information should be maximized with respect to the input probability $P(x)$, which is equivalent to maximization with respect to $P(y)$ according to (C.22). The distribution $P(y)$ satisfies two constraints, which is to be taken into account in maximization:

$$\int P(y)\, dy = 1, \quad \int y^2 P(y)\, dy = J^2 + J_0^2 \tag{C.26}$$

as can be verified from (C.22) and (C.20). By using Lagrange multipliers to reflect the constraints (C.26), the extremization condition

$$\frac{\delta}{\delta P(y)} \left\{ -\int P(y) \log_2 P(y)\, dy \right.$$
$$\left. -\lambda_1 \left(\int P(y)\, dy - 1 \right) - \lambda_2 \left(\int y^2 P(y)\, dy - J^2 - J_0^2 \right) \right\} = 0 \tag{C.27}$$

reads

$$-\log_2 P(y) - \lambda_2 y^2 - \text{const} = 0. \tag{C.28}$$

The solution is

$$P(y) = \frac{1}{\sqrt{2\pi(J^2 + J_0^2)}} \exp\left\{ -\frac{y^2}{2(J^2 + J_0^2)} \right\}, \tag{C.29}$$

where the constants in (C.28) have been fixed so that the result satisfies (C.26). Insertion of this formula into (C.25) immediately yields the capacity as

$$C = \frac{1}{2} \log_2 \left(1 + \frac{J_0^2}{J^2} \right). \tag{C.30}$$

C.4   Typical sequence and random coding

We continue to discuss the properties of a sequence of symbols with the input and output of a noisy channel in mind. Let us consider a sequence of symbols of length M in which the symbol a_i appears m_i times $(i = 1, 2, \ldots, L; M = \sum_i m_i)$.

If M is very large and symbols are generated one by one independently, m_i is approximately equal to Mp_i, where p_i is the probability that a_i appears in a single event. More precisely, according to the *weak law of large numbers*, the inequality

$$\left| \frac{m_i}{M} - p_i \right| < \epsilon \tag{C.31}$$

holds for any positive ϵ if one takes sufficiently large M.

Then the probability p_{typ} that a_i appears m_i times $(i = 1, \ldots, L)$ in the sequence is

$$
\begin{aligned}
p_{\text{typ}} &= p_1^{m_1} \cdots p_L^{m_L} \\
&\approx p_1^{Mp_1} \cdots p_L^{Mp_L} \\
&= 2^{M(p_1 \log_2 p_1 + \cdots + p_L \log_2 p_L)} \\
&\equiv 2^{-MH(U)}.
\end{aligned}
\tag{C.32}
$$

A sequence with a_i appearing Mp_i times $(i = 1, \ldots, L)$ is called the *typical sequence*. All typical sequences appear with the same probability $2^{-MH(U)}$.

The number of typical sequences is the number of ways to distribute m_i of the a_i among the M symbols $(i = 1, \ldots, L)$:

$$N_{\text{typ}} = \frac{M!}{m_1! \ldots m_L!}. \tag{C.33}$$

For sufficiently large M and $m_1, m_2, \ldots, m_L$, we find from the Stirling formula and $m_i = Mp_i$,

$$
\begin{aligned}
\log_2 N_{\text{typ}} &= M(\log_2 M - 1) - \sum_i m_i(\log_2 m_i - 1) \\
&= -M \sum_i p_i \log_2 p_i \\
&= MH(U).
\end{aligned}
\tag{C.34}
$$

Thus N_{typ} is the inverse of p_{typ},

$$N_{\text{typ}} = 2^{MH(U)} = (p_{\text{typ}})^{-1}. \tag{C.35}$$

This result is quite natural as all sequences in the set of typical sequences appear with the same probability. Equation (C.35) also confirms that $H(U)$ is the uncertainty about the outcome from U.

We restrict ourselves to binary symbols (0 or 1, for example) for simplicity from now on (i.e. $L = 2$). The set of inputs to the channel is denoted by X and that of outputs by Y. Both are composed of sequences of length M. The original source message has the length N. *Random coding* is a method of channel coding in which one randomly chooses code words from typical sequences in X. More precisely, a source message has the length N and the total number of messages

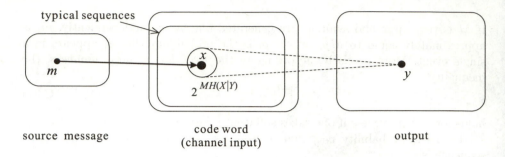

source message   code word   output
                              (channel input)

FIG. C.2. The original message m has a one-to-one correspondence with a code
word x. There are $2^{MH(X|Y)}$ possible inputs corresponding to an output of
the channel. Only one of these $2^{MH(X|Y)}$ code words (marked as a dot) should
be the code word assigned to an original message.

is 2^N. We assign a code word for each of the source messages by randomly
choosing a typical sequence of length $M\,(> N)$ from the set X. Note that there
are $2^{MH(X)}$ typical sequences in X, and only 2^N of them are chosen as code
words. The code rate is $R = N/M$. This random coding enables us to decode
the message without errors if the code rate is smaller than the channel capacity
$R < C$ as shown below.

C.5   Channel coding theorem

Our goal is to show that the probability of correct decoding can be made arbitrarily close to one in the limit of infinite length of code word. Such an ideal
decoding is possible if only a single code word $x\,(\in X)$ corresponds to a given
output of the channel $y\,(\in Y)$. However, we know that there are $2^{MH(X|Y)}$ pos-
sibilities as the input corresponding to a given output. The only way out is that
none of these $2^{MH(X|Y)}$ sequences are code words in our random coding except
a single one, the correct input (see Fig. C.2). To estimate the probability of such
a case, we first note that the probability that a typical sequence of length M is
chosen as a code word for an original message of length N is

$$\frac{2^N}{2^{MH(X)}} = 2^{-M[H(X)-R]} \tag{C.36}$$

because 2^N sequences are chosen from $2^{MH(X)}$. Thus the probability that an
arbitrary typical sequence of length M is not a code word is

$$1 - 2^{-M[H(X)-R]}. \tag{C.37}$$

Now, we require that the $2^{MH(X|Y)}$ sequences of length M (corresponding to
a given output of the channel) are not code words except the single correct one.
Such a probability is clearly

$$p_{\text{correct}} = \left[1 - 2^{-M[H(X)-R]}\right]^{2^{MH(X|Y)}-1}$$

$$\approx 1 - 2^{-M[H(X)-R-H(X|Y)]}. \tag{C.38}$$

The maximum possible value of $p_{correct}$ is given by replacing $H(X) - H(X|Y)$ by its largest value C, the channel capacity:

$$\max\{p_{correct}\} = 1 - 2^{-M(C-R)}, \tag{C.39}$$

which tends to one as $M \to \infty$ if $R < C$. This completes the argument for the channel coding theorem.

APPENDIX D

DISTRIBUTION AND FREE ENERGY OF K-SAT

In this appendix we derive the self-consistent equation (9.53) and the equilibrium free energy (9.55) of K-SAT from the variational free energy (9.51) under the RS ansatz (9.52).

The function $c(\boldsymbol{\sigma})$ depends on $\boldsymbol{\sigma}$ only through the number of down spins j in the set $\boldsymbol{\sigma} = (\sigma_1, \ldots, \sigma_n)$ if we assume symmetry between replicas; we thus sometimes use the notation $c(j)$ for $c(\boldsymbol{\sigma})$. The free energy (9.51) is then expressed as

$$-\frac{\beta F}{N} = -\sum_{j=0}^{n} \binom{n}{j} c(j) \log c(j) + \alpha \log \left\{ \sum_{j_1=0}^{n} \cdots \sum_{j_K=0}^{n} c(j_1) \ldots c(j_K) \right.$$

$$\left. \cdot \sum_{\boldsymbol{\sigma}_1(j_1)} \cdots \sum_{\boldsymbol{\sigma}_K(j_K)} \prod_{\alpha=1}^{n} \left(1 + (e^{-\beta} - 1) \prod_{k=1}^{K} \delta(\sigma_k^\alpha, 1) \right) \right\}, \qquad (D.1)$$

where the sum over $\boldsymbol{\sigma}_i(j_i)$ is for the $\boldsymbol{\sigma}_i$ with j_i down spins. Variation of (D.1) with respect to $c(j)$ yields

$$\frac{\delta}{\delta c(j)} \left(-\frac{\beta F}{N} \right) = -\binom{n}{j} (\log c(j) + 1) + \frac{K \alpha g}{f}, \qquad (D.2)$$

where

$$f = \sum_{j_1=0}^{n} \cdots \sum_{j_K=0}^{n} c(j_1) \ldots c(j_K)$$

$$\cdot \sum_{\boldsymbol{\sigma}_1(j_1)} \cdots \sum_{\boldsymbol{\sigma}_K(j_K)} \prod_{\alpha=1}^{n} \left(1 + (e^{-\beta} - 1) \prod_{k=1}^{K} \delta(\sigma_k^\alpha, 1) \right) \qquad (D.3)$$

$$g = \sum_{j_1=0}^{n} \cdots \sum_{j_{K-1}=0}^{n} c(j_1) \ldots c(j_{K-1}) \sum_{\boldsymbol{\sigma}_1(j_1)} \cdots \sum_{\boldsymbol{\sigma}_{K-1}(j_{K-1})}$$

$$\cdot \sum_{\boldsymbol{\sigma}(j)} \prod_{\alpha=1}^{n} \left(1 + (e^{-\beta} - 1)\delta(\sigma^\alpha, 1) \prod_{k=1}^{K-1} \delta(\sigma_k^\alpha, 1) \right). \qquad (D.4)$$

These functions f and g are expressed in terms of the local magnetization density $P(m)$ defined by

$$c(\boldsymbol{\sigma}) = \int_{-1}^{1} dm \, P(m) \prod_{\alpha=1}^{n} \frac{1 + m\sigma^{\alpha}}{2} \tag{D.5}$$

as

$$f = \int_{-1}^{1} \prod_{k=1}^{K} dm_k \, P(m_k)(A_K)^n \tag{D.6}$$

$$g = \binom{n}{j} \int_{-1}^{1} \prod_{k=1}^{K-1} dm_k \, P(m_k)(A_{K-1})^{n-j}, \tag{D.7}$$

where

$$A_K = 1 + (e^{-\beta} - 1) \prod_{k=1}^{K} \frac{1 + m_k}{2}. \tag{D.8}$$

Equations (D.6) and (D.7) are derived as follows.

Recalling that the sum over $\boldsymbol{\sigma}(j)$ in g appearing in (D.4) is for the $\boldsymbol{\sigma}$ with j down spins (for which $\delta(\sigma^{\alpha}, 1) = 0$), we find

$$g = \sum_{j_1=0}^{n} \cdots \sum_{j_{K-1}=0}^{n} c(j_1) \ldots c(j_{K-1})$$

$$\cdot \sum_{\boldsymbol{\sigma}_1(j_1)} \cdots \sum_{\boldsymbol{\sigma}_{K-1}(j_{K-1})} \sum_{\boldsymbol{\sigma}(j)} \prod_{\alpha=1}^{n}{}' \left(1 + (e^{-\beta} - 1) \prod_{k=1}^{K-1} \delta(\sigma_k^{\alpha}, 1) \right)$$

$$= \sum_{\boldsymbol{\sigma}(j)} \sum_{\boldsymbol{\sigma}_1} \cdots \sum_{\boldsymbol{\sigma}_{K-1}} c(\boldsymbol{\sigma}_1) \ldots c(\boldsymbol{\sigma}_{K-1})$$

$$\cdot \prod_{\alpha=1}^{n}{}' \left(1 + (e^{-\beta} - 1) \prod_{k=1}^{K-1} \delta(\sigma_k^{\alpha}, 1) \right), \tag{D.9}$$

where the product is over the replicas with $\sigma^{\alpha} = 1$. If we insert (D.5) into this equation and carry out the sums over $\boldsymbol{\sigma}_1$ to $\boldsymbol{\sigma}_{K-1}$, we find

$$g = \sum_{\boldsymbol{\sigma}(j)} \int_{-1}^{1} \prod_{k=1}^{K-1} dm_k \, P(m_k) \prod_{\alpha=1}^{n}{}' A_{K-1}$$

$$= \binom{n}{j} \int_{-1}^{1} \prod_{k=1}^{K-1} dm_k \, P(m_k)(A_{K-1})^{n-j}, \tag{D.10}$$

proving (D.7). Similar manipulations lead to (D.6).

In the extremization of F with respect to $c(j)$, we should take into account the symmetry $c(j) = c(n-j)$ coming from $c(\boldsymbol{\sigma}) = c(-\boldsymbol{\sigma})$ as well as the normalization condition $\sum_{j=0}^{n} \binom{n}{j} c(j) = 1$. Using a Lagrange multiplier for the latter and from (D.2), the extremization condition is

$$-2\left(\log c(j) + 1\right) + K\alpha \int_{-1}^{1} \prod_{k=1}^{K-1} dm_k P(m_k) f^{-1}\{(A_{K-1})^{n-j} + (A_{K-1})^j\} - 2\lambda = 0,$$

(D.11)

from which we find

$$c(j) = \exp\left\{-\lambda - 1 + \frac{K\alpha}{2f} \int_{-1}^{1} \prod_{k=1}^{K-1} dm_k P(m_k)((A_{K-1})^{n-j} + (A_{K-1})^k)\right\}.$$

(D.12)

The number of replicas n has so far been arbitrary. Letting $n \to 0$, we obtain the self-consistent equation for $P(m)$. The value of the Lagrange multiplier λ in the limit $n \to 0$ is evaluated from (D.12) for $j = 0$ using $c(0) = 1$. The result is $\lambda = K\alpha - 1$, which is to be used in (D.12) to erase λ. The distribution $P(m)$ is now derived from the inverse relation of

$$c(j) = \int_{-1}^{1} dm\, P(m) \left(\frac{1+m}{2}\right)^{n-j} \left(\frac{1-m}{2}\right)^{j}$$

(D.13)

in the limit $n \to 0$; that is,

$$P(m) = \frac{1}{\pi(1-m^2)} \int_{-\infty}^{\infty} dy\, c(iy) \exp\left(-iy \log \frac{1-y}{1+y}\right).$$

(D.14)

Inserting (D.12) (in the limit $n \to 0$) with λ replaced by $K\alpha - 1$ into the right hand side of the above equation, we finally arrive at the desired relation (9.53) for $P(m)$.

It is necessary to consider the $\mathcal{O}(n)$ terms to derive the free energy (9.55) expressed in terms of $P(m)$. Let us start from (D.1):

$$-\frac{\beta F}{N} = -\sum_{j=0}^{n} \binom{n}{j} c(j) \log c(j) + \alpha \log f.$$

(D.15)

The expression (D.6) for f implies that f is expanded in n as

$$f = 1 + na + \mathcal{O}(n^2)$$

(D.16)

$$a = \int_{-1}^{1} \prod_{k=1}^{K} dm_k P(m_k) \log A_K.$$

(D.17)

The first term on the right hand side of (D.15) is, using (D.12),

$$-\sum_{j=0}^{n} \binom{n}{j} c(j) \log c(j) = \lambda + 1 - \frac{K\alpha}{2f} \sum_{j=0}^{n} \binom{n}{j} c(j)$$

$$\cdot \int_{-1}^1 \prod_{k=1}^{K-1} dm_k P(m_k)((A_{K-1})^{n-j} + (A_{K-1})^j). \qquad \text{(D.18)}$$

We should therefore expand λ to $\mathcal{O}(n)$. For this purpose, we equate (D.12) and (D.13) to get

$$e^{\lambda+1} = \frac{\exp\left\{(K\alpha/2f)\int_{-1}^1 \prod_{k=1}^{K-1} dm_k P(m_k)((A_{K-1})^{n-j} + (A_{K-1})^j)\right\}}{\int_{-1}^1 dm\, P(m) \left(\frac{1+m}{2}\right)^{n-j} \left(\frac{1-m}{2}\right)^j}. \qquad \text{(D.19)}$$

Since the left hand side is independent of j, we may set $j = 0$ on the right hand side. We then expand the right hand side to $\mathcal{O}(n)$ to obtain

$$\lambda + 1 = K\alpha + n\left\{K\alpha\left(-a + \frac{b}{2}\right) + \log 2 - \int_{-1}^1 dm P(m) \log(1-m^2)\right\} + \mathcal{O}(n^2), \qquad \text{(D.20)}$$

where

$$b = \int_{-1}^1 \prod_{k=1}^{K-1} dm_k\, P(m_k) \log A_{K-1}. \qquad \text{(D.21)}$$

The final term on the right hand side of (D.18) is evaluated as

$$\sum_{j=0}^n \binom{n}{j} \int_{-1}^1 dm_k\, P(m_k) \left(\frac{1+m_k}{2}\right)^{n-j} \left(\frac{1-m_k}{2}\right)^j$$

$$\cdot \int_{-1}^1 \prod_{k=1}^{K-1} dm_k\, P(m_k)((A_{K-1})^{n-j} + (A_{K-1})^j)$$

$$= 2 \int_{-1}^1 \prod_{k=1}^K dm_k\, P(m_k) \left(\frac{1+m_k}{2} A_{K-1} + \frac{1-m_k}{2}\right)^n$$

$$= 2f. \qquad \text{(D.22)}$$

Combining (D.15), (D.16), (D.18), (D.20), and (D.22), we find

$$-\frac{\beta F}{Nn} = \log 2 + \alpha(1-K)a + \frac{\alpha K b}{2} - \frac{1}{2}\int_{-1}^1 dm\, P(m) \log(1-m^2) + \mathcal{O}(n), \qquad \text{(D.23)}$$

which gives the final answer (9.55) for the equilibrium free energy.

REFERENCES

Note: A reference in the form '*cond-mat/yymmnnn*' refers to a preprint at the Los Alamos e-print archive (http://xxx.lanl.gov/), in the condensed matter section, in the year *yy*, month *mm*, and number *nnn*.

Aarão Reis, F. D. A., de Queiroz, S. L. A., and dos Santos, R. R. (1999). *Physical Review B*, **60**, 6740–8.

Aarts, E. and Korst, J. (1989). *Simulated annealing and Boltzmann machines*. Wiley, Chichester.

Abe, S. and Okamoto, Y. (eds) (2001). *Nonextensive statistical mechanics and its applications*, Lecture Notes in Physics. Springer, New York.

Abu-Mostafa, Y. S. (1989). *Neural Computation*, **1**, 312–17.

Amari, S. (1997). In *Theoretical aspects of neural computation* (ed. K. Y. M. Wong, I. King, and D.-y. Yeung), pp. 1–15. Springer, Singapore.

Amari, S. and Maginu, K. (1988). *Neural Networks*, **1**, 63–73.

Amit, D. J. (1989). *Modeling brain function*. Cambridge University Press.

Amit, D. J., Gutfreund, H., and Sompolinsky, H. (1985). *Physical Review A*, **32**, 1007–18.

Amit, D. J., Gutfreund, H., and Sompolinsky, H. (1987). *Annals of Physics*, **173**, 30–67.

Arazi, B. (1988). *A commonsense approach to the theory of error-correcting codes*. MIT Press, Cambridge, Massachusetts.

Ash, R. (1990). *Information theory*. Dover, New York.

Aspvall, B., Plass, M. F., and Tarjan, R. E. (1979). *Information Processing Letters*, **8**, 121–3.

Barber, D. and Sollich, P. (1998). In *On-line learning in neural networks* (ed. D. Saad), pp. 279–302. Cambridge University Press.

Barkai, N., Seung, H. S., and Sompolinsky, H. (1995). *Physical Review Letters*, **75**, 1415–18.

Besag, J. (1986). *Journal of the Royal Statistical Society B*, **48**, 259–302.

Biehl, M. and Riegler, P. (1994). *Europhysics Letters*, **28**, 525–30.

Biehl, M., Riegler, P., and Stechert, M. (1995). *Physical Review E*, **52**, 4624–7.

Bilbro, G. L., Snyder, W. E., Garnier, S. J., and Gault, J. W. (1992). *IEEE Transactions on Neural Networks*, **3**, 131–8.

Binder, K. and Young, A. P. (1986). *Reviews of Modern Physics*, **58**, 801–976.

Bishop, C. M. (1995). *Neural networks for pattern recognition*. Oxford University Press.

Chellappa, R. and Jain, A. (eds) (1993). *Markov random fields: theory and applications*. Academic Press, New York.

Cho, S. and Fisher, M. P. A. (1997). *Physical Review B*, **55**, 1025–31.

Clark, Jr, G. C. and Cain, J. B. (1981). *Error-correction coding for digital communications*. Plenum, New York.

Coolen, A. C. C. (2001). In *Handbook of brain theory and neural networks* 2nd edn (ed. M. A. Arbib). MIT Press, Cambridge, Massachusetts. (In press); In *Handbook of biological physics IV* (ed. F. Moss and S. Gielen). Elsevier, Amsterdam. (In press).

Coolen, A. C. C. and Ruijgrok, Th. W. (1988). *Physical Review A*, **38**, 4253–5.

Coolen, A. C. C. and Saad, D. (2000). *Physical Review E*, **62**, 5444–87.

Coolen, A. C. C. and Sherrington, D. (1994). *Physical Review E*, **49**, 1921–34 and 5906–6.

Coolen, A. C. C. and Sherrington, D. (2001). *Statistical physics of neural networks*. Cambridge University Press. (In press).

de Almeida, J. R. L. and Thouless, D. J. (1978). *Journal of Physics A*, **11**, 983–990.

Derin, H., Elliott, H., Cristi, H. R., and Geman, D. (1984). *IEEE Transactions on Pattern Analysis and Machine Intelligence*, **6**, 707–20.

Derrida, B. (1981). *Physical Review B*, **24**, 2613–26.

Domany, E., van Hemmen, J. L., and Schulten, K. (eds) (1991). *Models of neural networks*. Springer, Berlin.

Domany, E., van Hemmen, J. L., and Schulten, K. (eds) (1995). *Models of neural networks III*. Springer, New York.

Dotsenko, V. (2001). *Introduction to the replica theory of disordered statistical systems*. Cambridge University Press.

Duplantier, B. (1981). *Journal of Physics A*, **14**, 283–5.

Edwards, S. F. and Anderson, P. W. (1975). *Journal of Physics F*, **5**, 965–74.

Ferreira, F. F. and Fontanari, J. F. (1998). *Journal of Physics A*, **31**, 3417–28.

Fischer, K. H. and Hertz, J. (1991). *Spin glasses*. Cambridge University Press.

Fontanari, J. F. (1995). *Journal of Physics A*, **28**, 4751–9.

Fontanari, J. F. and Köberle, R. (1987). *Physical Review A*, **36**, 2475–7.

Fu, Y. and Anderson, P. W. (1986). *Journal of Physics A*, **19**, 1605–20.

Gardner, E. (1985). *Nuclear Physics B*, **257** [**FS14**], 747–65.

Gardner, E. (1987). *Europhysics Letters*, **4**, 481–5.

Gardner, E. (1988). *Journal of Physics A*, **21**, 257–70.

Gardner, E. Derrida, B., and Mottishaw, P. (1987). *Journal de Physique*, **48**, 741–55.

Garey, M. R. and Johnson, D. S. (1979). *Computers and intractability: a guide to the theory of NP-completeness*. Freeman, San Francisco.

Geiger, D. and Girosi, F. (1991). *IEEE Transactions on Pattern Analysis and Machine Intelligence*, **13**, 401–12.

Geman, S. and Geman, D. (1984). *IEEE Transactions on Pattern Analysis and Machine Intelligence*, **6**, 721–41.

Gent, I. P. and Walsh, T. (1996). In *Proceedings of the 12th European conference on artificial intelligence* (ed. W. Wahlster), pp. 170–4. Wiley, New York.

Georges, A., Hansel, D., Le Doussal, P., and Bouchaud, J.-P. (1985). *Journal de Physique*, **46**, 1827–36.

Gillin, P., Nishimori, H., and Sherrington, D. (2001). *Journal of Physics A*, **34**, 2949–64.

Gingras, M. J. P. and Sørensen, E. S. (1998). *Physical Review B*, **57**, 10 264–7.

Goerdt, A. (1996). *Journal of Computer and System Sciences*, **53**, 469–86.

Gross, D. J. and Mézard, M. (1984). *Nuclear Physics B*, **240** [**FS12**], 431–52.

Gruzberg, I. A., Read, N., and Ludwig, A. W. W. (2001). *Physical Review B*, **63**, 104422-1–27.

Györgyi, G. and Tishby, N. (1990). In *Neural networks and spin glasses* (ed. K. Thuemann and R. Köberle), pp. 3–36. World Scientific, Singapore.

Haussler, D., Kearns, M., and Schapire, R. (1991). In *IVth annual workshop on computational learning theory (COLT 21)*, pp. 61–74. Morgan-Kaufmann, Santa Cruz, California.

Heegard, C. and Wicker, S. B. (1999). *Turbo coding*. Kluwer, Boston.

Hertz, J., Krogh, A., and Palmer, R. G. (1991). *Introduction to the theory of neural computation*. Perseus Books, Reading, Massachusetts.

Heskes, T. and Wiegerinck, W. (1998). In *On-line learning in neural networks* (ed. D. Saad), pp. 251–78. Cambridge University Press.

Hogg, T., Hubermann, B. A., and Williams, C. (eds) (1996). *Frontiers in problem solving: phase transitions and complexity. Artificial Intelligence*, **81** (1–2).

Honecker, A., Picco, M., and Pujol, P. (2000). *cond-mat/0010143*.

Hopfield, J. J. (1982). *Proceedings of the National Academy of Sciences of the United States of America*, **79**, 2554–8.

Horiguchi, T. (1981). *Physics Letters*, **81A**, 530–2.

Horiguchi, T. and Morita, T. (1981). *Journal of Physics A*, **14**, 2715–31.

Horiguchi, T. and Morita, T. (1982*a*). *Journal of Physics A*, **15**, L75–80.

Horiguchi, T. and Morita, T. (1982*b*). *Journal of Physics A*, **15**, 3511–60.

Horner, H., Bormann, D., Frick, M., Kinzelbach, H., and Schmidt, A. (1989). *Zeitschrift für Physik B*, **76**, 381–98.

Hukushima, K. (2000). *Journal of the Physical Society of Japan*, **69**, 631–4.

Iba, Y. (1999). *Journal of Physics A*, **32**, 3875–88.

Inoue, J. (1997). *Journal of Physics A*, **30**, 1047–58.

Inoue, J. (2001). *Physical Review E*, **63**, 046114-1–10.

Inoue, J. and Carlucci, D. M. (2000). *cond-mat/0006389*.

Inoue, J. and Nishimori, H. (1997). *Physical Review E*, **55**, 4544–51.

Inoue, J., Nishimori, H., and Kabashima, Y. (1997). *Journal of Physics A*, **30**, 3795–816. **31**, 123–44.

Kabashima, Y. and Saad, D. (1998). *Europhysics Letters*, **44**, 668–74.

Kabashima, Y. and Saad, D. (1999). *Europhysics Letters*, **45**, 97–103.

Kabashima, Y. and Saad, D. (2001). In *Advanced mean field methods – theory and practice* (ed. M. Opper and D. Saad), pp. 65–106. MIT Press, Cambridge, Massachusetts.

Kabashima, Y., Murayama, T., and Saad, D. (2000a). *Physical Review Letters*, **84**, 1355–8.

Kabashima, Y., Murayama, T., and Saad, D. (2000b). *Physical Review Letters*, **84**, 2030–3.

Kabashima, Y., Sazuka, N., Nakamura, K., and Saad, D. (2000c). *cond-mat/* 0010173.

Kanter, I. and Saad, D. (2000). *Physical Review E*, **61**, 2137–40.

Kawamura, H. and Li, M. S. (1996). *Physical Review*, **B54**, 619–36.

Kawashima, N. and Aoki, T. (2000). In *Frontiers in magnetism* (ed. Y. Miyako, H. Takayama, and S. Miyashita), pp. 169–77. *Journal of the Physical Society of Japan*, **69**, Supplement A.

Kinouchi, O. and Caticha, N. (1992). *Journal of Physics A*, **25**, 6243–50.

Kinzel, W. and Ruján, P. (1990). *Europhysics Letters*, **13**, 473–7.

Kirkpatrick, S. and Selman, B. (1994). *Science*, **264**, 1297–301.

Kirkpatrick, S., Gelatt, Jr, C. D., and Vecchi, M. P. (1983). *Science*, **220**, 671–80.

Kitatani, H. (1992). *Journal of the Physical Society of Japan*, **61**, 4049–55.

Kitatani, H. and Oguchi, T. (1990). *Journal of the Physical Society of Japan*, **59**, 3823–6.

Kitatani, H. and Oguchi, T. (1992). *Journal of the Physical Society of Japan*, **61**, 1598–605.

Korutcheva, E., Opper, M., and Lòpez, L. (1994). *Journal of Physics A*, **27**, L645–50.

Lakshmanan, S. and Derin, H. (1989). *IEEE Transactions on Pattern Analysis and Machine Intelligence*, **11**, 799–813.

Le Doussal, P. and Harris, A. B. (1988). *Physical Review Letters*, **61**, 625–8.

Le Doussal, P. and Harris, A. B. (1989). *Physical Review B*, **40**, 9249–52.

Lin, S. and Costello, Jr, D. J. (1983). *Error control coding: fundamentals and applications*. Prentice Hall, Upper Saddle River, New Jersey.

MacKay, D. J. C. (1999). *IEEE Transactions on Information Theory*, **45**, 399–431.

MacKay, D. J. C. and Neal, R. M. (1997). *Electronics Letters*, **33**, 457–8.

Marroquin, J., Mitter, S., and Poggio, T. (1987). *Journal of the American Statistical Association*, **82**, 76–89.

McEliece, R. J. (1977). *The theory of information and coding*. Addison-Wesley, San Francisco.

Mélin, R. and Peysson, S. (2000). *European Physical Journal B*, **14**, 169–76.

Mertens, S. (1998). *Physical Review Letters*, **81**, 4281–4.

Mézard, M., Parisi, G., and Virasoro, M. A. (1986). *Europhysics Letters*, **1**, 77–82.

Mézard, M., Parisi, G., and Virasoro, M. A. (1987). *Spin glass theory and beyond*. World Scientific, Singapore.

Miyajima, T., Hasegawa, T., and Haneishi, M. (1993). *IEICE Transactions on Communications*, **E76-B**, 961–8.

Miyako, Y., Takayama, H., and Miyashita, S. (eds) (2000). *Frontiers in magnetism*, *Journal of the Physical Society of Japan*, **69**, Supplement A.

Miyashita, S. and Shiba, H. (1984). *Journal of the Physical Society of Japan*, **53**, 1145–54.

Molina, R. A., Katsaggelos, A. K., and Mateos, J. (1999). *IEEE Transactions on Image Processing*, **8**, 231–46.

Monasson, R. and Zecchina, R. (1996). *Physical Review Letters*, **76**, 3881–4.

Monasson, R. and Zecchina, R. (1997). *Physical Review E*, **56**, 1357–70.

Monasson, R. and Zecchina, R. (1998). *Journal of Physics A*, **31**, 9209–17.

Monasson, R., Zecchina, R., Kirkpatrick, S., Selman, B., and Troyansky, L. (1999). *Nature*, **400**, 133–7.

Monasson, R., Zecchina, R., Kirkpatrick, S., Selman, B., and Troyansky, L. (2000). *cond-mat*/9910080.

Montanari, A. (2000). *European Physical Journal B*, **18**, 121–36.

Montanari, A. and Sourlas, N. (2000). *European Physical Journal B*, **18**, 107–119.

Morita, T. and Horiguchi, T. (1980). *Physics Letters*, **76A**, 424–6.

Morita, T. and Tanaka, T. (1996). *Physica A*, **223**, 244–62.

Morita, T. and Tanaka, T. (1997). *Pattern Recognition Letters*, **18**, 1479–93.

Müller, K.-R., Ziehe, A., Murata, N., and Amari, S. (1998). In *On-line learning in neural networks* (ed. D. Saad), pp. 93–110. Cambridge University Press.

Murayama, T., Kabashima, Y., Saad, D., and Vicente, R. (2000). *Physical Review E*, **62**, 1577–91.

Nakamura, K., Kabashima, Y., and Saad, D. (2000). *cond-mat*/0010073.

Nemoto, K. and Takayama, H. (1985). *Journal of Physics C*, **18**, L529–35.

Nishimori, H. (1980). *Journal of Physics C*, **13**, 4071–6.

Nishimori, H. (1981). *Progress of Theoretical Physics*, **66**, 1169–81.

Nishimori, H. (1986*a*). *Progress of Theoretical Physics*, **76**, 305–6.

Nishimori, H. (1986*b*). *Journal of the Physical Society of Japan*, **55**, 3305–7.

Nishimori, H. (1992). *Journal of the Physical Society of Japan*, **61**, 1011–12.

Nishimori, H. (1993). *Journal of the Physical Society of Japan*, **62**, 2973–5.

Nishimori, H. (1994). *Physica A*, **205**, 1–14.

Nishimori, H. and Inoue, J. (1998). *Journal of Physics A*, **31**, 5661–72.

Nishimori, H. and Ozeki, T. (1993). *Journal of Physics A*, **26**, 859–71.

Nishimori, H. and Sherrington, D. (2001). In *Disordered and complex systems* (ed. P. Sollich, A. C. C. Coolen, L. P. Hughston, and R. F. Streater), pp. 67–72. American Institute of Physics, Melville, New York.

Nishimori, H. and Stephen, M. J. (1983). *Physical Review B*, **27**, 5644–52.

Nishimori, H. and Wong, K. Y. M. (1999). *Physical Review E*, **60**, 132–44.

Okada, M. (1995). *Neural Networks*, **8**, 833–8.

Okada, M., Doya, K., Yoshioka, T., and Kawato, M. (1999). *Technical report of IEICE*, **NC98**-184, 239-246. [In Japanese].

Opper, M. and Haussler, D. (1991). *Physical Review Letters*, **66**, 2677–80.

Opper, M. and Kinzel, W. (1995). In *Models of neural networks III* (ed. E.

Domany, J. L. van Hemmen, and K. Schulten), pp. 151–209. Springer, New York.

Opper, M. and Saad, D. (eds) (2001). *Advanced mean field methods – theory and practice*. MIT Press, Cambridge, Massachusetts.

Opper, M. and Winther, O. (2001). In *Advanced mean field methods – theory and practice* (ed. M. Opper and D. Saad), pp. 9–26. MIT Press, Cambridge, Massachusetts.

Opper, M., Kinzel, W., Kleinz, J., and Nehl, R. (1990). *Journal of Physics A*, **23**, L581–6.

Ozeki, Y. (1990). *Journal of the Physical Society of Japan*, **59**, 3531–41.

Ozeki, Y. (1995). *Journal of Physics A*, **28**, 3645–55.

Ozeki, Y. (1997). *Journal of Physics: Condensed Matter*, **9**, 11 171–7.

Ozeki, Y. and Ito, N. (1998). *Journal of Physics A*, **31**, 5451–65.

Ozeki, Y. and Nishimori, H. (1987). *Journal of the Physical Society of Japan*, **56**, 1568–76 and 3265–9.

Ozeki, Y. and Nishimori, H. (1993). *Journal of Physics A*, **26**, 3399–429.

Parisi, G. (1979). *Physics Letters*, **73A**, 203–5.

Parisi, G. (1980). *Journal of Physics A*, **13**, L115–21, 1101–12, and 1887–95.

Plefka, T. (1982). *Journal of Physics A*, **15**, 1971–8.

Pryce, J. M. and Bruce, A. D. (1995). *Journal of Physics A*, **28**, 511–32.

Read, N. and Ludwig, W. W. (2000). *Physical Review B*, **63**, 024404-1–12.

Rhee, M. Y. (1989). *Error-correcting coding theory*. McGraw-Hill, New York.

Riedel, U., Kühn, R., and van Hemmen, J. L. (1988). *Physical Review A*, **38**, 1105–8.

Rieger, H., Schreckenberg, M., and Zittartz, J. (1989). *Zeitschrift für Physik B*, **74**, 527–38.

Ruján, P. (1993). *Physical Review Letters*, **70**, 2968–71.

Saad, D. (ed.) (1998). *On-line learning in neural networks*. Cambridge University Press.

Saad, D., Kabashima, Y., and Vicente, R. (2001). In *Advanced mean field methods – theory and practice* (ed. M. Opper and D. Saad), pp. 85–106. MIT Press, Cambridge, Massachusetts.

Sasamoto, T., Toyoizumi, T. and Nishimori, H. (2001). *Journal of Physics A*. (Submitted).

Senthil, T. and Fisher, M. P. A. (2000). *Physical Review B*, **61**, 9690–8.

Seung, H. S., Sompolinsky, H., and Tishby, N. (1992). *Physical Review A*, **45**, 6056–91.

Sherrington, D. and Kirkpatrick, S. (1975). *Physical Review Letters*, **35**, 1792–6.

Shiino, M. and Fukai, T. (1993). *Physical Review E*, **48**, 867–97.

Shiino, M., Nishimori, H., and Ono, M. (1989). *Journal of the Physical Society of Japan*, **58**, 763–6.

Shinomoto, S. and Kabashima, Y. (1991). *Journal of Physics A*, **24**, L141–4.

Simon, M. K., Omura, J. K., Scholotz, R. A., and Levitt, B. K. (1994). *Spread spectrum communications handbook*. McGraw-Hill, New York.

Singh, R. R. P. (1991). *Physical Review Letters*, **67**, 899–902.

Singh, R. R. P. and Adler, J. (1996). *Physical Review B*, **54**, 364–7.

Sollich, P. (1994). *Physical Review E*, **49**, 4637–51.

Sollich, P. and Barber, D. (1997). *Europhysics Letters*, **38**, 477–82.

Sørensen, E. S., Gingras, M. J. P., and Huse, D. A. (1998). *Europhysics Letters*, **44**, 504–10.

Sourlas, N. (1989). *Nature*, **339**, 693–5.

Sourlas, N. (1994). *Europhysics Letters*, **25**, 159–64.

Stanley, H. E. (1987). *Introduction to phase transitions and critical phenomena*. Oxford University Press.

Steffan, H. and Kühn, R. (1994). *Zeitschrift für Physik B*, **95**, 249–60.

Tanaka, K. (1999). *Butsuri*, **54**, 25–33. [In Japanese].

Tanaka, K. (2001*a*). *Transactions of the Japanese Society for Artificial Intelligence*, **16**, 246–58.

Tanaka, K. (2001*b*). *Transactions of the Japanese Society for Artificial Intelligence*, **16**, 259–67. [In Japanese].

Tanaka, K. and Horiguchi, T. (2000). *Electronics Communications in Japan*, **3-83**, 84–94.

Tanaka, K. and Inoue, J. (2000). *Technical report of IEICE*, **100**, 41-8. [In Japanese].

Tanaka, K. and Morita, T. (1995). *Physics Letters*, **203A**, 122–8.

Tanaka, K. and Morita, T. (1996). In *Theory and applications of the cluster variation and path probability methods* (ed. J. L. Morán-López and J. M. Sanchez), pp. 357–73. Plenum, New York.

Tanaka, K. and Morita, T. (1997). *Transactions of IEICE*, **J80-A**, 1033–7. [In Japanese].

Tanaka, T. (2001). In *Advances in Neural Information Processing Systems* (ed. T. K. Leen, T. G. Dietterich, and V. Tresp), Vol. 13. MIT Press, Cambridge, Massachusetts; *Europhysics Letters*, **54**, 504–6.

Thouless, D. J., Anderson, P. W., and Palmer, R. G. (1977). *Philosophical Magazine*, **35**, 593–601.

Toulouse, G. (1980). *Journal de Physique*, **41**, L447–9.

Ueno, Y. and Ozeki, Y. (1991). *Journal of Statistical Physics*, **64**, 227–49.

Vallet, F. (1989). *Europhysics Letters*, **8**, 747–51.

van Hemmen, J. L. and Morgenstern, I. (eds) (1987). *Heidelberg colloquium on glassy dynamics*, Lecture Notes in Physics. Springer, Berlin.

Vicente, R., Saad, D., and Kabashima, Y. (1999). *Physical Review E*, **60**, 5352–66.

Vicente, R., Saad, D., and Kabashima, Y. (2000). *Europhysics Letters*, **51**, 698–704.

Villain, J. (1977). *Journal of Physics C*, **10**, 4793–803.

Viterbi, A. J. (1995). *CDMA: Principles of spread spectrum communication*. Addison-Wesley, Reading, Massachusetts.

Watkin, T. L. H. and Rau, A. (1992). *Physical Review A*, **45**, 4102–10.

Watkin, T. L. H. and Rau, A. (1993). *Reviews of Modern Physics*, **65**, 499–556.

Wicker, S. B. (1995). *Error control systems for digital communications and storage*. Prentice Hall, Upper Saddle River, New Jersey.

Wong, K. Y. M. and Sherrington, D. (1988). *Journal of Physics A*, **21**, L459–66.

Wong, K. Y. M., King, I., and Yeung, D.-y. (eds) (1997). *Theoretical aspects of neural computation*. Springer, Singapore.

Yeomans, J. M. (1992). *Statistical mechanics of phase transitions*. Oxford University Press.

Young, A. P. (ed.) (1997). *Spin glasses and random fields*. World Scientific, Singapore.

Zerubia, J. and Chellappa, R. (1993). *IEEE Transactions on Neural Networks*, **4**, 703–9.

Zhang, J. (1992). *IEEE Transactions on Signal Processing*, **40**, 2570–83.

Zhang, J. (1996). *IEEE Transactions on Image Processing*, **5**, 1208–14.

Zhou, Z., Leahy, R. M., and Qi, J. (1997). *IEEE Transactions on Image Processing*, **6**, 844–61.

INDEX